D1072100

Control Engineering

Guillermo J. Silva
Aniruddha Datta
S.P. Bhattacharyya

PID Controllers for Time-Delay Systems

Birkhäuser
Boston • Basel • Berlin

Guillermo J. Silva
IBM
11400 Burnet Road
Austin, TX 78758
USA

Aniruddha Datta
Department of Electrical Engineering
Texas A&M University
College Station, TX 77843
USA

S.P. Bhattacharyya
Department of Electrical Engineering
Texas A&M University
College Station, TX 77843
USA

AMS Subject Classifications: 30-02, 37F10, 65-02, 93D99

Library of Congress Cataloging-in-Publication Data
Silva, G. J., 1973-
PID controllers for time-delay systems / G.J. Silva, A. Datta, S.P. Bhattacharyya.
p. cm. – (Control engineering)
ISBN 0-8176-4266-8 (alk. paper)
1. PID controllers–Design and construction. 2. Time delay systems. I. Datta, Aniruddha, 1963- II. Bhattacharyya, S. P. (Shankar P.), 1946- III. Title. IV. Control engineering (Birkhäuser)

TJ223.P55S55 2004
629.8'3–dc22 2004062387

ISBN 0-8176-4266-8 Printed on acid-free paper.

Printed in the United States of America. (SB)

9 8 7 6 5 4 3 2 1 SPIN 10855839

www.birkhauser.com

THIS BOOK IS DEDICATED TO

My wife Seziş for her loving support and endless patience, and my parents Guillermo and Elvia.
G. J. Silva

My wife Anindita and my daughters Aparna and Anisha.
A. Datta

The memory of my friend and mentor, the late Yakov Z. Tsypkin, Russian control theorist and academician whose many contributions include the first results, in 1946, analyzing the stability of time-delay systems.
S. P. Bhattacharyya

Contents

Preface

This monograph presents our recent results on the proportional-integral-derivative (PID) controller and its design, analysis, and synthesis. The focus is on linear time-invariant plants that may contain a time delay in the feedback loop. This setting captures many real-world practical and industrial situations. The results given here include and complement those published in *Structure and Synthesis of PID Controllers* by Datta, Ho, and Bhattacharyya [10]. In [10] we mainly dealt with the delay-free case.

The main contribution described here is the efficient computation of the *entire set* of PID controllers achieving stability and various performance specifications. The performance specifications that can be handled within our machinery are classical ones such as gain and phase margin as well as modern ones such as H_∞ norms of closed-loop transfer functions. Finding the entire set is the key enabling step to realistic design with several design criteria. The computation is efficient because it reduces most often to linear programming with a sweeping parameter, which is typically the proportional gain. This is achieved by developing some preliminary results on root counting, which generalize the classical Hermite-Biehler Theorem, and also by exploiting some fundamental results of Pontryagin on quasi-polynomials to extract useful information for controller synthesis. The efficiency is important for developing software design packages, which we are sure will be forthcoming in the near future, as well as the development of further capabilities such as adaptive PID design and online implementation. It is also important for creating a realistic interactive design environment where multiple performance specifications that are appropriately prioritized can be overlaid and intersected to telescope down to a small and satisfactory controller set. Within this set further design choices must be made that reflect concerns such as cost, size, packaging, and other intangibles beyond the scope of the theory given here.

The PID controller is very important in control engineering applications and is widely used in many industries. Thus any improvement in design methodology has the potential to have a significant engineering and economic impact. An excellent account of many practical aspects of PID control is given in *PID Controllers: Theory, Design and Tuning* by Astrom and Hagglund [2], to which we refer the interested reader; we have chosen to not repeat these considerations here. At the other end of the spectrum there is a vast mathematical literature on the analysis of stability of time-delay systems which we have also not included. We refer the reader to the excellent and comprehensive recent work *Stability of Time-Delay Systems* by Gu, Kharitonov, and Chen [15] for these results. In other respects our work is self-contained in the sense that we present proofs and justfications of all results and algorithms developed by us.

We believe that these results are timely and in phase with the resurgence of interest in the PID controller and the general rekindling of interest in fixed and low-order controller design. As we know there are hardly any results in modern and postmodern control theory in this regard while such controllers are the ones of choice in applications. Classical control theory approaches, on the other hand, generally produce a single controller based on *ad hoc* loop-shaping techniques and are also inadequate for the kind of computer-aided multiple performance specifications design applications advocated here. Thus we hope that our monograph acts as a catalyst to bridge the theory-practice gap in the control field as well as the classical-modern gap.

The results reported here were derived in the Ph.D. theses of Ming-Tzu Ho, Guillermo Silva, and Hao Xu at Texas A&M University and we thank the Electrical Engineering Department for its logistical support. We also acknowledge the financial support of the National Science Foundation's Engineering Systems Program under the directorship of R. K. Baheti and the support of National Instruments, Austin, Texas.

Austin, Texas — *G. J. Silva*
College Station, Texas — *A. Datta*
College Station, Texas — *S. P. Bhattacharyya*

October 2004

PID Controllers
for Time-Delay Systems

1
Introduction

In this chapter we give a quick overview of control theory, explaining why integral feedback control works, describing PID controllers, and summarizing some of the currently available techniques for PID controller design. This background will serve to motivate our results on PID control, presented in the subsequent chapters.

1.1 Introduction to Control

Control theory and control engineering deal with dynamic systems such as aircraft, spacecraft, ships, trains, and automobiles, chemical and industrial processes such as distillation columns and rolling mills, electrical systems such as motors, generators, and power systems, and machines such as numerically controlled lathes and robots. In each case the *setting* of the control problem is

1. There are certain dependent variables, called *outputs*, to be controlled, which must be made to behave in a prescribed way. For instance it may be necessary to *assign* the temperature and pressure at various points in a process, or the position and velocity of a vehicle, or the voltage and frequency in a power system, to given desired fixed values, despite uncontrolled and unknown variations at other points in the system.

2. Certain independent variables, called *inputs*, such as voltage applied to the motor terminals, or valve position, are available to regulate

and control the behavior of the system. Other dependent variables, such as position, velocity, or temperature, are accessible as dynamic *measurements* on the system.

3. There are unknown and unpredictable *disturbances* impacting the system. These could be, for example, the fluctuations of load in a power system, disturbances such as wind gusts acting on a vehicle, external weather conditions acting on an air conditioning plant, or the fluctuating load torque on an elevator motor, as passengers enter and exit.

4. The equations describing the plant dynamics, and the parameters contained in these equations, are not known at all or at best known imprecisely. This uncertainty can arise even when the physical laws and equations governing a process are known well, for instance, because these equations were obtained by linearizing a nonlinear system about an operating point. As the operating point changes so do the system parameters.

These considerations suggest the following general representation of the *plant* or system to be controlled.

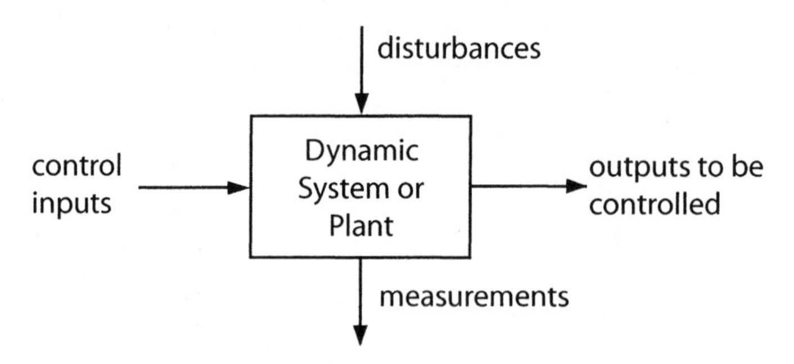

FIGURE 1.1. A general plant.

In Fig. 1.1 the inputs or outputs shown could actually be representing a vector of signals. In such cases the plant is said to be a *multivariable plant* as opposed to the case where the signals are scalar, in which case the plant is said to be a *scalar or monovariable plant.*

Control is exercised by feedback, which means that the corrective control input to the plant is generated by a device that is driven by the available measurements. Thus the controlled system can be represented by the *feedback* or *closed-loop system* shown in Fig. 1.2.

The control design problem is to determine the characteristics of the controller so that the controlled outputs can be

1. Set to prescribed values called *references*;

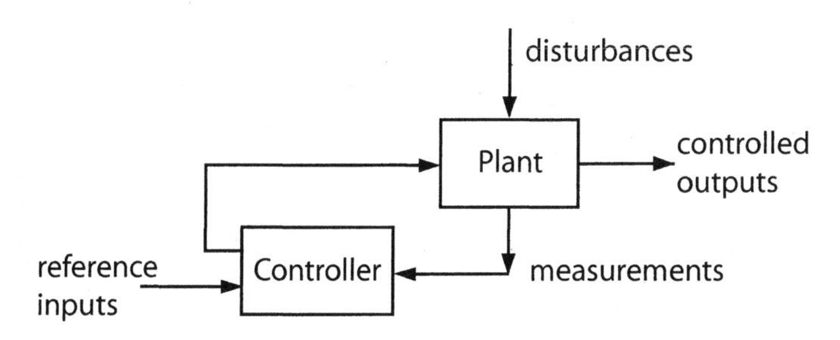

FIGURE 1.2. A feedback control system.

2. Maintained at the reference values despite the unknown disturbances;
3. Conditions (1) and (2) are met despite the inherent uncertainties and changes in the plant dynamic characteristics.

The first condition above is called *tracking*, the second, *disturbance rejection*, and the third, *robustness* of the system. The simultaneous satisfaction of (1), (2), and (3) is called *robust tracking and disturbance rejection* and control systems designed to achieve this are called *robust servomechanisms.*

In the next section we discuss how integral and PID control are useful in the design of robust servomechanisms.

1.2 The Magic of Integral Control

Integral control is used almost universally in the control industry to design robust servomechanisms. Integral action is most easily implemented by computer control. It turns out that hydraulic, pneumatic, electronic, and mechanical integrators are also commonly used elements in control systems. In this section we explain how integral control works in general to achieve robust tracking and disturbance rejection.

Let us first consider an integrator as shown in Fig. 1.3.

FIGURE 1.3. An integrator.

The input-output relationship is

$$y(t) = K\int_0^t u(\tau)d\tau + y(0) \tag{1.1}$$

or

$$\frac{dy}{dt} = Ku(t) \tag{1.2}$$

where K is the integrator gain.

Now *suppose* that the output $y(t)$ is a *constant*. It follows from (1.2) that

$$\frac{dy}{dt} = 0 = Ku(t) \, \forall \, t > 0. \tag{1.3}$$

Equation (1.3) proves the following important facts about the operation of an integrator:

1. If the output of an integrator is *constant* over a segment of time, then the input must be identically *zero* over that same segment.

2. The output of an integrator changes as long as the input is nonzero.

The simple fact stated above suggests how an integrator can be used to solve the servomechanism problem. If a plant output $y(t)$ is to track a constant reference value r, despite the presence of unknown constant disturbances, it is enough to

a. attach an integrator to the plant and make the error

$$e(t) = r - y(t)$$

the input to the integrator;

b. ensure that the closed-loop system is asymptotically stable so that under constant reference and disturbance inputs, all signals, including the integrator output, reach constant steady-state values.

This is depicted in the block diagram shown in Fig. 1.4. If the system

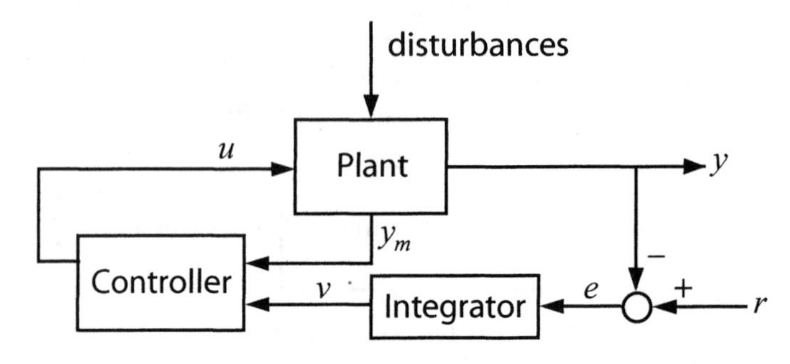

FIGURE 1.4. Servomechanism.

shown in Fig. 1.4 is asymptotically stable, and the inputs r and d (disturbances) are constant, it follows that all signals in the closed loop will tend to constant values. In particular the integrator output $v(t)$ tends to a constant value. Therefore by the fundamental fact about the operation of an integrator established above, it follows that the integrator input tends to *zero.* Since we have arranged that this input is the tracking error it follows that $e(t) = r - y(t)$ tends to zero and hence $y(t)$ tracks r as $t \to \infty$.

We emphasize that the steady-state tracking property established above is *very robust.* It holds as long as the closed loop is asymptotically stable and is (1) independent of the particular values of the constant disturbances or references, (2) independent of the initial conditions of the plant and controller, and (3) independent of whether the plant and controller are linear or nonlinear. Thus the tracking problem is reduced to guaranteeing that stability is assured. In many practical systems stability of the closed-loop system can even be ensured without detailed and exact knowledge of the plant characteristics and parameters; this is known as *robust stability.*

We next discuss how several plant outputs $y_1(t), y_2(t), \ldots, y_m(t)$ can be pinned down to prescribed but arbitrary constant reference values $r_1, r_2, \ldots, r_m$ in the presence of unknown but constant disturbances $d_1, d_2, \ldots, d_q$. The previous argument can be extended to this multivariable case by attaching m integrators to the plant and driving each integrator with its corresponding error input $e_i(t) = r_i - y_i(t), i = 1, \ldots, m$. This is shown in the configuration in Fig. 1.5.

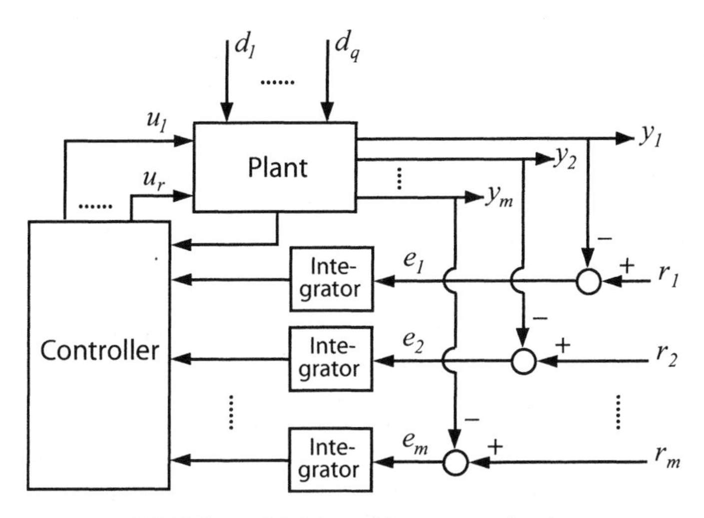

FIGURE 1.5. Multivariable servomechanism.

Once again it follows that as long as the closed-loop system is stable, all signals in the system must tend to constant values and integral action forces the $e_i(t), i = 1, \ldots, m$ to tend to zero asymptotically, regardless of the actual values of the disturbances $d_j, j = 1, \ldots, q$. The existence of steady-state inputs $u_1, u_2, \ldots, u_r$ that make $y_i = r_i, i = 1, \ldots, m$ for arbitrary $r_i, i = 1, \ldots, m$ requires that the plant equations relating $y_i, i = 1, \ldots, m$ to $u_j, j = 1, \ldots, r$ be invertible for constant inputs. In the case of linear time-invariant systems this is equivalent to the requirement that the corresponding transfer matrix have rank equal to m at $s = 0$. Sometimes this is restated as two conditions: (1) $r \geq m$ or at least as many control inputs as outputs to be controlled and (2) $G(s)$ has no transmission zero at $s = 0$.

In general, the addition of an integrator to the plant tends to make the system less stable. This is because the integrator is an inherently unstable device; for instance, its response to a step input, a bounded signal, is a ramp, an unbounded signal. Therefore the problem of stabilizing the closed loop becomes a critical issue even when the plant is stable to begin with.

Since integral action and thus the attainment of zero steady-state error is *independent* of the particular value of the integrator gain K, we can see that this gain can be used to try to stabilize the system. This single degree of freedom is sometimes insufficient for attaining stability and an acceptable transient response, and additional gains are introduced as explained in the next section. This leads naturally to the PID controller structure commonly used in industry.

1.3 PID Controllers

In the last section we saw that when an integrator is part of an asymptotically stable system and constant inputs are applied to the system, the integrator input is forced to become zero. This simple and powerful principle is the basis for the design of linear, nonlinear, single-input single-output, and multivariable servomechanisms. All we have to do is (1) attach as many integrators as outputs to be regulated, (2) drive the integrators with the tracking errors required to be zeroed, and (3) stabilize the closed-loop system by using any adjustable parameters.

As argued in the last section the input zeroing property is independent of the gain cascaded to the integrator. Therefore this gain can be freely used to attempt to stabilize the closed-loop system. Additional free parameters for stabilization can be obtained, without destroying the input zeroing property, by adding parallel branches to the controller, processing in addition to the integral of the error, the error itself and its derivative, when it can be obtained. This leads to the PID controller structure shown in Fig. 1.6.

As long as the closed loop is stable it is clear that the input to the integrator will be driven to zero independent of the values of the gains. Thus the function of the gains k_p, k_i, and k_d is to stabilize the closed-loop system if possible and to adjust the transient response of the system.

In general the derivative can be computed or obtained if the error is varying slowly. Since the response of the derivative to high-frequency inputs is much higher than its response to slowly varying signals (see Fig. 1.7), the derivative term is usually omitted if the error signal is corrupted by high-frequency noise.

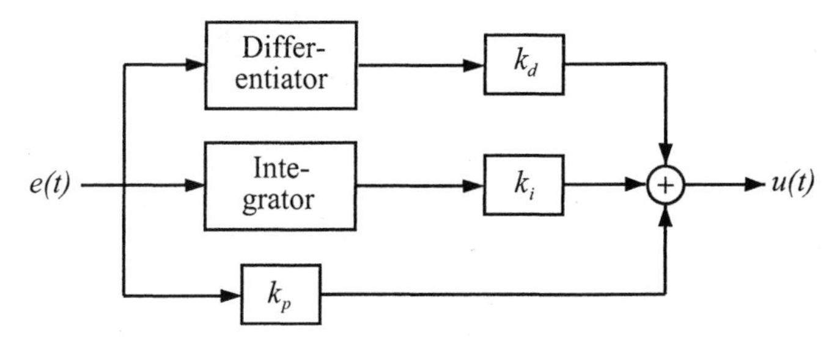

FIGURE 1.6. PID controller.

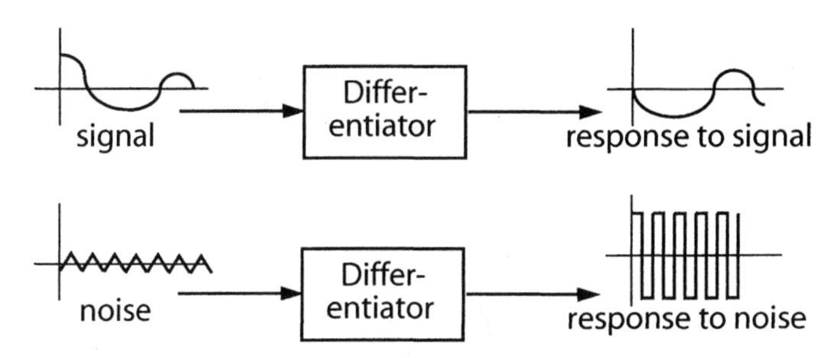

FIGURE 1.7. Response of derivative to signal and noise.

In such cases the derivative gain k_d is set to zero or equivalently the differentiator is switched off and the controller is a proportional-integral or PI controller. Such controllers are most common in industry.

In subsequent chapters of the book we solve the problem of stabilization of a linear time-invariant plant by a PID controller. Both delay-free systems and systems with time delay are considered. Our solutions uncover the entire set of stabilizing controllers in a computationally tractable way.

In the rest of this introductory chapter we briefly discuss the currently available techniques for PID controller design. Many of them are based on empirical observations. For a comprehensive survey on tuning methods for PID controllers, we refer the reader to [2].

1.4 Some Current Techniques for PID Controller Design

1.4.1 The Ziegler-Nichols Step Response Method

The PID controller we are concerned with is implemented as follows:

$$C(s) = k_p + \frac{k_i}{s} + k_d s \tag{1.4}$$

where k_p is the proportional gain, k_i is the integral gain, and k_d is the derivative gain. In real life, the derivative term is often replaced by $\frac{k_d s}{1+T_d s}$, where T_d is a small positive value that is usually fixed. This circumvents the problem of pure differentiation when the error signals are contaminated by noise.

The Ziegler-Nichols step response method is an experimental open-loop tuning method and is only applicable to open-loop *stable* plants. This method first characterizes the plant by two parameters A and L obtained from its step response. A and L can be determined graphically from a measurement of the step response of the plant as illustrated in Fig. 1.8. First, the point on the step response curve with the maximum slope is determined and the tangent is drawn. The intersection of the tangent with the vertical axis gives A, while the intersection of the tangent with the horizontal axis gives L. Once A and L are determined, the PID controller parameters are

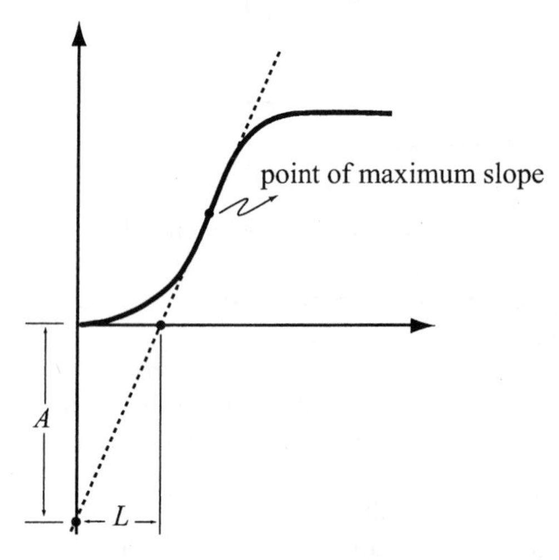

FIGURE 1.8. Graphical determination of parameters A and L.

then given in terms of A and L by the following formulas:

$$\begin{aligned} k_p &= \frac{1.2}{A} \\ k_i &= \frac{0.6}{AL} \\ k_d &= \frac{0.6L}{A}. \end{aligned}$$

These formulas for the controller parameters were selected to obtain an amplitude decay ratio of 0.25, which means that the first overshoot decays to $\frac{1}{4}$th of its original value after one oscillation. Intense experimentation showed that this criterion gives a small settling time.

1.4.2 The Ziegler-Nichols Frequency Response Method

The Ziegler-Nichols frequency response method is a closed-loop tuning method. This method first determines the point where the Nyquist curve of the plant $G(s)$ intersects the negative real axis. It can be obtained experimentally in the following way: Turn the integral and differential actions off and set the controller to be in the proportional mode only and close the loop as shown in Fig. 1.9. Slowly increase the proportional gain k_p until a periodic oscillation in the output is observed. This critical value of k_p is called the *ultimate gain* (k_u). The resulting period of oscillation is referred to as the *ultimate period* (T_u). Based on k_u and T_u, the Ziegler-Nichols frequency response method gives the following simple formulas for setting PID controller parameters:

$$\begin{aligned} k_p &= 0.6k_u \\ k_i &= \frac{1.2k_u}{T_u} \\ k_d &= 0.075k_uT_u\,. \end{aligned} \tag{1.5}$$

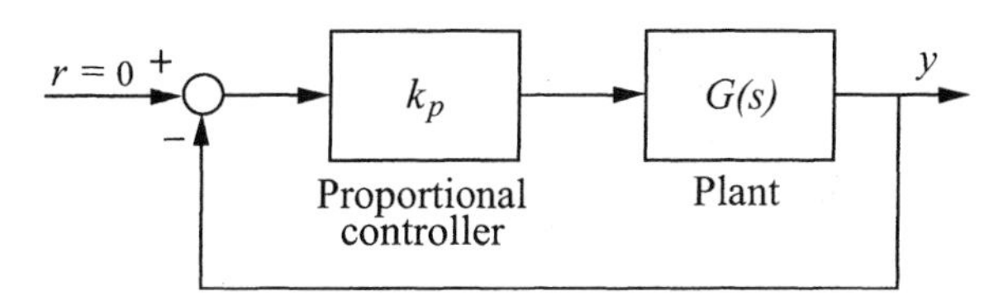

FIGURE 1.9. The closed-loop system with the proportional controller.

This method can be interpreted in terms of the Nyquist plot. Using PID control it is possible to move a given point on the Nyquist curve to an arbitrary position in the complex plane. Now, the first step in the frequency response method is to determine the point $(-\frac{1}{k_u}, 0)$ where the Nyquist curve of the open-loop transfer function intersects the negative real axis. We will study how this point is changed by the PID controller. Using (1.5) in (1.4), the frequency response of the controller at the ultimate frequency w_u is

$$\begin{aligned} C(jw_u) &= 0.6k_u - j\left(\frac{1.2k_u}{T_uw_u}\right) + j(0.075k_uT_uw_u) \\ &= 0.6k_u(1+j0.4671)\ \ [\text{since } T_uw_u = 2\pi]\,. \end{aligned}$$

From this we see that the controller gives a phase advance of 25 degrees at the ultimate frequency. The loop transfer function is then

$$G_{loop}(jw_u) = G(jw_u)C(jw_u) = -0.6(1+j0.4671) = -0.6 - j0.28\,.$$

Thus the point $(-\frac{1}{k_u}, 0)$ is moved to the point (-0.6, -0.28). The distance from this point to the critical point is almost 0.5. This means that the frequency response method gives a sensitivity greater than 2.

The procedure described above for measuring the ultimate gain and ultimate period requires that the closed-loop system be operated close to instability. To avoid damaging the physical system, this procedure needs to be executed carefully. Without bringing the system to the verge of instability, an alternative method was proposed by Astrom and Hagglund using relay to generate a relay oscillation for measuring the ultimate gain and ultimate period. This is done by using the relay feedback configuration shown in Fig. 1.10. In Fig. 1.10, the relay is adjusted to induce a self-sustaining oscillation in the loop.

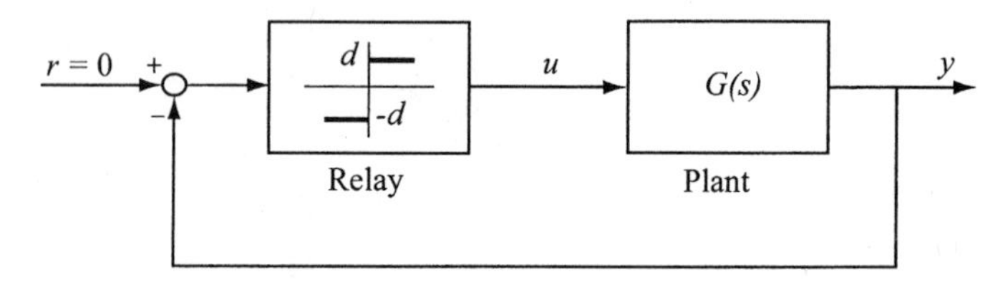

FIGURE 1.10. Block diagram of relay feedback.

Now we explain why this relay feedback can be used to determine the ultimate gain and ultimate period. The relay block is a nonlinear element that can be represented by a *describing function.* This describing function is obtained by applying a sinusoidal signal $asin(wt)$ at the input of the nonlinearity and calculating the ratio of the Fourier coefficient of the first harmonic at the output to a. This function can be thought of as an equivalent gain of the nonlinear system. For the case of the relay its describing function is given by

$$N(a) = \frac{4d}{a\pi}$$

where a is the amplitude of the sinusoidal input signal and d is the relay amplitude. The conditions for the presence of limit cycle oscillations can be derived by investigating the propagation of a sinusoidal signal around the loop. Since the plant $G(s)$ acts as a low pass filter, the higher harmonics produced by the nonlinear relay will be attenuated at the output of the plant. Hence, the condition for oscillation is that the fundamental sine waveform comes back with the same amplitude and phase after traversing through the loop. This means that for sustained oscillations at a frequency of ω, we must have

$$G(j\omega)N(a) = -1 \,. \tag{1.6}$$

This equation can be solved by plotting the Nyquist plot of $G(s)$ and the line $-\frac{1}{N(a)}$. As shown in Fig. 1.11, the plot of $-\frac{1}{N(a)}$ is the negative real axis, so the solution to (1.6) is given by the two conditions:

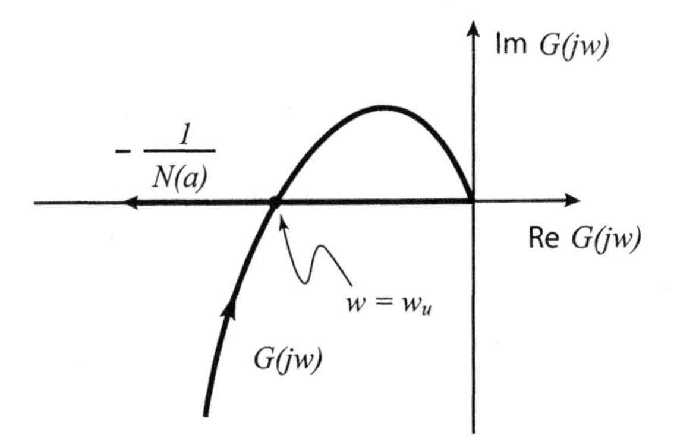

FIGURE 1.11. Nyquist plots of the plant $G(j\omega)$ and the describing function $-\frac{1}{N(a)}$.

$$
\begin{aligned}
|G(j\omega_u)| &= \frac{a\pi}{4d} \\
&\triangleq \frac{1}{k_u} \\
\text{and } arg\, G(j\omega_u) &= -\pi\,.
\end{aligned}
$$

The ultimate gain and ultimate period can now be determined by measuring the amplitude and period of the oscillations. This relay feedback technique is widely used in automatic PID tuning.

Remark 1.1 *Both Ziegler-Nichols tuning methods require very little knowledge of the plants and simple formulas are given for controller parameter settings. These formulas are obtained by extensive simulations of many stable and simple plants. The main design criterion of these methods is to obtain a quarter amplitude decay ratio for the load disturbance response. As pointed out by Astrom and Hagglund [2], little emphasis is given to measurement noise, sensitivity to process variations, and setpoint response. Even though these methods provide good rejection of load disturbance, the resulting closed-loop system can be poorly damped and sometimes can have poor stability margins.*

1.4.3 PID Settings using the Internal Model Controller Design Technique

The internal model controller (IMC) structure has become popular in process control applications. This structure, in which the controller includes an explicit model of the plant, is particularly appropriate for the design and implementation of controllers for open-loop stable systems. The fact that many of the plants encountered in process control happen to be open-loop stable possibly accounts for the popularity of IMC among practicing engineers. In this section, we consider the IMC configuration for a stable plant

$G(s)$ as shown in Fig. 1.12. The IMC controller consists of a stable *IMC parameter* $Q(s)$ and a model of the plant $\hat{G}(s)$, which is usually referred to as the *internal model.* $F(s)$ is the IMC filter chosen to enhance robustness with respect to the modelling error and to make the overall IMC parameter $Q(s)F(s)$ proper. From Fig. 1.12 the equivalent feedback controller $C(s)$ is

$$C(s) = \frac{F(s)Q(s)}{1 - F(s)Q(s)\hat{G}(s)}.$$

The IMC design objective considered in this section is to choose $Q(s)$ which

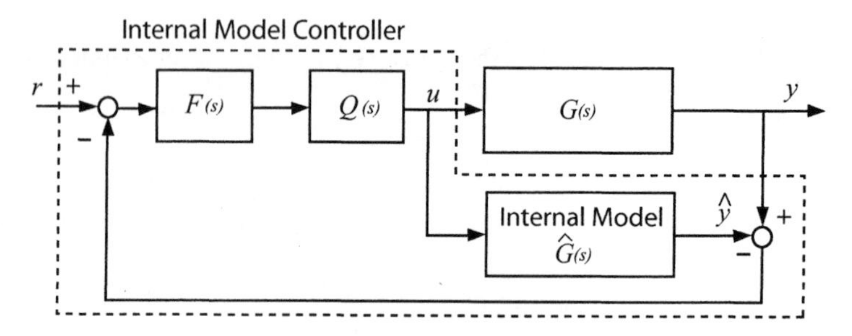

FIGURE 1.12. The IMC configuration.

minimizes the L_2 norm of the tracking error $r - y$, i.e., achieves an H_2-optimal control design. In general, complex models lead to complex IMC H_2-optimal controllers. However, it has been shown that, for first-order plants with deadtime and a step command signal, the IMC H_2-optimal design results in a controller with a PID structure. This will be clearly borne out by the following discussion.

Assume that the plant to be controlled is a first-order model with deadtime:

$$G(s) = \frac{k}{1 + Ts} e^{-Ls}.$$

The control objective is to minimize the L_2 norm of the tracking error due to setpoint changes. Using Parseval's Theorem, this is equivalent to choosing $Q(s)$ for which $\min \|[1 - \hat{G}(s)Q(s)]R(s)\|_2$ is achieved, where $R(s) = \frac{1}{s}$ is the Laplace transform of the unit step command.

Approximating the deadtime with a first-order Padé approximation, we have

$$e^{-Ls} \cong \frac{1 - \frac{L}{2}s}{1 + \frac{L}{2}s}.$$

The resulting rational transfer function of the internal model $\hat{G}(s)$ is given by

$$\hat{G}(s) = \frac{k}{(1 + Ts)} \frac{1 - \frac{L}{2}s}{1 + \frac{L}{2}s}.$$

Choosing $Q(s)$ to minimize the H_2 norm of $[1 - \hat{G}(s)Q(s)]R(s)$, we obtain

$$Q(s) = \frac{1 + Ts}{k}.$$

Since this $Q(s)$ is improper, we choose

$$F(s) = \frac{1}{1 + \lambda s}$$

where $\lambda > 0$ is a small number. The equivalent feedback controller becomes

$$\begin{aligned} C(s) &= \frac{F(s)Q(s)}{1 - F(s)Q(s)\hat{G}(s)} \\ &= \frac{(1 + Ts)(1 + \frac{L}{2}s)}{ks(L + \lambda + \frac{L\lambda}{2}s)} \\ &\cong \frac{(1 + Ts)(1 + \frac{L}{2}s)}{ks(L + \lambda)}. \end{aligned} \tag{1.7}$$

From (1.7) we can extract the following parameters for a standard PID controller:

$$\begin{aligned} k_p &= \frac{2T + L}{2k(L + \lambda)} \\ k_i &= \frac{1}{k(L + \lambda)} \\ k_d &= \frac{TL}{2k(L + \lambda)}. \end{aligned}$$

Since a first-order Padé *approximation* was used for the time delay, ensuring the robustness of the design to modelling errors is all the more important. This can be done by properly selecting the design variable λ to achieve the appropriate compromise between performance and robustness. Morari and Zafiriou [31] have proposed that a suitable choice for λ is $\lambda > 0.2T$ and $\lambda > 0.25L$.

Remark 1.2 *The IMC PID design procedure minimizes the L_2 norm of the tracking error due to setpoint changes. Therefore, as expected, this design method gives good response to setpoint changes. However, for lag dominant plants the method gives poor load disturbance response because of the pole-zero cancellation inherent in the design methodology.*

1.4.4 Dominant Pole Design: The Cohen-Coon Method

Dominant pole design attempts to position a few poles to achieve certain control performance specifications. The Cohen-Coon method is a dominant pole design method. This tuning method is based on the first-order plant model with deadtime:

$$G(s) = \frac{k}{1+Ts}e^{-Ls}.$$

The key feature of this tuning method is to attempt to locate three dominant poles, a pair of complex poles and one real pole, such that the amplitude decay ratio for load disturbance response is 0.25 and the integrated error $\int_0^\infty e(t)dt$ is minimized. Thus, the Cohen-Coon method gives good load disturbance rejection. Based on analytical and numerical computation, Cohen and Coon gave the following PID controller parameters in terms of k, T, and L:

$$\begin{aligned} k_p &= \frac{1.35(1-0.82b)}{a(1-b)} \\ k_i &= \frac{1.35(1-0.82b)(1-0.39b)}{aL(1-b)(2.5-2b)} \\ k_d &= \frac{1.35L(0.37-0.37b)}{a(1-b)} \end{aligned}$$

where

$$\begin{aligned} a &= \frac{kL}{T} \\ b &= \frac{L}{L+T}. \end{aligned}$$

Note that for small b, the controller parameters given by the above formulas are close to the parameters obtained by the Ziegler-Nichols step response method.

1.4.5 New Tuning Approaches

The tuning methods described in the previous subsections are easy to use and require very little information about the plant to be controlled. However, since they do not capture all aspects of desirable PID performance, many other new approaches have been developed. These methods can be classified into three categories.

Time Domain Optimization Methods

The idea behind these methods is to choose the PID controller parameters to minimize an integral cost functional. Zhuang and Atherton [53] used an integral criterion with data from a relay experiment. The time-weighted system error integral criterion was chosen as

$$J_n(\theta) = \int_0^\infty t^n e^2(\theta,t)dt$$

where θ is a vector containing the controller parameters and $e(\theta, t)$ represents the error signal. Experimentation showed that for $n = 1$, the controller obtained produced a step response of desirable form. This gave birth to the integral square time error (ISTE) criterion. Another contribution is due to Pessen [36], who used the integral absolute error (IAE) criterion:

$$J(\theta) = \int_0^\infty |e(\theta, t)| dt \ .$$

In order to minimize the above integral cost functions, Parseval's Theorem can be invoked to express the time functions in terms of their Laplace transforms. Definite integrals of the form encountered in this approach have been evaluated in terms of the coefficients of the numerator and denominator of the Laplace transforms (see [32]). Once the integration is carried out, the parameters of the PID controller are adjusted in such a way as to minimize the integral cost function. Recently Atherton and Majhi [3] proposed a modified form of the PID controller (see Fig. 1.13). In this structure an internal proportional-derivative (PD) feedback is used to change the poles of the plant transfer function to more desirable locations and then a PI controller is used in the forward loop. The parameters of the controller are obtained by minimization of the ISTE criterion.

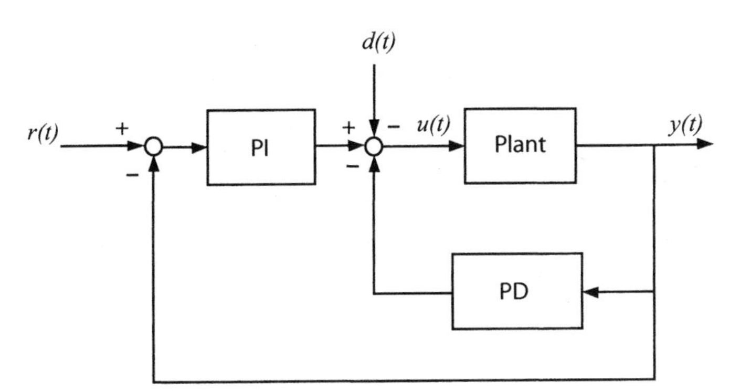

FIGURE 1.13. PI-PD feedback control structure.

Frequency Domain Shaping

These methods seek a set of controller parameters that give a desired frequency response. Astrom and Hagglund [2] proposed the idea of using a set of rules to achieve a desired phase margin specification. In the same spirit, Ho, Hang, and Zhou [24] developed a PID self-tuning method with specifications on the gain and phase margins. Another contribution by Voda and Landau [48] presented a method to shape the compensated system frequency response.

Optimal Control Methods

This new trend has been motivated by the desire to incorporate several control system performance objectives such as reference tracking, disturbance rejection, and measurement noise rejection. Grimble and Johnson [14] incorporated all these objectives into an LQG optimal control problem. They proposed an algorithm to minimize an LQG-cost function where the controller structure is fixed to a particular PID industrial form. In a similar fashion, Panagopoulos, Astrom, and Hagglund [35] presented a method to design PID controllers that captures demands on load disturbance rejection, set point response, measurement noise, and model uncertainty. Good load disturbance rejection was obtained by minimization of the integral control error. Good set point response was obtained by using a structure with two degrees of freedom. Measurement noise was dealt with by filtering. Robustness was achieved by requiring a maximum sensitivity of less than a specified value.

1.5 Integrator Windup

An important element of the control strategy discussed in Section 1.2 is the actuator, which applies the control signal u to the plant. However, all actuators have limitations that make them nonlinear elements. For instance, a valve cannot be more than fully opened or fully closed. During the regular operation of a control system, it can very well happen that the control variable reaches the actuator limits. When this situation arises, the feedback loop is broken and the system runs as an open loop because the actuator will remain at its limit independently of the process output. In this scenario, if the controller is of the PID type, the error will continue to be integrated. This in turn means that the integral term may become very large, which is commonly referred to as *windup*. In order to return to a normal state, the error signal needs to have an opposite sign for a long period of time. As a consequence of all this, a system with a PID controller may give large transients when the actuator saturates.

The phenomenon of windup has been known for a long time. It may occur in connection with large setpoint changes or it may be caused by large disturbances or equipment malfunction. Several techniques are available to avoid windup when using an integral term in the controller. We describe some of these techniques in this section.

1.5.1 Setpoint Limitation

The easiest way to avoid integrator windup is to introduce limiters on the setpoint variations so that the controller output will never reach the actuator bounds. However, this approach has several disadvantages: (a) it leads to conservative bounds; (b) it imposes limitations on the controller performance; and (c) it does not prevent windup caused by disturbances.

1.5.2 Back-Calculation and Tracking

This technique is illustrated in Fig. 1.14. If we compare this figure to Fig. 1.6, we see that the controller has an extra feedback path. This path is generated by measuring the actual actuator output $u(t)$ and forming the error signal $e_s(t)$ as the difference between the output of the controller $v(t)$ and the signal $u(t)$. This signal $e_s(t)$ is fed to the input of the integrator through a gain $1/T_t$.

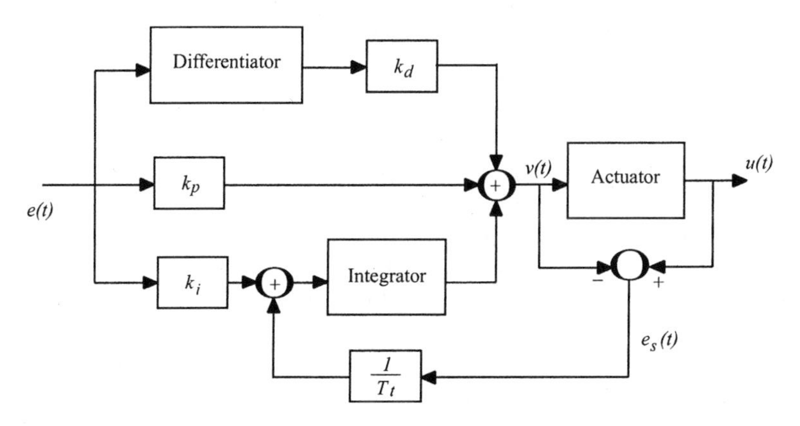

FIGURE 1.14. Controller with antiwindup.

When the actuator is within its operating range, the signal $e_s(t)$ is zero. Thus it will not have any effect on the normal operation of the controller. When the actuator saturates, the signal $e_s(t)$ is different from zero. The normal feedback path around the process is broken because the process input remains constant. However, there is a new feedback path around the integrator due to $e_s(t) \neq 0$ and this prevents the integrator from winding up. The rate at which the controller output is reset is governed by the feedback gain $1/T_t$. The parameter T_t can thus be interpreted as the time constant that determines how quickly the integral action is reset. In general, the smaller the value of T_t, the faster the integrator is reset. However, if the parameter T_t is chosen too small, spurious errors can cause saturation of the output, which accidentally resets the integrator. Astrom and Hagglund [2] recommend T_t to be larger than $\frac{k_d}{k}$ and smaller than $\frac{k}{k_i}$.

1.5.3 Conditional Integration

Conditional integration is an alternative to the back-calculation technique. It simply consists of switching off the integral action when the control is far from the steady state. This means that the integral action is only used when certain conditions are fulfilled, otherwise the integral term is kept constant. We now consider a couple of these switching conditions.

One simple approach is to switch off the integral action when the control error $e(t)$ is large. Another one is to switch off the integral action when the actuator saturates. However, both approaches have a disadvantage: the controller may get stuck at a nonzero control error $e(t)$ if the integral term has a large value at the time of switch off.

Because of the previous disadvantage, a better approach is to switch off integration when the controller is saturated and the integrator update is such that it causes the control signal to become more saturated. For example, consider that the controller becomes saturated at its upper bound. Integration is then switched off if the control error is positive, but not if it is negative.

1.6 Contribution of this Book

In concluding this chapter, it is important to point out that in addition to the approaches discussed above, there are many other approaches for tuning PID controllers [2]. Despite this, for plants having order higher than two, *there is no approach that can be used to determine the set of all stabilizing PID gain values.* The principal contribution of this book to the PID literature is the development of a methodology that provides a complete answer to this long-standing open problem for both delay-free plants as well as for plants with time delay. For the former class of plants, the results were first reported in [10]. In this book, we give results for determining, in a computationally efficient way, the complete set of PID controllers that stabilize a given linear time-invariant plant and achieve prescribed levels of performance. These results apply to plants with and without time delay. In going from delay-free plants to plants with time delays, one has to transition from the realm of polynomials to that of *quasi-polynomials.* When considering the latter, the early results of Pontryagin are very useful.

1.7 Notes and References

The Ziegler-Nichols methods were first presented in [54]. The alternative method using relay feedback is described in [1]. The relay feedback technique in Section 1.4.2 and its applications to automatic PID tuning can be found in the works of Astrom and Hagglund [1, 2]. For a better understanding of describing functions, the book by Khalil [27] is recommended. For a more detailed explanation of the IMC structure and its applications in process control, the reader is referred to [31]. The Cohen-Coon method can be found in [9]. A complete description of antiwindup techniques can be found in [2]. Needless to say there is an extensive literature covering all aspects of PID control. We have not attempted to be complete in citing this literature. Instead, we have tried to cite all relevant publications related to the new results given in this book.

2

The Hermite-Biehler Theorem and its Generalization

In this chapter we introduce the classical Hermite-Biehler Theorem for Hurwitz polynomials. We also present several generalizations of this theorem that are useful for solving the problem of finding the set of proportional (P), PI, and PID controllers that stabilize a given finite-dimensional linear time-invariant system.

2.1 Introduction

The problem of determining conditions under which all of the roots of a given real polynomial lie in the open left half of the complex plane plays an important role in the theory of stability of linear time-invariant systems. A polynomial for which such a property holds is said to be Hurwitz. Many conditions have been proposed for ascertaining the Hurwitz stability of a given real polynomial without determining the actual roots. Results of this nature were first obtained by Routh, Hurwitz, and Hermite in the 19th century.

In this chapter, we introduce another well-known result: the classical Hermite-Biehler Theorem. This theorem states that a given real polynomial is Hurwitz if and only if it satisfies a certain interlacing property. This result has played an important role in studying the parametric robust stability problem. However, when a given polynomial is not Hurwitz stable, the Hermite-Biehler Theorem does not provide any information about the root distribution of the polynomial. Recent research has produced several generalizations of the Hermite-Biehler Theorem applicable to the case of real polynomials that are not necessarily Hurwitz. Some of these generalizations will be introduced in this chapter and used in later chapters to solve the important problem of finding the set of stabilizing PID controllers for a system described by a rational transfer function.

The chapter is organized as follows. In Section 2.2, we provide a statement of the Hermite-Biehler Theorem as well as a useful equivalent characterization. Section 2.3 contains important generalizations of the Hermite-Biehler Theorem. These generalizations, which are essentially root counting formulas, will be used in later chapters to solve the PID stabilization problem for finite-dimensional linear time-invariant systems.

2.2 The Hermite-Biehler Theorem for Hurwitz Polynomials

In this section, we state the classical Hermite-Biehler Theorem, which provides necessary and sufficient conditions for the Hurwitz stability of a given real polynomial. Before stating the theorem, we establish some notation.

Definition 2.1 *Let $\delta(s) = \delta_0 + \delta_1 s + \cdots + \delta_n s^n$ be a given real polynomial of degree n. Write*

$$\delta(s) = \delta_e(s^2) + s\delta_o(s^2)$$

where $\delta_e(s^2)$, $s\delta_o(s^2)$ are the components of $\delta(s)$ made up of the even and odd powers of s respectively. For every frequency $\omega \in \mathcal{R}$, denote

$$\delta(j\omega) = p(\omega) + jq(\omega)$$

where $p(\omega) = \delta_e(-\omega^2)$, $q(\omega) = \omega\delta_o(-\omega^2)$. Let $\omega_{e_1}, \omega_{e_2}, \ldots$ denote the non-negative real zeros of $\delta_e(-\omega^2)$ and let $\omega_{o_1}, \omega_{o_2}, \ldots$ denote the non-negative real zeros of $\delta_o(-\omega^2)$, both arranged in ascending order of magnitude.

Theorem 2.1 (Hermite-Biehler Theorem) *Let $\delta(s) = \delta_0 + \delta_1 s + \cdots + \delta_n s^n$ be a given real polynomial of degree n. Then $\delta(s)$ is Hurwitz stable if and only if all the zeros of $\delta_e(-\omega^2)$, $\delta_o(-\omega^2)$ are real and distinct, δ_n and δ_{n-1} are of the same sign, and the non-negative real zeros satisfy the following interlacing property*

$$0 < \omega_{e_1} < \omega_{o_1} < \omega_{e_2} < \omega_{o_2} < \cdots \tag{2.1}$$

This important theorem is based on the fact that a Hurwitz polynomial $\delta(s)$ satisfies the *monotonic phase increase property*, that is, the phase of $\delta(j\omega)$ is a continuous and strictly increasing function of ω on $(-\infty, +\infty)$. Moreover, using this property, we can show that the parametric plot of $\delta(j\omega) = p(\omega) + jq(\omega)$ in the $\delta(j\omega)$-plane must move strictly counterclockwise and go through n quadrants in turn as ω increases from 0 to $+\infty$ [5].

Figure 2.1 illustrates the admissible plots of $\delta(j\omega)$ for a Hurwitz polynomial $\delta(s)$.

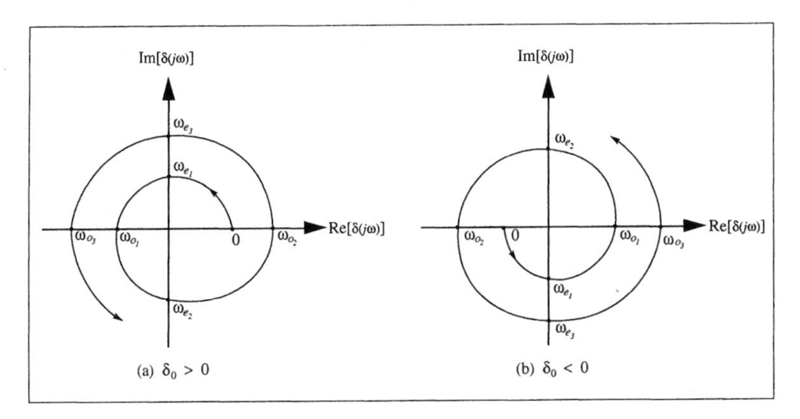

FIGURE 2.1. The monotonic phase increase property for a Hurwitz polynomial.

The following example illustrates the application of Theorem 2.1 to verify the Hurwitz stability of a real polynomial.

Example 2.1 *Consider the real polynomial*

$$\delta(s) = s^7 + 4s^6 + 11s^5 + 29s^4 + 36s^3 + 61s^2 + 34s + 36 .$$

Then

$$\delta(j\omega) = p(\omega) + jq(\omega)$$

where

$$\begin{aligned} p(\omega) &= -4\omega^6 + 29\omega^4 - 61\omega^2 + 36 \\ q(\omega) &= \omega(-\omega^6 + 11\omega^4 - 36\omega^2 + 34) . \end{aligned}$$

The plots of $p(\omega)$ and $q(\omega)$ are shown in Fig. 2.2. They show that the polynomial $\delta(s)$ satisfies the interlacing property. To verify that $\delta(s)$ is indeed a Hurwitz polynomial, we solve for the roots of $\delta(s)$:

$$\begin{array}{ll} -0.0477 \pm 1.9883j & -0.2008 \pm 1.4200j \\ -0.2898 \pm 1.1957j & -2.9233 . \end{array}$$

We see that all the roots of $\delta(s)$ are in the left half plane so that $\delta(s)$ is Hurwitz. $\triangle$

We now present some alternative characterizations of the Hermite-Biehler Theorem that will be used subsequently. We first introduce the standard signum function sgn : $\mathcal{R} \rightarrow \{-1, 0, 1\}$ defined by

$$\text{sgn}[x] = \begin{cases} -1 \text{ if } x < 0 \\ 0 \text{ if } x = 0 \\ 1 \text{ if } x > 0 \end{cases} .$$

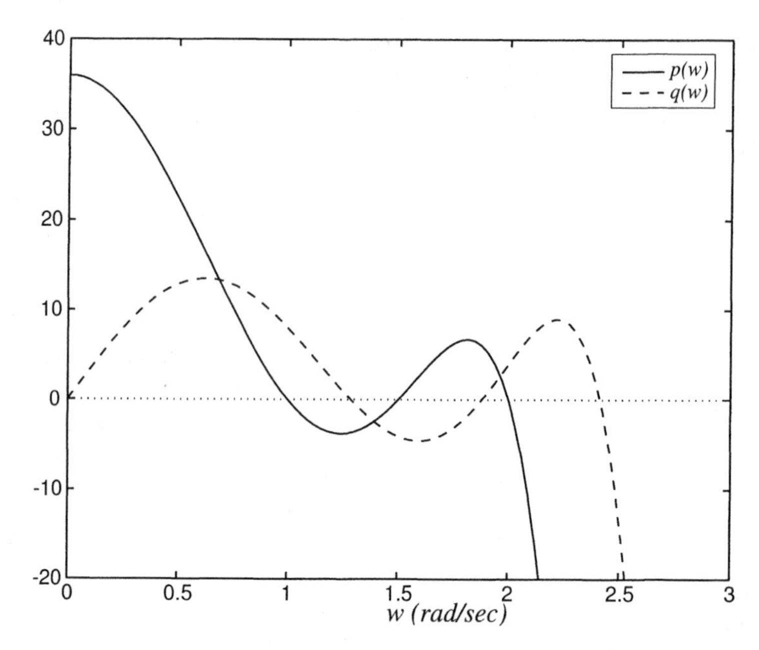

FIGURE 2.2. The interlacing property for a Hurwitz polynomial.

Lemma 2.1 *Let $\delta(s) = \delta_0 + \delta_1 s + \cdots + \delta_n s^n$ be a given real polynomial of degree n. Let $\omega_{e_1}, \omega_{e_2}, \ldots$ denote the non-negative real zeros of the even part of $\delta(j\omega)$, and let $\omega_{o_1}, \omega_{o_2}, \ldots$ denote the non-negative real zeros of the odd part of $\delta(j\omega)$, both arranged in ascending order of magnitude. Then the following conditions are equivalent:*

(i) $\delta(s)$ is Hurwitz stable

(ii) δ_n and δ_{n-1} are of the same sign and

$$n = \begin{cases} \operatorname{sgn}[\delta_0] \cdot \{\operatorname{sgn}[p(0)] - 2\operatorname{sgn}[p(\omega_{o_1})] + 2\operatorname{sgn}[p(\omega_{o_2})] + \cdots \\ +(-1)^{m-1} \cdot 2\operatorname{sgn}[p(\omega_{o_{m-1}})] + (-1)^m \cdot \operatorname{sgn}[p(\infty)]\}, \\ \quad \textit{for } n = 2m \\ \\ \operatorname{sgn}[\delta_0] \cdot \{\operatorname{sgn}[p(0)] - 2\operatorname{sgn}[p(\omega_{o_1})] + 2\operatorname{sgn}[p(\omega_{o_2})] + \cdots \\ +(-1)^{m-1} \cdot 2\operatorname{sgn}[p(\omega_{o_{m-1}})] + (-1)^m \cdot 2\operatorname{sgn}[p(\omega_{o_m})]\}, \\ \quad \textit{for } n = 2m+1 \end{cases} \tag{2.2}$$

(iii) δ_n *and* δ_{n-1} *are of the same sign and*

$$n = \begin{cases} \text{sgn}[\delta_0] \cdot \{2\text{sgn}[q(\omega_{e_1})] - 2\text{sgn}[q(\omega_{e_2})] + 2\text{sgn}[q(\omega_{e_3})] + \cdots \\ +(-1)^{m-2} \cdot 2\text{sgn}[q(\omega_{e_{m-1}})] + (-1)^{m-1} \cdot 2\text{sgn}[q(\omega_{e_m})]\}, \\ \quad \textit{for } n = 2m \\ \text{sgn}[\delta_0] \cdot \{2\text{sgn}[q(\omega_{e_1})] - 2\text{sgn}[q(\omega_{e_2})] + 2\text{sgn}[q(\omega_{e_3})] + \cdots \\ +(-1)^{m-1} \cdot 2\text{sgn}[q(\omega_{e_m})] + (-1)^{m} \cdot \text{sgn}[q(\infty)]\}, \\ \quad \textit{for } n = 2m+1 \end{cases} \tag{2.3}$$

Proof.

(1) (i) $\Leftrightarrow$ (ii)

We first show that (i) $\Rightarrow$ (ii).
From Fig. 2.1, it is clear that

$$\begin{cases} \text{for } n = 2m \\ \text{sgn}[\delta_0] \cdot \text{sgn}[p(0)] > 0 \\ -\text{sgn}[\delta_0] \cdot \text{sgn}[p(\omega_{o_1})] > 0 \\ \quad \vdots \\ (-1)^{m-1}\text{sgn}[\delta_0] \cdot \text{sgn}[p(\omega_{o_{m-1}})] > 0 \\ (-1)^{m}\text{sgn}[\delta_0] \cdot \text{sgn}[p(\infty)] > 0 \end{cases} \tag{2.4}$$

and

$$\begin{cases} \text{for } n = 2m+1 \\ \text{sgn}[\delta_0] \cdot \text{sgn}[p(0)] > 0 \\ -\text{sgn}[\delta_0] \cdot \text{sgn}[p(\omega_{o_1})] > 0 \\ \quad \vdots \\ (-1)^{m-1}\text{sgn}[\delta_0] \cdot \text{sgn}[p(\omega_{o_{m-1}})] > 0 \\ (-1)^{m}\text{sgn}[\delta_0] \cdot \text{sgn}[p(\omega_{o_m})] > 0 \,. \end{cases} \tag{2.5}$$

From (2.4) and (2.5), it follows that (2.2) holds.

(ii) $\Rightarrow$ (i)
Let $\omega_{o_0} = 0$ and for $n = 2m$, define $\omega_{o_m} = \infty$. Equation (2.2) holds if and only if $[p(\omega_{o_{l-1}})]$ and $[p(\omega_{o_l})]$ are of opposite signs for $l = 1, 2, \cdots, m$. By the continuity of $p(\omega)$, there exists at least one $\omega_e \in \mathcal{R}$, $\omega_{o_{l-1}} < \omega_e < \omega_{o_l}$ such that $p(\omega_e) = 0$. Moreover, since the maximum possible number of non-negative real roots of $p(\cdot)$ is m, it follows that there exists one and only one $\omega_e \in (\omega_{o_{l-1}}, \omega_{o_l})$ such that $p(\omega_e) = 0$, thereby leading us to the interlacing property (2.1).

(2) (i) $\Leftrightarrow$ (iii)
The proof of (2) follows along the same lines as that of (1).

■

The interlacing property in Theorem 2.1 gives a graphical interpretation of the Hermite-Biehler Theorem while Lemma 2.1 gives an equivalent analytical characterization. Note that from Lemma 2.1 if $\delta(s)$ is Hurwitz stable then all the zeros of $p(\omega)$ and $q(\omega)$ must be real and distinct, otherwise (2.2) and (2.3) will fail to hold. Furthermore, the signs of $p(\omega)$ at the successive zero crossings of $q(\omega)$ must alternate. This is also true for the signs of $q(\omega)$ at the successive zero crossings of $p(\omega)$.

Example 2.2 *Consider the same polynomial as in Example 2.1:*

$$\delta(s) = s^7 + 4s^6 + 11s^5 + 29s^4 + 36s^3 + 61s^2 + 34s + 36 .$$

Then

$$\delta(j\omega) = p(\omega) + jq(\omega)$$

where

$$\begin{aligned} p(\omega) &= -4\omega^6 + 29\omega^4 - 61\omega^2 + 36 \\ q(\omega) &= \omega(-\omega^6 + 11\omega^4 - 36\omega^2 + 34) . \end{aligned}$$

We have

$$\omega_{e_1} = 1, \ \omega_{e_2} = 1.5, \ \omega_{e_3} = 2$$

$$\omega_{o_1} = 1.2873, \ \omega_{o_2} = 1.8786, \ \omega_{o_3} = 2.4111$$

and

$$\text{sgn}[p(0)] = 1, \ \text{sgn}[p(\omega_{o_1})] = -1, \ \text{sgn}[p(\omega_{o_2})] = 1, \ \text{sgn}[p(\omega_{o_3})] = -1.$$

Now $\delta(s)$ is of degree $n = 7$ which is odd and

$$\text{sgn}[\delta_0] \cdot [\text{sgn}[p(0)] - 2\text{sgn}[p(\omega_{o_1})] + 2\text{sgn}[p(\omega_{o_2})] - 2\text{sgn}[p(\omega_{o_3})]] = 7$$

which shows that (2.2) holds.
Also, we have

$$\text{sgn}[q(\omega_{e_1})] = 1, \ \text{sgn}[q(\omega_{e_2})] = -1, \ \text{sgn}[q(\omega_{e_3})] = 1, \ \text{sgn}[q(\infty)] = -1$$

so that

$$\text{sgn}[\delta_0] \cdot [2\text{sgn}[q(\omega_{e_1})] - 2\text{sgn}[q(\omega_{e_2})] + 2\text{sgn}[q(\omega_{e_3})] - \text{sgn}[q(\infty)]] = 7.$$

Once again, this checks with (2.3). △

2.3 Generalizations of the Hermite-Biehler Theorem

Consider $\delta(j\omega) = p(\omega) + jq(\omega)$ where $p(\omega)$ and $q(\omega)$ are as illustrated in Fig. 2.3. From this figure we know that the polynomial $\delta(s)$ is not a Hurwitz polynomial because it fails to satisfy the interlacing property since ω_{o_1}, ω_{o_2}, ω_{o_3}, ω_{o_4} are successive roots of $q(\omega)$ between 0 and ω_{e_1}. However, we

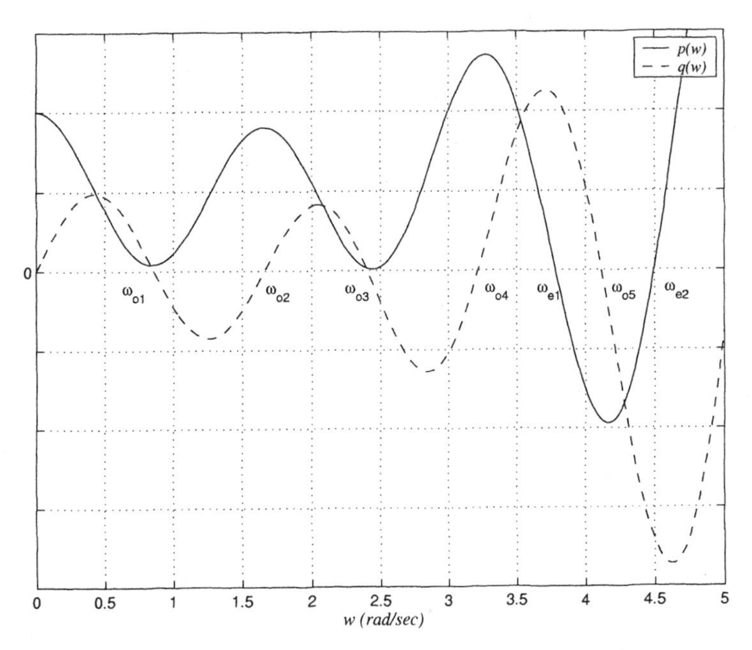

FIGURE 2.3. Interlacing property fails for non-Hurwitz polynomials.

would like to extract, if possible, more information from these plots beyond whether or not $\delta(s)$ is Hurwitz. This has motivated research with the goal of obtaining generalized versions of the Hermite-Biehler Theorem for not necessarily Hurwitz polynomials. In this section, we present some of these generalizations, which are useful for solving special cases of the fixed-order stabilization problem. As a preliminary step to the Generalized Hermite-Biehler Theorems, we introduce some notation and definitions. To this end, let $\mathcal{C}$ denote the complex plane, $\mathcal{C}^-$ the open left half plane, and $\mathcal{C}^+$ the open right half plane.

Consider a real polynomial $\delta(s)$ of degree n:

$$\begin{aligned} \delta(s) \;&=\; \delta_0 + \delta_1 s + \delta_2 s^2 + \cdots + \delta_n s^n,\ \delta_i \in \mathcal{R},\ i = 0, 1, \ldots, n,\ \delta_n \neq 0 \\ &\quad \text{such that } \delta(j\omega) \neq 0,\ \forall \omega \in (-\infty,\ \infty)\,. \end{aligned}$$

Let $p(\omega)$ and $q(\omega)$ be two functions defined pointwise by $p(\omega) = Re[\delta(j\omega)]$, $q(\omega) = Im[\delta(j\omega)]$. With this definition, we have

$$\delta(j\omega) = p(\omega) + jq(\omega) \ \forall\omega .$$

Furthermore,

$$\theta(\omega) \triangleq \angle\delta(j\omega) = \arctan\left[\frac{q(\omega)}{p(\omega)}\right] .$$

Let $\Delta_0^\infty\theta$ denote the net change in the argument $\theta(\omega)$ as ω increases from 0 to ∞ and let $l(\delta)$ and $r(\delta)$ denote the numbers of roots of $\delta(s)$ in $\mathcal{C}^-$ and $\mathcal{C}^+$ respectively. The following lemma shows a fundamental relationship betweeen the net accumulated phase of $\delta(j\omega)$ and the difference between the numbers of roots of the polynomial in $\mathcal{C}^-$ and $\mathcal{C}^+$.

Lemma 2.2 *Let $\delta(s)$ be a real polynomial with no imaginary axis roots. Then*

$$\Delta_0^\infty\theta = \frac{\pi}{2}(l(\delta) - r(\delta)) .$$

Proof. Each $\mathcal{C}^-$ root contributes $+\frac{\pi}{2}$ and each $\mathcal{C}^+$ root contributes $-\frac{\pi}{2}$ to the net change in argument. ■

We now define, mainly for notational convenience, the imaginary and real signatures associated with a real polynomial. These definitions are useful because they facilitate an elegant statement of the generalizations of the Hermite-Biehler Theorem.

Definition 2.2 *Let $\delta(s)$ be any given real polynomial of degree n with k denoting the multiplicity of a root at the origin. Define*

$$p_f(\omega) := \frac{p(\omega)}{(1+\omega^2)^{\frac{n}{2}}},\ q_f(\omega) := \frac{q(\omega)}{(1+\omega^2)^{\frac{n}{2}}}.$$

Let $0 = \omega_0 < \omega_1 < \omega_2 < \cdots < \omega_{m-1}$ be the real, non-negative, distinct finite zeros of $q_f(\omega)$ with odd multiplicities, and define $\omega_m = \infty$. Then the imaginary signature *$\sigma_i(\delta)$ of $\delta(s)$ is defined by*

$$\sigma_i(\delta) := \begin{cases} \{\mathrm{sgn}[p_f^{(k)}(\omega_0)] - 2\mathrm{sgn}[p_f(\omega_1)] + 2\mathrm{sgn}[p_f(\omega_2)] + \cdots \\ +(-1)^{m-1}\cdot 2\mathrm{sgn}[p_f(\omega_{m-1})] + (-1)^m\mathrm{sgn}[p_f(\omega_m)]\} \\ \cdot(-1)^{m-1}\mathrm{sgn}[q(\infty)] \\ \qquad \textit{if } n \textit{ is even} \\ \\ \{\mathrm{sgn}[p_f^{(k)}(\omega_0)] - 2\mathrm{sgn}[p_f(\omega_1)] + 2\mathrm{sgn}[p_f(\omega_2)] + \cdots \\ +(-1)^{m-1}\cdot 2\mathrm{sgn}[p_f(\omega_{m-1})]\}\cdot(-1)^{m-1}\mathrm{sgn}[q(\infty)] \\ \qquad \textit{if } n \textit{ is odd} \end{cases} \tag{2.6}$$

where $p_f^{(k)}(\omega_0) := \frac{d^k}{d\omega^k}[p_f(\omega)]|_{\omega=\omega_0}$.

Definition 2.3 *Let $\delta(s)$ be any given real polynomial of degree n with k denoting the multiplicity of a root at the origin. Let $p_f(\omega)$, $q_f(\omega)$ be as in the last definition, and let $0 < \omega_1 < \omega_2 < \cdots < \omega_{m-1}$ be the real, non-negative, distinct finite zeros of $p_f(\omega)$ with odd multiplicities, and define $\omega_0 = 0$, $\omega_m = \infty$. Then the* real signature $\sigma_r(\delta)$ *of $\delta(s)$ is defined by*

$$\sigma_r(\delta) := \begin{cases} \{\mathrm{sgn}[q_f^{(k)}(\omega_0)] - 2\mathrm{sgn}[q_f(\omega_1)] + 2\mathrm{sgn}[q_f(\omega_2)] + \cdots \\ +(-1)^{m-1}2 \cdot \mathrm{sgn}[q_f(\omega_{m-1})]\} \cdot (-1)^m \mathrm{sgn}[p(\infty)] \\ \qquad\qquad \textit{if } n \textit{ is even} \\ \\ \{\mathrm{sgn}[q_f^{(k)}(\omega_0)] - 2\mathrm{sgn}[q_f(\omega_1)] + 2\mathrm{sgn}[q_f(\omega_2)] + \cdots \\ +(-1)^{m-1}2 \cdot \mathrm{sgn}[q_f(\omega_{m-1})] + (-1)^m \mathrm{sgn}[q_f(\omega_m)]\} \cdot (-1)^m \\ \cdot\mathrm{sgn}[p(\infty)] \qquad \textit{if } n \textit{ is odd} \end{cases} \tag{2.7}$$

where $q_f^{(k)}(\omega_0) := \frac{d^k}{d\omega^k}[q_f(\omega)]|_{\omega=\omega_0}$.

With these definitions, we now present several generalizations of Theorem 2.1 applicable under progressively less restrictive conditions.

2.3.1 No Imaginary Axis Roots

In this subsection, we focus on real polynomials with no imaginary axis roots. Let $p(\omega)$, $q(\omega)$, $p_f(\omega)$, $q_f(\omega)$ be as already defined and let

$$0 = \omega_0 < \omega_1 < \omega_2 < \cdots < \omega_{m-1}$$

be the real, non-negative distinct finite zeros of $q_f(\omega)$ with odd multiplicities, and define $\omega_m = \infty$.

Then we can make the following simple observations:

1. If ω_i, ω_{i+1} are both zeros of $q_f(\omega)$ then

$$\Delta_{\omega_i}^{\omega_{i+1}}\theta = \frac{\pi}{2}\left[\mathrm{sgn}[p_f(\omega_i)] - \mathrm{sgn}[p_f(\omega_{i+1})]\right] \cdot \mathrm{sgn}[q_f(\omega_i^+)] \,. \tag{2.8}$$

2. If ω_i is a zero of $q_f(\omega)$ while ω_{i+1} is not a zero of $q_f(\omega)$, a situation possible only when $\omega_{i+1} = \infty$ is a zero of $p_f(\omega)$ and n is odd, then

$$\Delta_{\omega_i}^{\omega_{i+1}}\theta = \frac{\pi}{2}\mathrm{sgn}[p_f(\omega_i)] \cdot \mathrm{sgn}[q_f(\omega_i^+)] \,. \tag{2.9}$$

3. We also have

$$\mathrm{sgn}[q_f(\omega_{i+1}^+)] = -\mathrm{sgn}[q_f(\omega_i^+)], \; i = 0, 1, 2, \ldots, m-2 \,. \tag{2.10}$$

Equation (2.8) above is obvious while (2.10) simply states that $q_f(\omega)$ changes sign when it passes through a zero of odd multiplicity. Equation (2.9), on the other hand, is a special case of (2.8) for the case where $p_f(\omega_{i+1}) = 0$.

Using (2.10) repeatedly, we obtain

$$\begin{aligned}\operatorname{sgn}[q_f(\omega_i^+)] &= (-1)^{m-1-i} \cdot \operatorname{sgn}[q_f(\omega_{m-1}^+)], \\ & \qquad i = 0,\ 1, \ldots,\ m-1\ . \end{aligned} \tag{2.11}$$

Substituting (2.11) into (2.8), we see that if ω_i, ω_{i+1} are both zeros of $q_f(\omega)$ then

$$\begin{aligned}\Delta_{\omega_i}^{\omega_{i+1}}\theta &= \frac{\pi}{2}\left[\operatorname{sgn}[p_f(\omega_i)] - \operatorname{sgn}[p_f(\omega_{i+1})]\right] \\ &\quad \cdot(-1)^{m-1-i} \cdot \operatorname{sgn}[q_f(\omega_{m-1}^+)]\ . \end{aligned} \tag{2.12}$$

The above observations enable us to state and prove the following.

Theorem 2.2 *Let $\delta(s)$ be a given real polynomial of degree n with no roots on the $j\omega$ axis, that is, the normalized plot $\delta_f(j\omega) = p_f(\omega) + jq_f(\omega)$ does not pass through the origin. Then*

$$l(\delta) - r(\delta) = \sigma_i(\delta) \tag{2.13}$$

and

$$l(\delta) - r(\delta) = \sigma_r(\delta)\ . \tag{2.14}$$

Proof. We note that under the conditions of this theorem, $k = 0$ in Definition 2.2 so that $p_f^{(k)}(\omega_0) = p_f(\omega_0)$. First let us suppose that n is even. Then $\omega_m = \infty$ is a zero of $q_f(\omega)$. By repeatedly using (2.12) to determine $\Delta_0^\infty\theta$, applying Lemma 2.2, and then using the fact that $\operatorname{sgn}[q_f(\omega_{m-1}^+)] = \operatorname{sgn}[q(\infty)]$, it follows that $l(\delta) - r(\delta)$ is equal to the first expression in (2.6). Hence (2.13) holds for n even.

Next let us consider the case in which n is odd. Then $\omega_m = \infty$ is not a zero of $q_f(\omega)$. Hence,

$$\begin{aligned}\Delta_0^\infty\theta &= \sum_{i=0}^{m-2} \Delta_{\omega_i}^{\omega_{i+1}}\theta + \Delta_{\omega_{m-1}}^{\infty}\theta \\ &= \sum_{i=0}^{m-2} \frac{\pi}{2}\left[\operatorname{sgn}[p_f(\omega_i)] - \operatorname{sgn}[p_f(\omega_{i+1})]\right] \cdot (-1)^{m-1-i}\operatorname{sgn}[q_f(\omega_{m-1}^+)] \\ &\quad + \frac{\pi}{2}\operatorname{sgn}[p_f(\omega_{m-1})] \cdot \operatorname{sgn}[q_f(\omega_{m-1}^+)] \\ &\quad \text{(using (2.12) and (2.9))}\ . \end{aligned} \tag{2.15}$$

Applying Lemma 2.2, and then using the fact that $\operatorname{sgn}[q_f(\omega_{m-1}^+)] = \operatorname{sgn}[q(\infty)]$, it follows that $l(\delta) - r(\delta)$ is equal to the second expression in (2.6). Hence (2.13) also holds for n odd.

The proof for (2.14) is omitted here since it follows along the same lines as that of (2.13). Notice that in (2.14), the expression for $l(\delta) - r(\delta)$ is determined using the values of the frequencies where $\delta_f(j\omega)$ crosses the imaginary axis. ■

Remark 2.1 *Theorem 2.2 generalizes Lemma 2.1, parts (ii) and (iii) to the case of not necessarily Hurwitz polynomials. It is precisely in this sense that Theorem 2.2 is a generalization of the Hermite-Biehler Theorem.*

2.3.2 Roots Allowed on the Imaginary Axis Except at the Origin

In this subsection, we extend Theorem 2.2 so that $\delta(s)$ is now allowed to have *nonzero* imaginary axis roots.

Theorem 2.3 *Let $\delta(s)$ be a given real polynomial of degree n with no roots at the origin. Then*

$$l(\delta) - r(\delta) = \sigma_i(\delta) \tag{2.16}$$

and

$$l(\delta) - r(\delta) = \sigma_r(\delta) . \tag{2.17}$$

Proof. We present here the proof of (2.16) since that of (2.17) follows along similar lines. $\delta(s)$ can be factored as

$$\delta(s) = \bar{\delta_o}(s)\bar{\delta_e}(s)\delta^*(s)$$

where $\bar{\delta_o}(s)$ contains all the $j\omega$ axis roots of $\delta(s)$ with odd multiplicities, $\bar{\delta_e}(s)$ contains all the $j\omega$ axis roots of $\delta(s)$ with even multiplicities, while $\delta^*(s)$ has no $j\omega$ axis roots. $\bar{\delta_o}(s)$ and $\bar{\delta_e}(s)$ must necessarily be of the form

$$\bar{\delta_o}(s) = \prod_{i_o=1,2,3,\cdots} (s^2 + \alpha_{i_o}^2)^{n_{i_o}},$$

where $\alpha_{i_o} > 0$, $n_{i_o} \geq 0$, n_{i_o} is odd, $\alpha_1 < \alpha_2 < \cdots$, and

$$\bar{\delta_e}(s) = \prod_{i_e=1,2,3,\cdots} (s^2 + \beta_{i_e}^2)^{n_{i_e}},$$

where $\beta_{i_e} > 0$, $n_{i_e} \geq 0$, n_{i_e} is even.

The proof is carried out in two steps. First we show that multiplying $\delta^*(s)$ by $\bar{\delta_e}(s)$ has no effect on (2.13). Thereafter, we use an inductive argument to show that multiplying $\bar{\delta_e}(s)\delta^*(s)$ by $\bar{\delta_o}(s)$ also does not affect (2.13).

Step I. Define[1]

$$\begin{aligned} \delta_0(s) &= \bar{\delta_e}(s)\delta^*(s) \\ &= \prod_{i_e}(s^2 + \beta_{i_e}^2)^{n_{i_e}}\delta^*(s). \end{aligned}$$

---

[1]Note that in this proof $\delta_0(s), \delta_1(s), \ldots, \delta_k(s)$, etc., represent particular polynomials that should not be confused with the coefficients of $\delta(s)$ defined earlier.

We want to show that $\delta_0(s)$ satisfies (2.16).

Define

$$\begin{aligned}\delta^*(j\omega) &:= p^*(\omega) + jq^*(\omega)\\ \delta_0(j\omega) &:= p_0(\omega) + jq_0(\omega)\end{aligned}$$

so that $p^*(\omega)$, $p_0(\omega)$, $q^*(\omega)$, $q_0(\omega)$ are related by

$$p_0(\omega) = \prod_{i_e}(-\omega^2 + \beta_{i_e}^2)^{n_{ie}} p^*(\omega) \tag{2.18}$$

$$q_0(\omega) = \prod_{i_e}(-\omega^2 + \beta_{i_e}^2)^{n_{ie}} q^*(\omega). \tag{2.19}$$

Let $0 = \omega_0 < \omega_1 < \omega_2 < \cdots < \omega_{m-1}$ be the real, non-negative, distinct finite zeros of $q_f^*(\omega)$ with odd multiplicities. Also define $\omega_m = \infty$. First let us assume that $\delta^*(s)$ is of even degree. Then, from Theorem 2.2, we have

$$\begin{aligned}l(\delta^*) - r(\delta^*) &= \sigma_i(\delta^*)\\ &= \{\text{sgn}[p_f^*(\omega_0)] - 2\text{sgn}[p_f^*(\omega_1)] + 2\text{sgn}[p_f^*(\omega_2)] + \cdots\\ &\quad +(-1)^{m-1}2\text{sgn}[p_f^*(\omega_{m-1})] + (-1)^m\text{sgn}[p_f^*(\omega_m)]\}\\ &\quad \cdot(-1)^{m-1}\text{sgn}[q^*(\infty)].\end{aligned}$$

From (2.19), it follows that ω_i, $i = 0, 1, \ldots, m-1$ are also the real, non-negative, distinct finite zeros of $q_{0_f}(\omega)$ with odd multiplicities. Furthermore, from (2.18) and (2.19), we have

$$\begin{aligned}\text{sgn}[p_f^*(\omega_i)] &= \text{sgn}[p_{0_f}(\omega_i)],\ i = 0, 1, \ldots, m\\ \text{sgn}[q^*(\infty)] &= \text{sgn}[q_0(\infty)].\end{aligned}$$

Since

$$l(\delta_0) - r(\delta_0) = l(\delta^*) - r(\delta^*)$$

it follows that (2.16) is true for $\delta_0(s)$ of even degree. The fact that (2.16) is also true for $\delta_0(s)$ of odd degree can be verified by proceeding along exactly the same lines.

Step II: Proof by Induction. Let the induction index j be equal to 1 and consider

$$\begin{aligned}\delta_1(s) &= (s^2 + \alpha_1^2)^{n_1} \prod_{i_e}(s^2 + \beta_{i_e}^2)^{n_{ie}} \delta^*(s)\\ &= (s^2 + \alpha_1^2)^{n_1} \delta_0(s).\end{aligned} \tag{2.20}$$

Define

$$\delta_1(j\omega) = p_1(\omega) + jq_1(\omega)$$

so that $p_1(\omega)$, $p_0(\omega)$, $q_1(\omega)$, $q_0(\omega)$ are related by

$$\begin{aligned} p_1(\omega) &= (-\omega^2+\alpha_1^2)^{n_1} p_0(\omega) && (2.21)\\ q_1(\omega) &= (-\omega^2+\alpha_1^2)^{n_1} q_0(\omega). && (2.22)\end{aligned}$$

Let $0=\omega_0 < \omega_1 < \omega_2 < \cdots < \omega_{m-1}$ be the real, non-negative, distinct finite zeros of $q_{0_f}(\omega)$ with odd multiplicities. Also define $\omega_m=\infty$. First let us assume that $\delta_0(s)$ has even degree. Then from Step I

$$\begin{aligned} l(\delta_0)-r(\delta_0) &= \sigma_i(\delta_0)\\ &= \{\mathrm{sgn}[p_{0_f}(\omega_0)] - 2\mathrm{sgn}[p_{0_f}(\omega_1)] + 2\mathrm{sgn}[p_{0_f}(\omega_2)] + \cdots\\ &\quad +(-1)^{m-1}2\mathrm{sgn}[p_{0_f}(\omega_{m-1})] + (-1)^m\mathrm{sgn}[p_{0_f}(\omega_m)]\}\\ &\quad \cdot(-1)^{m-1}\mathrm{sgn}[q_0(\infty)]. && (2.23)\end{aligned}$$

From (2.22), it follows that ω_i, $i=0,1,\ldots,m-1$; α_1 are the real, non-negative, distinct finite zeros of $q_{1_f}(\omega)$ with odd multiplicities. Let us assume that $\omega_l < \alpha_1 < \omega_{l+1}$. Then from (2.21) and (2.22), we have

$$\left.\begin{aligned} \mathrm{sgn}[p_{0_f}(\omega_i)] &= \mathrm{sgn}[p_{1_f}(\omega_i)],\ i=0,1,\ldots,l\\ \mathrm{sgn}[p_{0_f}(\omega_i)] &= -\mathrm{sgn}[p_{1_f}(\omega_i)],\ i=l+1,l+2,\ldots,m\\ \mathrm{sgn}[p_{1_f}(\alpha_1)] &= 0\\ \mathrm{sgn}[q_0(\infty)] &= -\mathrm{sgn}[q_1(\infty)].\end{aligned}\right\} \quad (2.24)$$

Since $l(\delta_1)-r(\delta_1)=l(\delta_0)-r(\delta_0)$, using (2.23), (2.24), we obtain

$$\begin{aligned} l(\delta_1)-r(\delta_1) &= \{\mathrm{sgn}[p_{1_f}(\omega_0)] - 2\mathrm{sgn}[p_{1_f}(\omega_1)] + 2\mathrm{sgn}[p_{1_f}(\omega_2)]\\ &\quad + \cdots + (-1)^l 2\mathrm{sgn}[p_{1_f}(\omega_l)] + (-1)^{l+1}2\mathrm{sgn}[p_{1_f}(\alpha_1)]\\ &\quad +(-1)^{l+2}2\mathrm{sgn}[p_{1_f}(\omega_{l+1})] + \cdots\\ &\quad +(-1)^m 2\mathrm{sgn}[p_{1_f}(\omega_{m-1})] + (-1)^{m+1}\mathrm{sgn}[p_{1_f}(\omega_m)]\}\\ &\quad \cdot(-1)^m\mathrm{sgn}[q_1(\infty)]\\ &= \sigma_i(\delta_1)\end{aligned}$$

which shows that (2.16) is true for $\delta_1(s)$ of even degree. The fact that (2.16) holds for $\delta_1(s)$ of odd degree can be verified likewise. This completes the first step of the induction argument.

Now let $j=k$ and consider

$$\delta_k(s) = \prod_{i_o=1}^{k}(s^2+\alpha_{i_o}^2)^{n_{i_o}} \prod_{i_e}(s^2+\beta_{i_e}^2)^{n_{i_e}}\delta^*(s). \quad (2.25)$$

Assume that (2.16) is true for $\delta_k(s)$ (inductive assumption). Then

$$\begin{aligned} \delta_{k+1}(s) &= \prod_{i_o=1}^{k+1}(s^2+\alpha_{i_o}^2)^{n_{i_o}} \prod_{i_e}(s^2+\beta_{i_e}^2)^{n_{i_e}}\delta^*(s)\\ &= (s^2+\alpha_{k+1}^2)^{n_{k+1}}\delta_k(s). && (2.26)\end{aligned}$$

Define

$$\begin{aligned}\delta_k(j\omega) &= p_k(\omega) + jq_k(\omega)\\ \delta_{k+1}(j\omega) &= p_{k+1}(\omega) + jq_{k+1}(\omega)\end{aligned}$$

so that $p_{k+1}(\omega)$, $p_k(\omega)$, $q_{k+1}(\omega)$, $q_k(\omega)$ are related by

$$p_{k+1}(\omega) = (-\omega^2 + \alpha_{k+1}^2)^{n_{k+1}} p_k(\omega) \tag{2.27}$$
$$q_{k+1}(\omega) = (-\omega^2 + \alpha_{k+1}^2)^{n_{k+1}} q_k(\omega). \tag{2.28}$$

Let $0 = \omega_0 < \omega_1 < \omega_2 < \cdots < \omega_{m-1}$ be the real, non-negative, distinct finite zeros of $q_{k_f}(\omega)$ with odd multiplicities. Also define $\omega_m = \infty$. First let us assume that $\delta_k(s)$ is of even degree. Then from the inductive assumption, we have

$$\begin{aligned}l(\delta_k) - r(\delta_k) &= \sigma_i(\delta_k)\\ &= \{\mathrm{sgn}[p_{k_f}(\omega_0)] - 2\mathrm{sgn}[p_{k_f}(\omega_1)] + 2\mathrm{sgn}[p_{k_f}(\omega_2)]\\ &\quad + \cdots + (-1)^{m-1} 2\mathrm{sgn}[p_{k_f}(\omega_{m-1})] + (-1)^m\\ &\quad \mathrm{sgn}[p_{k_f}(\omega_m)]\} \cdot (-1)^{m-1}\mathrm{sgn}[q_k(\infty)].\end{aligned} \tag{2.29}$$

Now from (2.28), it follows that ω_i, $i = 0, 1, \ldots, m-1$; α_{k+1} are the real, non-negative, distinct finite zeros of $q_{k+1_f}(\omega)$ with odd multiplicities. Let us assume that $\omega_l < \alpha_{k+1} < \omega_{l+1}$. Then from (2.27) and (2.28), we have

$$\left.\begin{aligned}\mathrm{sgn}[p_{k_f}(\omega_i)] &= \mathrm{sgn}[p_{k+1_f}(\omega_i)],\ i = 0, 1, \ldots, l\\ \mathrm{sgn}[p_{k_f}(\omega_i)] &= -\mathrm{sgn}[p_{k+1_f}(\omega_i)],\ i = l+1, \ldots, m\\ \mathrm{sgn}[p_{k+1_f}(\alpha_{k+1})] &= 0\\ \mathrm{sgn}[q_k(\infty)] &= -\mathrm{sgn}[q_{k+1}(\infty)].\end{aligned}\right\} \tag{2.30}$$

Since $l(\delta_{k+1}) - r(\delta_{k+1}) = l(\delta_k) - r(\delta_k)$, using (2.29), (2.30), we obtain

$$\begin{aligned}l(\delta_{k+1}) - r(\delta_{k+1}) &= \{\mathrm{sgn}[p_{k+1_f}(\omega_0)] - 2\mathrm{sgn}[p_{k+1_f}(\omega_1)]\\ &\quad +2\mathrm{sgn}[p_{k+1_f}(\omega_2)] + \cdots + (-1)^l 2\mathrm{sgn}[p_{k+1_f}(\omega_l)]\\ &\quad +(-1)^{l+1} \cdot 2\mathrm{sgn}[p_{k+1_f}(\alpha_{k+1})]\\ &\quad +(-1)^{l+2} 2\mathrm{sgn}[p_{k+1_f}(\omega_{l+1})] + \cdots\\ &\quad +(-1)^m 2\mathrm{sgn}[p_{k+1_f}(\omega_{m-1})] + (-1)^{m+1}\\ &\quad \cdot\mathrm{sgn}[p_{k+1_f}(\omega_m)]\} \cdot (-1)^m \mathrm{sgn}[q_{k+1}(\infty)]\\ &= \sigma_i(\delta_{k+1})\end{aligned} \tag{2.31}$$

which shows that (2.16) is true for $\delta_{k+1}(s)$ of even degree. The fact that (2.16) is true for $\delta_{k+1}(s)$ of odd degree can be similarly verified. This completes the induction argument and hence the proof. ∎

2.3.3 No Restriction on Root Locations

Theorem 2.3 presented in the last subsection requires that the polynomial $\delta(s)$ have no roots at the origin. In this subsection, we show that such restrictions can be removed.

Theorem 2.4 *Let $\delta(s)$ be a given real polynomial of degree n. Then*

$$l(\delta) - r(\delta) = \sigma_i(\delta) \tag{2.32}$$

and

$$l(\delta) - r(\delta) = \sigma_r(\delta). \tag{2.33}$$

Moreover, $\delta(s)$ is Hurwitz if and only if $\sigma_i(\delta) = \sigma_r(\delta) = n$.

Proof. A proof is presented only for the first part of the theorem. The proof for the second part is similar and is therefore omitted. If $\delta(s)$ has no roots at the origin, then (2.32) follows from Theorem 2.3. Let us assume that $\delta(s)$ has a root of multiplicity k at the origin. Then we can write

$$\delta(s) = s^k \delta^*(s)$$

where $\delta^*(s)$ is a real polynomial of degree n^* with no roots at the origin. Define

$$\begin{aligned} \delta^*(j\omega) &:= p^*(\omega) + jq^*(\omega) \\ \delta(j\omega) &:= p(\omega) + jq(\omega). \end{aligned}$$

The proof can be completed by considering four different cases, namely $k = 4l$, $k = 4l + 1$, $k = 4l + 2$, and $k = 4l + 3$. These four cases correspond to the four different ways in which multiplication by $(j\omega)^k$ affects the real and imaginary parts of $\delta^*(j\omega)$. Due to the fact that each of these cases is handled by proceeding along similar lines, we do not treat all of the cases here. Instead, we focus on a representative case, say $k = 4l+1$, and provide a detailed treatment for it.

For $k = 4l + 1$, we have

$$\begin{aligned} \delta(j\omega) &= p(\omega) + jq(\omega) \\ &= -\omega^{4l+1} q^*(\omega) + j\omega^{4l+1} p^*(\omega). \end{aligned}$$

First let us assume that n^* is even. Then, from Theorem 2.3, we have

$$\begin{aligned} l(\delta^*) - r(\delta^*) &= \sigma_r(\delta^*) \\ &= -\{2\mathrm{sgn}[q_f^*(\omega_1)] - 2\mathrm{sgn}[q_f^*(\omega_2)] + \cdots \\ &\quad +(-1)^{m-2} 2\mathrm{sgn}[q_f^*(\omega_{m-1})]\} \\ &\quad \cdot(-1)^m \mathrm{sgn}[p^*(\infty)] \end{aligned} \tag{2.34}$$

where $0 < \omega_1 < \omega_2 < \cdots < \omega_{m-1}$ are the real, non-negative, distinct finite zeros of $p_f^*(\omega)$ with odd multiplicities.

Define $w_0 := 0$. Since

$$\begin{aligned} p(\omega) &= -\omega^{4l+1}q^*(\omega) \\ p^{(4l+1)}(\omega_0) &= -(4l+1)!\,q^*(\omega_0) = 0 \\ q(\omega) &= \omega^{4l+1}p^*(\omega) \text{ and} \\ p_f^{(4l+1)}(\omega_0) &= p^{(4l+1)}(\omega_0) \end{aligned}$$

we have

$$\begin{aligned} \operatorname{sgn}[p_f^{(4l+1)}(\omega_0)] &= 0 && (2.35) \\ \operatorname{sgn}[q_f^*(\omega_i)] &= -\operatorname{sgn}[p_f(\omega_i)],\ i = 1, 2, \ldots, m && (2.36) \\ \operatorname{sgn}[p^*(\infty)] &= \operatorname{sgn}[q(\infty)]. && (2.37) \end{aligned}$$

Since n^* is even and $k = 4l+1$, it follows that n is odd. Moreover, since

$$l(\delta) - r(\delta) = l(\delta^*) - r(\delta^*),$$

using (2.34), (2.35), (2.36), (2.37), we have

$$\begin{aligned} l(\delta) - r(\delta) &= \{\operatorname{sgn}[p_f^{(k)}(\omega_0)] - 2\operatorname{sgn}[p_f(\omega_1)] + 2\operatorname{sgn}[p_f(\omega_2)] + \cdots \\ &\quad + (-1)^{m-1}2\operatorname{sgn}[p_f(\omega_{m-1})]\} \cdot (-1)^{m-1}\operatorname{sgn}[q(\infty)] \\ &= \sigma_i(\delta) && (2.38) \end{aligned}$$

which shows that (2.32) holds for $\delta(s)$ of odd degree. The fact that (2.32) also holds for $\delta(s)$ of even degree or equivalently n^* odd can be verified by proceeding along exactly the same lines. ■

Remark 2.2 *In view of the above theorem, it is clear that for any real polynomial $\delta(s)$, the value of the real signature $\sigma_r(\delta)$ is equal to the value of the imaginary signature $\sigma_i(\delta)$ and each is in fact equal to $l(\delta) - r(\delta)$. Thus either of them could be referred to as the signature $\sigma(\delta)$ of $\delta(s)$ with the subscript "r" or "i" indicating the formula that is being used in a particular situation to compute the value.*

Example 2.3 *Consider the real polynomial*

$$\delta(s) = s^3(s^2+1)^2(s^2+5)(s-3)(s^2+s+1)$$

of degree $n = 12$. Substituting $s = j\omega$, we have

$$\delta(j\omega) = p(\omega) + jq(\omega)$$

where

$$p(\omega) = \omega^{12} - 5\omega^{10} - 3\omega^8 + 17\omega^6 - 10\omega^4$$

and

$$q(\omega) = 2\omega^{11} - 17\omega^9 + 43\omega^7 - 43\omega^5 + 15\omega^3.$$

The real, positive finite zeros of $q_f(\omega)$ *with odd multiplicities are* $\omega_1 = 1.2247$ *and* $\omega_2 = 2.2361$. *Also define* $\omega_0 = 0$ *and* $\omega_3 = \infty$. *Since* $\delta(s)$ *is of even degree and with a root at the origin of multiplicity* 3, *from formula (2.6), it follows that*

$$\begin{aligned}\sigma_i(\delta) &= \{\mathrm{sgn}[p_f^{(3)}(\omega_0)] - 2\mathrm{sgn}[p_f(\omega_1)] + 2\mathrm{sgn}[p_f(\omega_2)] - \mathrm{sgn}[p_f(\omega_3)]\} \\ &\quad \cdot(-1)^2\mathrm{sgn}[q(\infty)] \ .\end{aligned}$$

Then, we have

$$\mathrm{sgn}[p_f^{(3)}(\omega_0)] = 0, \mathrm{sgn}[p_f(\omega_1)] = -1, \mathrm{sgn}[p_f(\omega_2)] = 0 \text{ and } \mathrm{sgn}[p_f(\omega_3)] = 1.$$

Hence,

$$\sigma_i(\delta) = 0 + 2 + 0 - 1 = 1 \ .$$

This agrees with the value for $l(\delta) - r(\delta) = 2 - 1$ *obtained from visual inspection of the factored form of* $\delta(s)$, *so Theorem 2.4 is verified. Finally, since* $\sigma_i(\delta) \neq 12$, *the polynomial is non-Hurwitz.* $\triangle$

2.4 Notes and References

The proof of the classical Hermite-Biehler Theorem can be found in [13]. For an alternative proof using the Boundary Crossing Theorem the reader is referred to [5]. The generalizations of the Hermite-Biehler Theorem presented in Section 2.3 are due to Ho, Datta, and Bhattacharyya [17]. A formula for $l(\delta) - r(\delta)$ appeared first in [34], for the case of polynomials with no roots on the imaginary axis and at most one root at the origin. Most of the material presented in this chapter is based on [10].

3

PI Stabilization of Delay-Free Linear Time-Invariant Systems

In this chapter we utilize the Generalized Hermite-Biehler Theorem presented in the previous chapter to give a solution to the problem of feedback stabilization of a given finite-dimensional linear-time invariant plant by a constant gain controller and by a PI controller. In each case the complete set of stabilizing solutions is found.

3.1 Introduction

In this chapter we first provide a complete analytical solution to the problem of stabilizing a given plant described by a rational transfer function using a constant gain (or zeroth order) controller. The solution derived in this chapter is based on the Generalized Hermite-Biehler Theorem developed in Chapter 2. This solution along with some illustrative examples are presented in Section 3.2. In Section 3.3 we derive a computational characterization of all stabilizing PI controllers. This characterization is in a quasi-closed form and can be used to optimize various performance criteria when the controller structure is constrained to be of the PI type.

The results of this chapter fill an important gap in control theory. In part, they are motivated by the limitations of modern optimal control techniques, which cannot accommodate constraints on the controller order or structure into their design methods. Because of this fact, techniques such as H_2 or H_∞ cannot currently be used for designing optimal or robust PI controllers.

3.2 A Characterization of All Stabilizing Feedback Gains

In this section we utilize the Generalized Hermite-Biehler Theorem to give a solution to the problem of feedback stabilization of a given linear time-invariant plant by a constant gain controller. Even though this problem can be solved using classical approaches such as the Nyquist stability criterion and the Routh-Hurwitz criterion, it is not clear how to extend these methods to the more complicated cases where PI or PID controllers are involved. By using the Generalized Hermite-Biehler Theorem, an elegant procedure is developed that can be extended to the aforementioned cases.

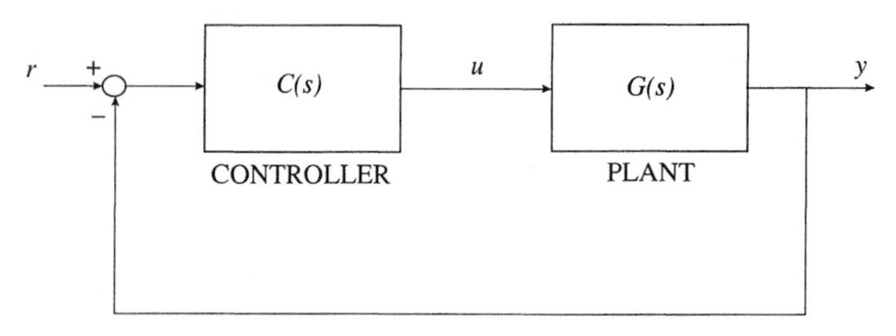

FIGURE 3.1. Feedback control system.

To this end, consider the feedback system shown in Fig. 3.1. Here r is the command signal, y is the output,

$$G(s) = \frac{N(s)}{D(s)}$$

is the plant to be controlled, $N(s)$ and $D(s)$ are coprime polynomials, and $C(s)$ is the controller to be designed. In the case of constant gain stabilization,

$$C(s) = k$$

so that the closed-loop characteristic polynomial $\delta(s,k)$ is given by

$$\delta(s,k) = D(s) + kN(s)\,. \tag{3.1}$$

Our objective is to determine those values of k, if any, for which the closed-loop system is stable, that is, $\delta(s,k)$ is Hurwitz.

If we now consider the even-odd decompositions of $N(s)$ and $D(s)$

$$\begin{aligned} N(s) &= N_e(s^2) + sN_o(s^2) \\ D(s) &= D_e(s^2) + sD_o(s^2) \end{aligned}$$

then, (3.1) can be rewritten as

$$\delta(s,k) = [kN_e(s^2) + D_e(s^2)] + s[kN_o(s^2) + D_o(s^2)]\,.$$

It is clear from this expression that both the even as well as the odd parts of $\delta(s,k)$ depend on k. This creates difficulties when trying to use Lemma 2.1 to ensure the Hurwitz stability of $\delta(s,k)$. To overcome this problem, we will now construct a polynomial for which only the even part depends on k, and to which Theorem 2.4 is applicable.

Suppose that the degree of $D(s)$ is n while the degree of $N(s)$ is m and $m \leq n$. Define

$$N^*(s) := N(-s) = N_e(s^2) - sN_o(s^2).$$

Multiplying $\delta(s,k)$ by $N^*(s)$ we obtain the following result.

Lemma 3.1 *$\delta(s,k)$ is Hurwitz if and only if*

$$\sigma_i(\delta(s,k)N^*(s)) = n - (l(N(s)) - r(N(s))). \tag{3.2}$$

Proof. Since $l(a(s) \cdot b(s)) = l(a(s)) + l(b(s))$ and $r(a(s) \cdot b(s)) = r(a(s)) + r(b(s))$, we have

$$\begin{aligned} l(\delta(s,k)N^*(s)) - r(\delta(s,k)N^*(s)) &= l(\delta(s,k)) - r(\delta(s,k)) \\ &\quad + l(N^*(s)) - r(N^*(s)) \\ &= l(\delta(s,k)) - r(\delta(s,k)) \\ &\quad + l(N(-s)) - r(N(-s)) \\ &= l(\delta(s,k)) - r(\delta(s,k)) \\ &\quad - (l(N(s)) - r(N(s))). \end{aligned}$$

Now, $\delta(s,k)$ of degree n is Hurwitz if and only if $l(\delta(s,k)) = n$ and $r(\delta(s,k)) = 0$. Furthermore, from Theorem 2.4

$$\sigma_i(\delta(s,k)N^*(s)) = l(\delta(s,k)N^*(s)) - r(\delta(s,k)N^*(s)).$$

Thus

$$\sigma_i(\delta(s,k)N^*(s)) = n - (l(N(s)) - r(N(s))) .$$

■

In order to solve our stabilization problem, we need to determine those values of k, if any, for which (3.2) holds. Notice that in this expression, the values of n and $l(N(s)) - r(N(s))$ are known and fixed.

Using the even-odd decompositions of $N(s)$ and $D(s)$ we have

$$\delta(s,k)N^*(s) = h_1(s^2) + kh_2(s^2) + sg_1(s^2)$$

where

$$\begin{aligned} h_1(s^2) &= D_e(s^2)N_e(s^2) - s^2D_o(s^2)N_o(s^2) \\ h_2(s^2) &= N_e(s^2)N_e(s^2) - s^2N_o(s^2)N_o(s^2) \\ g_1(s^2) &= N_e(s^2)D_o(s^2) - D_e(s^2)N_o(s^2). \end{aligned}$$

Substituting $s = j\omega$, we obtain

$$\delta(j\omega, k)N^*(j\omega) = p(\omega, k) + jq(\omega) \tag{3.3}$$

where

$$\begin{aligned} p(\omega, k) &= p_1(\omega) + kp_2(\omega) && (3.4)\\ p_1(\omega) &= [D_e(-\omega^2)N_e(-\omega^2) + \omega^2 D_o(-\omega^2)N_o(-\omega^2)] && (3.5)\\ p_2(\omega) &= [N_e(-\omega^2)N_e(-\omega^2) + \omega^2 N_o(-\omega^2)N_o(-\omega^2)] && (3.6)\\ q(\omega) &= \omega[N_e(-\omega^2)D_o(-\omega^2) - D_e(-\omega^2)N_o(-\omega^2)]\,. && (3.7) \end{aligned}$$

Also, define

$$\begin{aligned} p_f(\omega, k) &= \frac{p(\omega, k)}{(1+\omega^2)^{\frac{m+n}{2}}} \\ q_f(\omega) &= \frac{q(\omega)}{(1+\omega^2)^{\frac{m+n}{2}}}\,. \end{aligned}$$

Note that the zeros of the imaginary part $q(\omega)$ are independent of k. For clarity of presentation, we first introduce some definitions before formally stating the main result of this section.

Definition 3.1 *Let the integers m, n and the function $q_f(\omega)$ be as already defined. Let $0 = \omega_0 < \omega_1 < \omega_2 < \cdots < \omega_{l-1}$ be the real, non-negative, distinct finite zeros of $q_f(\omega)$ with odd multiplicities. Define a sequence of numbers $i_0, i_1, i_2, \cdots, i_l$ as follows:*

(i)

$$i_0 = \begin{cases} \operatorname{sgn}[p_{1_f}^{(k_n)}(0)] & \text{if } N^*(s) \text{ has a zero of multiplicity } k_n \text{ at the origin} \\ \alpha & \text{otherwise} \end{cases}$$

where $\alpha \in \{-1, 1\}$ and

$$p_{1_f}(\omega) := \frac{p_1(\omega)}{(1+\omega^2)^{\frac{(m+n)}{2}}}\,,$$

(ii) For $t = 1, 2, \ldots, l-1$:

$$i_t = \begin{cases} 0 & \text{if } N^*(j\omega_t) = 0 \\ \alpha & \text{otherwise} \end{cases}\,,$$

(iii)

$$i_l = \begin{cases} \alpha & \text{if } n+m \text{ is even} \\ 0 & \text{if } n+m \text{ is odd} \end{cases}\,.$$

With $i_0, i_1, \ldots$ defined in this way, we define the string $\mathcal{I} : N \rightarrow R$ as the following sequence of numbers:

$$\mathcal{I} := \{i_0, i_1, \ldots, i_l\} .$$

Define A to be the set of all possible strings $\mathcal{I}$ that can be generated to satisfy the preceding requirements.

Next we introduce the "imaginary signature" $\gamma(\mathcal{I})$ associated with any element $\mathcal{I} \in A$. This definition is motivated by Theorem 3.1 to follow.

Definition 3.2 *Let the integers m, n and the functions $q(\omega)$, $q_f(\omega)$ be as already defined. Let $0 = \omega_0 < \omega_1 < \omega_2 < \cdots < \omega_{l-1}$ be the real, non-negative, distinct finite zeros of $q_f(\omega)$ with odd multiplicities. Also define $\omega_l = \infty$. For each string $\mathcal{I} = \{i_0, i_1, \ldots\}$ in A, let $\gamma(\mathcal{I})$ denote the "imaginary signature" associated with the string $\mathcal{I}$ defined by*

$$\gamma(\mathcal{I}) := [i_0 - 2i_1 + 2i_2 + \cdots + (-1)^{l-1} 2i_{l-1} + (-1)^l i_l] \cdot (-1)^{l-1} \mathrm{sgn}[q(\infty)] . \tag{3.8}$$

Remark 3.1 *Note that if we make the identification $i_0 = \mathrm{sgn}[p_f^{(k_n)}(0, k)]$, $i_t = \mathrm{sgn}[p_f(\omega_t, k)]$ for $t \neq 0$, then the imaginary signature of $\delta(s, k)N^*(s)$ as determined from (2.6) is the same as the quantity $\gamma(\mathcal{I})$ defined above. Hence, referring to $\gamma(\mathcal{I})$ as the "imaginary signature" of $\mathcal{I}$ is appropriate terminology.*

Definition 3.3 *The set F^* of feasible strings for the constant gain stabilization problem is defined as*

$$F^* = \{\mathcal{I} \in A | \gamma(\mathcal{I}) = n - (l(N(s)) - r(N(s)))\} .$$

The following example illustrates these definitions.

Example 3.1 *From (3.3), we have*

$$\delta(j\omega, k)N^*(j\omega) = p(\omega, k) + jq(\omega)$$

where

$$p(\omega, k) = p_1(\omega) + kp_2(\omega).$$

Now suppose, for example, that $\delta(s, k)$ is of degree $n = 6$, the degree of $N^*(s)$ is $m = 4$, and $l(N(s)) - r(N(s)) = 2$. Let $q(\omega)$ have three real, non-negative, distinct finite zeros ω_0, ω_1, ω_2 with odd multiplicities, let $\omega_3 = \infty$ and $\mathrm{sgn}[(q(\infty)] = 1$. Also let $N^*(j\omega_i) \neq 0$ for $i = 0, 1, 2$, so that $N^*(s)$ has no zeros on the imaginary axis.

Since $m+n=10$ is even, the strings $\mathcal{I}$ will have the structure $\{i_0, i_1, i_2, i_3\}$, where $i_t \in \{-1, 1\}$, $t = 0, 1, 2, 3$. The set A of all possible strings $\mathcal{I}$ with this structure is

$$A = \left\{ \begin{array}{ll} \{-1,-1,-1,-1\} & \{1,-1,-1,-1\} \\ \{-1,-1,-1,1\} & \{1,-1,-1,1\} \\ \{-1,-1,1,-1\} & \{1,-1,1,-1\} \\ \{-1,-1,1,1\} & \{1,-1,1,1\} \\ \{-1,1,-1,-1\} & \{1,1,-1,-1\} \\ \{-1,1,-1,1\} & \{1,1,-1,1\} \\ \{-1,1,1,-1\} & \{1,1,1,-1\} \\ \{-1,1,1,1\} & \{1,1,1,1\} \end{array} \right\}.$$

Since

$$(-1)^{(l-1)}\mathrm{sgn}[q(\infty)] = 1,$$

it follows using Definition 3.2 that the imaginary signature of every string $\mathcal{I}$ is given by

$$\gamma(\mathcal{I}) = i_0 - 2i_1 + 2i_2 - i_3 \,.$$

From Lemma 3.1, we have $\delta(s,k)$ is Hurwitz if and only if

$$\sigma_i(\delta(s,k)N^*(s)) = n - (l(N(s)) - r(N(s))) = 4.$$

Thus the set F^* of strings that have $\gamma(\mathcal{I}) = 4$ is given by

$$F^* = \{\{-1,-1,1,-1\},\{1,-1,1,1\}\}.$$

Therefore, the constant gain stabilization problem now reduces to the problem of determining the values of k, if any, such that $\mathrm{sgn}[p_f(\omega_j,k)] = i_j$, $j = 0,\ 1,\ 2,\ 3$ and $\{i_0, i_1, i_2, i_3\} \in F^*$. △

We are now ready to state the main result of this section.

Theorem 3.1 (Constant Gain Stabilization) *The constant gain feedback stabilization problem is solvable for a given plant with rational transfer function $G(s)$ if and only if the following conditions hold:*

(i) F^* is not empty, i.e., at least one feasible string exists and

(ii) there exists a string $\mathcal{I} = \{i_0, i_1, \ldots\} \in F^*$ such that

$$\max_{\{t:i_t>0\}} (L_t) < \min_{\{t:i_t<0\}} (U_t)$$

where

$$L_t := -\frac{p_1(\omega_t)}{p_2(\omega_t)} \quad \text{for } i_t \in \mathcal{I},\ i_t > 0$$

$$U_t := -\frac{p_1(\omega_t)}{p_2(\omega_t)} \quad \text{for } i_t \in \mathcal{I},\ i_t < 0$$

$p_1(\omega_t)$, $p_2(\omega_t)$ are given by (3.5) and (3.6), respectively, and $\omega_0, \omega_1, \omega_2, \ldots$ are as already defined.

Furthermore, if the above conditions are satisfied by the feasible strings $\mathcal{I}_1, \mathcal{I}_2, \ldots, \mathcal{I}_s \in F^*$, then the set of all stabilizing gains is given by $K = \cup_{r=1}^{s} K_r$ where

$$K_r = \left(\max_{\{t: i_t > 0, i_t \in \mathcal{I}_r\}} (L_t), \min_{\{t: i_t < 0, i_t \in \mathcal{I}_r\}} (U_t) \right) \quad r = 1, 2, \ldots, s \, .$$

Proof. From (3.2), we know that $\delta(s,k)$ is Hurwitz if and only if

$$\sigma_i(\delta(s,k)N^*(s)) = n - (l(N(s)) - r(N(s))).$$

Thus $\delta(s,k)$ is Hurwitz if and only if $\mathcal{I} \in F^*$, where (see Definition 3.3)

$$F^* = \{\mathcal{I} \in A/\gamma(\mathcal{I}) = n - (l(N(s)) - r(N(s)))\}$$

and

$$\begin{aligned} \mathcal{I} &= \{i_0, i_1, \ldots\} \\ i_0 &= \operatorname{sgn}[p_f^{(k_n)}(\omega_0, k)] \\ i_j &= \operatorname{sgn}[p_f(\omega_j, k)], \quad \text{for } j = 1, 2, \ldots, l-1 \\ i_l &= \begin{cases} \operatorname{sgn}[p_f(\omega_l, k)] & \text{if } n+m \text{ is even} \\ 0 & \text{if } n+m \text{ is odd} \end{cases} . \end{aligned}$$

Let us now consider two different cases.

Case 1: $N^*(s)$ does not have any zeros on the imaginary axis. In this case, for all stabilizing values of the gain k, $\delta(s,k)N^*(s)$ will also not have any zeros on the $j\omega$ axis so that $i_j \in \{-1, 1\}$ for $j = 0, 1, 2, \ldots, l-1$, and $i_l \in \{-1, 0, 1\}$. Next we consider the two different possibilities:

(a) If $i_j > 0$, then the stability requirement is

$$p_1(\omega_j) + kp_2(\omega_j) \; > \; 0.$$

From (3.6), we note that

$$p_2(\omega) = |N(j\omega)|^2.$$

Since $N^*(s)$ does not have any zeros on the $j\omega$ axis, it follows that $p_2(\omega_j) > 0$. Hence

$$k \; > \; -\frac{p_1(\omega_j)}{p_2(\omega_j)}. \tag{3.9}$$

(b) If $i_j < 0$, then the stability requirement is

$$p_1(\omega_j) + kp_2(\omega_j) \; < \; 0.$$

Once again, since $p_2(\omega_j) > 0$, it follows that

$$k < -\frac{p_1(\omega_j)}{p_2(\omega_j)}. \tag{3.10}$$

Case 2: $N^*(s)$ has one or more zeros on the $j\omega$ axis including a zero of multiplicity k_n at the origin. In this case, for all stabilizing values of the gain k, $\delta(s,k)N^*(s)$ will also have the same set of $j\omega$-axis zeros. Furthermore, it is clear that these zero locations will be a subset of $\{\omega_0, \omega_1, \ldots, \omega_{l-1}\}$. Since the location of these zeros depends on $N^*(s)$ and is independent of the gain k, it is reasonable to expect that such a zero, at ω_m say, will not impose any additional constraint on k. Instead it will only mandate that $i_m \in \mathcal{I}$ be constrained to a particular value. We next proceed to establish rigorously these facts. We consider two possibilities:

(a) $m \neq 0$. Here $N^*(s)$ has a zero at $j\omega_m$ where $\omega_m \neq 0$. This implies that

$$N_e(-\omega_m^2) = N_o(-\omega_m^2) = 0$$

so that from (3.5), (3.6) we obtain

$$p_1(\omega_m) = 0 \text{ and } p_2(\omega_m) = 0.$$

Thus from (3.4), it follows that

$$p(\omega_m, k) = 0.$$

Thus $i_m = 0$ independent of k and this constraint on $\mathcal{I}$ was already incorporated into the definition of A.

(b) $m = 0$. Here $N^*(s)$ has a zero at the origin of multiplicity k_n. Since

$$N^*(j\omega) = N_e(-\omega^2) - j\omega N_o(-\omega^2)$$

it follows that $N_e(-\omega^2)$ and $\omega N_o(-\omega^2)$ must each have zeros at the origin of multiplicity at least k_n. Thus from (3.6), we see that $p_2(\omega)$ will have a zero at the origin of multiplicity $2k_n$ so that for $k_n \geq 1$,

$$p_{2_f}^{(k_n)}(0) = 0.$$

Since

$$p_f^{(k_n)}(0,k) = p_{1_f}^{(k_n)}(0) + k p_{2_f}^{(k_n)}(0)$$

it follows that for $k_n \geq 1$

$$p_f^{(k_n)}(0,k) = p_{1_f}^{(k_n)}(0)$$

independent of k. Thus, although no constraints on k appear, we must have

$$i_0 = \mathrm{sgn}[p_{1_f}^{(k_n)}(0)].$$

Once again, we note that this condition has been explicitly incorporated into the definition of the set A.

Of the two cases discussed above, only Case 1 imposes constraints on k as given by (3.9) and (3.10). This leads us to the conclusion that each $i_j > 0$ in the string $\mathcal{I} \in F^*$ contributes a lower bound on k while each $i_j < 0$ contributes an upper bound on k. Thus, if the string $\mathcal{I} \in F^*$ is to correspond to a stabilizing k then we must have

$$\max_{i_t \in \mathcal{I}, i_t > 0} \left[-\frac{p_1(\omega_t)}{p_2(\omega_t)} \right] < \min_{i_t \in \mathcal{I}, i_t < 0} \left[-\frac{p_1(\omega_t)}{p_2(\omega_t)} \right] \tag{3.11}$$

which is condition (ii) in the theorem statement. This completes the proof of the necessary and sufficient conditions for the existence of a stabilizing k. The set of all stabilizing ks is now determined by taking the union of all ks that are obtained from all the feasible strings that satisfy (ii). ■

Remark 3.2 *Since*

$$G(s) = \frac{N(s)}{D(s)} = \frac{N_e(s^2) + sN_o(s^2)}{D_e(s^2) + sD_o(s^2)}$$

we have

$$\begin{aligned}
\frac{1}{G(j\omega)} &= \frac{D_e(-\omega^2) + j\omega D_o(-\omega^2)}{N_e(-\omega^2) + j\omega N_o(-\omega^2)} \\
&= \frac{[D_e(-\omega^2) + j\omega D_o(-\omega^2)][N_e(-\omega^2) - j\omega N_o(-\omega^2)]}{[N_e(-\omega^2) + j\omega N_o(-\omega^2)][N_e(-\omega^2) - j\omega N_o(-\omega^2)]} \\
&= \frac{[D_e(-\omega^2)N_e(-\omega^2) + \omega^2 D_o(-\omega^2)N_o(-\omega^2)]}{[N_e(-\omega^2)N_e(-\omega^2) + \omega^2 N_o(-\omega^2)N_o(-\omega^2)]} \\
&\quad + j\frac{\omega[N_e(-\omega^2)D_o(-\omega^2) - D_e(-\omega^2)N_o(-\omega^2)]}{[N_e(-\omega^2)N_e(-\omega^2) + \omega^2 N_o(-\omega^2)N_o(-\omega^2)]} \\
&= \frac{p_1(\omega) + jq(\omega)}{p_2(\omega)}.
\end{aligned}$$

Since $q(\omega_t) = 0$ for finite ω_t, it follows that for all such frequencies

$$-\frac{p_1(\omega_t)}{p_2(\omega_t)} = -\frac{1}{G(j\omega_t)}.$$

Remark 3.3 *It is appropriate to point out here that Theorem 3.1 parts (i) and (ii) do provide a characterization of all plants that are stabilizable by a* *constant gain. Also note that a necessary condition for F^* to be nonempty is that for $m+n$ even,*

$$l \geq \frac{|n - (l(N(s)) - r(N(s)))|}{2}$$

and for $m+n$ odd,

$$l \geq \frac{|n - (l(N(s)) - r(N(s)))| + 1}{2}.$$

The following examples illustrate the usefulness of Theorem 3.1 when solving the constant gain stabilization problem.

Example 3.2 *Consider a system described by*

$$\begin{aligned} D(s) &= s^4 + 5s^3 + 10s^2 + 4s + 6 \\ N(s) &= s^3 + 3s^2 + 2s - 2. \end{aligned}$$

The closed-loop characteristic polynomial is

$$\delta(s,k) = D(s) + kN(s).$$

Here $N_e(s^2) = 3s^2 - 2$ and $N_o(s^2) = s^2 + 2$, so that

$$N^*(s) = N(-s) = N_e(s^2) - sN_o(s^2).$$

Therefore

$$\begin{aligned} \delta(s,k)N^*(s) &= (-2s^6 + 14s^4 - 10s^2 - 12) \\ &\quad + k(-s^6 + 5s^4 - 16s^2 + 4) \\ &\quad + s(-s^6 + 3s^4 - 24s^2 - 20) \end{aligned}$$

so that

$$\delta(j\omega,k)N^*(j\omega) = p_1(\omega) + kp_2(\omega) + jq(\omega)$$

with

$$\begin{aligned} p_1(\omega) &:= 2\omega^6 + 14\omega^4 + 10\omega^2 - 12 \\ p_2(\omega) &:= \omega^6 + 5\omega^4 + 16\omega^2 + 4 \\ q(\omega) &:= \omega(\omega^6 + 3\omega^4 + 24\omega^2 - 20). \end{aligned}$$

The real, non-negative, distinct finite zeros of $q_f(\omega)$ with odd multiplicities are

$$\omega_0 = 0,\ \omega_1 = 0.8639.$$

Since $n+m=7$, which is odd, and $N^*(s)$ has no roots on the $j\omega$ axis, from Definition 3.1, the set A becomes

$$A = \left\{ \begin{array}{ll} \{-1,-1,0\} & \{1,-1,0\} \\ \{-1,1,0\} & \{1,1,0\} \end{array} \right\}.$$

Since $l(N(s)) - r(N(s)) = 1$ and $(-1)^{l-1}\mathrm{sgn}[q(\infty)] = -1$, it follows using Definition 3.3 that every string $\mathcal{I} = \{i_0,\ i_1,\ i_2\} \in F^*$ must satisfy

$$-(i_0 - 2i_1 + i_2) = 3\ .$$

Hence $F^* = \{\mathcal{I}_1\}$ where $\mathcal{I}_1 = \{-1,1,0\}$. Furthermore,

$$\begin{aligned} U_0 = -\frac{p_1(\omega_0)}{p_2(\omega_0)} &= 3, \\ L_1 = -\frac{p_1(\omega_1)}{p_2(\omega_1)} &= -0.2139. \end{aligned}$$

Hence from Theorem 3.1, we have

$$K_1 = (-0.2139, 3) \text{ for } \mathcal{I}_1\ .$$

Therefore $\delta(s,k)$ is Hurwitz for $k \in (-0.2139,\ 3)$. △

Example 3.3 *[10] Consider the constant gain stabilization problem with*

$$\begin{aligned} D(s) &= s^5 + 11s^4 + 22s^3 + 60s^2 + 47s + 25 \\ N(s) &= s^4 + 6s^3 + 12s^2 + 54s + 16. \end{aligned}$$

The closed-loop characteristic polynomial is

$$\delta(s,k) = D(s) + kN(s)\ .$$

Here $N_e(s^2) = s^4 + 12s^2 + 16$ and $N_o(s^2) = 6s^2 + 54$ so that

$$N^*(s) = N(-s) = N_e(s^2) - sN_o(s^2).$$

Therefore

$$\begin{aligned} \delta(s,k)N^*(s) &= (5s^8 + 6s^6 - 549s^4 - 1278s^2 + 400) \\ &\quad +k(s^8 - 12s^6 - 472s^4 - 2532s^2 + 256) \\ &\quad +s(s^8 - 32s^6 - 627s^4 - 2474s^2 - 598) \end{aligned}$$

so that

$$\delta(j\omega,k)N^*(j\omega) = p_1(\omega) + kp_2(\omega) + jq(\omega)$$

with

$$\begin{aligned} p_1(\omega) &:= 5\omega^8 - 6\omega^6 - 549\omega^4 + 1278\omega^2 + 400 \\ p_2(\omega) &:= \omega^8 + 12\omega^6 - 472\omega^4 + 2532\omega^2 + 256 \\ q(\omega) &:= \omega(\omega^8 + 32\omega^6 - 627\omega^4 + 2474\omega^2 - 598). \end{aligned}$$

The real, non-negative, distinct finite zeros of $q_f(\omega)$ with odd multiplicities are

$$\omega_0 = 0,\ \omega_1 = 0.50834,\ \omega_2 = 2.41735,\ \omega_3 = 2.91515.$$

Since $n + m = 9$, which is odd, and $N^*(s)$ has no roots on the $j\omega$ axis, from Definition 3.1, the set A becomes

$$A = \left\{ \begin{array}{ll} \{-1,-1,-1,-1,0\} & \{1,-1,-1,-1,0\} \\ \{-1,-1,-1,1,0\} & \{1,-1,-1,1,0\} \\ \{-1,-1,1,-1,0\} & \{1,-1,1,-1,0\} \\ \{-1,-1,1,1,0\} & \{1,-1,1,1,0\} \\ \{-1,1,-1,-1,0\} & \{1,1,-1,-1,0\} \\ \{-1,1,-1,1,0\} & \{1,1,-1,1,0\} \\ \{-1,1,1,-1,0\} & \{1,1,1,-1,0\} \\ \{-1,1,1,1,0\} & \{1,1,1,1,0\} \end{array} \right\}.$$

Since $l(N(s)) - r(N(s)) = 4$ and $(-1)^{l-1}\mathrm{sgn}[q(\infty)] = -1$, it follows using Definition 3.3 that every string $\mathcal{I} = \{i_0,\ i_1,\ i_2,\ i_3,\ i_4\} \in F^*$ must satisfy

$$-(i_0 - 2i_1 + 2i_2 - 2i_3 + i_4) = 1.$$

Hence $F^* = \{I_1, I_2, I_3\}$ where

$$\begin{aligned} \mathcal{I}_1 &= \{1,-1,-1,1,0\} \\ \mathcal{I}_2 &= \{1,1,1,1,0\} \\ \mathcal{I}_3 &= \{1,1,-1,-1,0\}. \end{aligned}$$

Furthermore,

$$\begin{aligned} -\frac{p_1(\omega_0)}{p_2(\omega_0)} &= -1.56250 \\ -\frac{p_1(\omega_1)}{p_2(\omega_1)} &= -0.78898 \\ -\frac{p_1(\omega_2)}{p_2(\omega_2)} &= 2.50345 \\ -\frac{p_1(\omega_3)}{p_2(\omega_3)} &= 22.49390. \end{aligned}$$

Hence from Theorem 3.1, we have

$$\begin{cases} K_1 = \emptyset \text{ for } \mathcal{I}_1 \\ K_2 = (22.49390,\ \infty) \text{ for } \mathcal{I}_2 \\ K_3 = (-0.78898,\ 2.50345) \text{ for } \mathcal{I}_3. \end{cases}$$

Therefore $\delta(s,k)$ is Hurwitz for $k \in (-0.78898,\ 2.50345) \cup (22.49390,\ \infty)$.

$\triangle$

3.3 Computation of All Stabilizing PI Controllers

In this section, we show how the results developed in Section 3.2 for the constant gain stabilization problem can be extended to solve the problem of PI stabilization. As in the previous case, we consider the feedback control system shown in Fig. 3.1. Now the controller being used is of the PI type so $C(s)$ is given by

$$C(s) = k_p + \frac{k_i}{s} = \frac{k_i + k_p s}{s}.$$

The closed-loop characteristic polynomial is then

$$\delta(s, k_p, k_i) = sD(s) + (k_i + k_p s)N(s).$$

Let n be the degree of $\delta(s, k_p, k_i)$ and m be the degree of $N(s)$. The problem of stabilization using a PI controller is to determine the values of k_p and k_i for which the closed-loop characteristic polynomial $\delta(s, k_p, k_i)$ is Hurwitz.

Clearly, k_p and k_i both affect the even and odd parts of $\delta(s, k_p, k_i)$. Motivated by the approach used in Section 3.2, we now proceed to construct a new polynomial whose even part depends on k_i and odd part depends on k_p. Consider the even-odd decompositions

$$\begin{aligned} N(s) &= N_e(s^2) + sN_o(s^2) \\ D(s) &= D_e(s^2) + sD_o(s^2). \end{aligned}$$

Define

$$N^*(s) = N(-s) = N_e(s^2) - sN_o(s^2).$$

Multiplying $\delta(s, k_p, k_i)$ by $N^*(s)$ and examining the resulting polynomial, we obtain

$$\begin{aligned} l(\delta(s,k_p,k_i)N^*(s)) - r(\delta(s,k_p,k_i)N^*(s)) &= l(\delta(s,k_p,k_i)) \\ &\quad -r(\delta(s,k_p,k_i)) \\ &\quad -(l(N(s)) - r(N(s))) \,. \end{aligned}$$

$\delta(s, k_p, k_i)$ of degree n is Hurwitz if and only if $l(\delta(s, k_p, k_i)) = n$ and $r(\delta(s, k_p, k_i)) = 0$. Therefore, in view of Theorem 2.4, we have the following result.

Lemma 3.2 *$\delta(s, k_p, k_i)$ is Hurwitz if and only if*

$$\sigma_i(\delta(s, k_p, k_i)N^*(s)) = n - (l(N(s)) - r(N(s))) \,. \tag{3.12}$$

Our task now is to determine those values of k_p, k_i for which (3.12) holds. It can be verified that

$$\begin{aligned}\delta(s, k_p, k_i)N^*(s) \;=\;& [s^2(N_e(s^2)D_o(s^2) - D_e(s^2)N_o(s^2)) \\ &+k_i(N_e(s^2)N_e(s^2) - s^2N_o(s^2)N_o(s^2))] \\ &+s[D_e(s^2)N_e(s^2) - s^2D_o(s^2)N_o(s^2) \\ &+k_p(N_e(s^2)N_e(s^2) - s^2N_o(s^2)N_o(s^2))] \,.\end{aligned} \tag{3.13}$$

Substituting $s = j\omega$, we obtain

$$\delta(j\omega, k_p, k_i)N^*(j\omega) \;=\; p(\omega,\, k_i) + jq(\omega,\, k_p)$$

where

$$\begin{aligned}p(\omega,\, k_i) &= p_1(\omega) + k_i p_2(\omega) \\ q(\omega,\, k_p) &= q_1(\omega) + k_p q_2(\omega) \\ p_1(\omega) &= -\omega^2(N_e(-\omega^2)D_o(-\omega^2) - D_e(-\omega^2)N_o(-\omega^2)) \\ p_2(\omega) &= N_e(-\omega^2)N_e(-\omega^2) + \omega^2N_o(-\omega^2)N_o(-\omega^2)) \\ q_1(\omega) &= \omega(D_e(-\omega^2)N_e(-\omega^2) + \omega^2D_o(-\omega^2)N_o(-\omega^2)) \\ q_2(\omega) &= \omega(N_e(-\omega^2)N_e(-\omega^2) + \omega^2N_o(-\omega^2)N_o(-\omega^2)).\end{aligned}$$

Also, define

$$\begin{aligned}p_f(\omega, k_i) &= \frac{p(\omega, k_i)}{(1+\omega^2)^{\frac{m+n}{2}}} \\ q_f(\omega, k_p) &= \frac{q(\omega, k_p)}{(1+\omega^2)^{\frac{m+n}{2}}}.\end{aligned}$$

From these expressions, we first note that k_i, k_p appear affinely in $p(\omega, k_i)$, $q(\omega, k_p)$ respectively. Moreover, for every fixed k_p, the zeros of $q(\omega, k_p)$ do not depend on k_i, and so the results of Section 3.2 are applicable in this case. Thus, by sweeping over all real k_p and solving a constant gain stabilization problem at each stage, we can determine the set of all stabilizing (k_p, k_i) values for the given plant.

However, there is no need to sweep over all real values of the parameter k_p. As will be shown briefly, the range of k_p values over which the sweeping needs to be carried out can be considerably reduced in many cases. Recall from Remark 3.3 that for a fixed k_p, a necessary condition for the existence of a stabilizing k_i value is that the number of real, non-negative, distinct finite zeros of odd multiplicities of $q(\omega, k_p)$ be at least

$$\frac{|n - (l(N(s)) - r(N(s)))|}{2} \quad \text{if } m+n \text{ is even}$$

or

$$\frac{|n - (l(N(s)) - r(N(s)))| + 1}{2} \quad \text{if } m + n \text{ is odd .}$$

Such a necessary condition can be checked by first rewriting $q(\omega,\ k_p)$ as follows:

$$q(\omega,\ k_p) \quad = \quad \omega[U(\omega) + k_p V(\omega)] \tag{3.14}$$

where

$$\begin{aligned} U(\omega) &= D_e(-\omega^2)N_e(-\omega^2) + \omega^2 D_o(-\omega^2)N_o(-\omega^2) \\ V(\omega) &= N_e(-\omega^2)N_e(-\omega^2) + \omega^2 N_o(-\omega^2)N_o(-\omega^2). \end{aligned}$$

From (3.14), we see that $q(\omega,\ k_p)$ has at least one real non-negative root at the origin. Now applying the root locus ideas that will be presented shortly, we can determine the real root distributions of $q(\omega,\ k_p)$ corresponding to different ranges of k_p. Then, using the fact that $n - (l(N(s)) - r(N(s)))$ is known, one can identify the ranges of k_p for which $q(\omega, k_p)$ does not satisfy the necessary condition stated above. Such k_p ranges do not need to be swept over and can, therefore, be safely discarded.

Before we proceed further, let us consider the problem of determining the root locus of $U(\omega) + k_p V(\omega) = 0$, where $U(\omega)$ and $V(\omega)$ are real and coprime polynomials and k_p varies from $-\infty$ to $+\infty$. We now make the following observations:

1. The real breakaway points on the root loci of $U(\omega) + k_p V(\omega) = 0$ correspond to a real multiple root and must, therefore, satisfy

$$\frac{d\left(\frac{V(\omega)}{U(\omega)}\right)}{d\omega} = 0 \Leftrightarrow \frac{U(\omega)\frac{dV(\omega)}{d\omega} - V(\omega)\frac{dU(\omega)}{d\omega}}{U^2(\omega)} = 0 \ .$$

   The real breakaway points are the real zeros of the above equation.

2. Let $k_1 < k_2 < \cdots < k_z$ be the distinct, finite, values of k_p corresponding to the real breakaway points ω_j, $j = 1, 2, \ldots, z$, on the root loci of $U(\omega) + k_p V(\omega) = 0$. Also define $k_0 = -\infty$ and $k_{z+1} = +\infty$. Then ω_j, $j = 1, 2, \ldots, z$, are the multiple real roots of $U(\omega) + k_p V(\omega) = 0$ and the corresponding values of k_p are k_j, $j = 1, 2, \ldots, z$. We note that for $k_p \in (k_j, k_{j+1})$, the real roots of $U(\omega) + k_p V(\omega) = 0$ are simple and the number of real roots is invariant.

3. If $U(0) + k_p V(0) \neq 0$, for all $k_p \in (k_j, k_{j+1})$, then the distribution of the real roots of $U(\omega) + k_p V(\omega) = 0$ with respect to the origin is invariant over this range of k_p values.

Using these root locus ideas we can narrow the sweeping range for the controller parameter k_p.

We now present a simple example to illustrate the detailed calculations involved in determining the stabilizing (k_p, k_i) values for a given plant.

Example 3.4 *Consider the problem of choosing stabilizing PI gains for the plant* $G(s) = \frac{N(s)}{D(s)}$ *where*

$$\begin{aligned} D(s) &= s^5 + 3s^4 + 29s^3 + 15s^2 - 3s + 60 \\ N(s) &= s^3 + 6s^2 - 2s + 1. \end{aligned}$$

The closed-loop characteristic polynomial is

$$\delta(s, k_p, k_i) = sD(s) + k_i N(s) + k_p s N(s).$$

The even-odd decompositions of the polynomials $N(s)$ *and* $D(s)$ *are given by*

$$\begin{aligned} D(s) &= D_e(s^2) + sD_o(s^2) \\ N(s) &= N_e(s^2) + sN_o(s^2) \end{aligned}$$

where

$$\begin{aligned} D_e(s^2) &= 3s^4 + 15s^2 + 60 \qquad & N_e(s^2) &= 6s^2 + 1 \\ D_o(s^2) &= s^4 + 29s^2 - 3 \qquad & N_o(s^2) &= s^2 - 2. \end{aligned}$$

Now

$$\begin{aligned} N^*(s) &= N(-s) = N_e(s^2) - sN_o(s^2) \\ &= (6s^2 + 1) - s(s^2 - 2). \end{aligned}$$

Therefore, from (3.13) we obtain

$$\begin{aligned} \delta(s, k_p, k_i)N^*(s) &= [s^2(3s^6 + 166s^4 - 19s^2 + 117) + k_i(-s^6 + 40s^4 \\ &\quad +8s^2 + 1)] + s[(-s^8 - 9s^6 + 154s^4 + 369s^2 + 60) \\ &\quad +k_p(-s^6 + 40s^4 + 8s^2 + 1)] \end{aligned}$$

so that

$$\delta(j\omega, k_p, k_i)N^*(j\omega) = [p_1(\omega) + k_i p_2(\omega)] + j[q_1(\omega) + k_p q_2(\omega)]$$

with

$$\begin{aligned} p_1(\omega) &= 3\omega^8 - 166\omega^6 - 19\omega^4 - 117\omega^2 \\ p_2(\omega) &= \omega^6 + 40\omega^4 - 8\omega^2 + 1 \\ q_1(\omega) &= -\omega^9 + 9\omega^7 + 154\omega^5 - 369\omega^3 + 60\omega \\ q_2(\omega) &= \omega^7 + 40\omega^5 - 8\omega^3 + \omega. \end{aligned}$$

We now use the root locus ideas introduced earlier to specify the range of k_p *values over which the sweeping should be carried out. From (3.14) we have*

$$q(\omega,\ k_p) = \omega[U(\omega) + k_p V(\omega)]$$

where

$$\begin{aligned} U(\omega) &= -\omega^8 + 9\omega^6 + 154\omega^4 - 369\omega^2 + 60 \\ V(\omega) &= \omega^6 + 40\omega^4 - 8\omega^2 + 1 \,. \end{aligned}$$

Now

$$\frac{U(\omega)\frac{dV(\omega)}{d\omega} - V(\omega)\frac{dU(\omega)}{d\omega}}{U^2(\omega)} = \frac{\alpha(\omega)}{\beta(\omega)}$$

where

$$\begin{aligned} \alpha(\omega) &= 2\omega^{13} + 160\omega^{11} - 460\omega^9 - 1180\omega^7 - 26750\omega^5 + 8984\omega^3 - 222\omega \\ \beta(\omega) &= (-\omega^8 + 9\omega^6 + 154\omega^4 - 369\omega^2 + 60)^2 \,. \end{aligned}$$

Then the distinct, finite k_p values producing either real breakaway points or a root at the origin are

$$k_1 = -61.67086,\ k_2 = -60,\ k_3 = -2.54119,\ k_4 = 16.44309$$

and the corresponding zeros ω_i are

$$\omega_1 = \pm 0.16390,\ \omega_2 = 0,\ \omega_3 = \pm 2.60928,\ \omega_4 = \pm 0.55140 \,.$$

The real root distributions of $U(\omega) + k_p V(\omega) = 0$ with respect to the origin, corresponding to the different ranges of k_p, are given below:

$k_p \in (-\infty, -61.67086)$: *no real roots*

$k_p \in (-61.67086, -60)$: *two positive simple real roots*
two negative simple real roots

$k_p \in (-60, -2.54119)$: *one positive simple real root*
one negative simple real root

$k_p \in (-2.54119, 16.44309)$: *three positive simple real roots*
three negative simple real roots

$k_p \in (16.44309, \infty)$: *one positive simple real root*
one negative simple real root.

For this example, $m + n$ is odd and

$$n - (l(N(s)) - r(N(s))) = 6 - (1 - 2) = 7.$$

Hence, for a given fixed k_p, a necessary condition for the existence of a stabilizing k_i value is that $q(\omega, k_p)$ must have at least four distinct, real non-negative roots of odd multiplicities. The root distributions presented above show that this is possible only for

$$k_p \in (-2.54119, 16.44309).$$

For each k_p in this range we can use the constant gain stabilization result of Section 3.2 to determine the exact ranges of stabilizing k_i. By sweeping over all $k_p \in (-2.54119,\ 16.44309)$ and using the constant gain stabilization result at each stage, we obtained the stabilizing region sketched in Fig. 3.2.

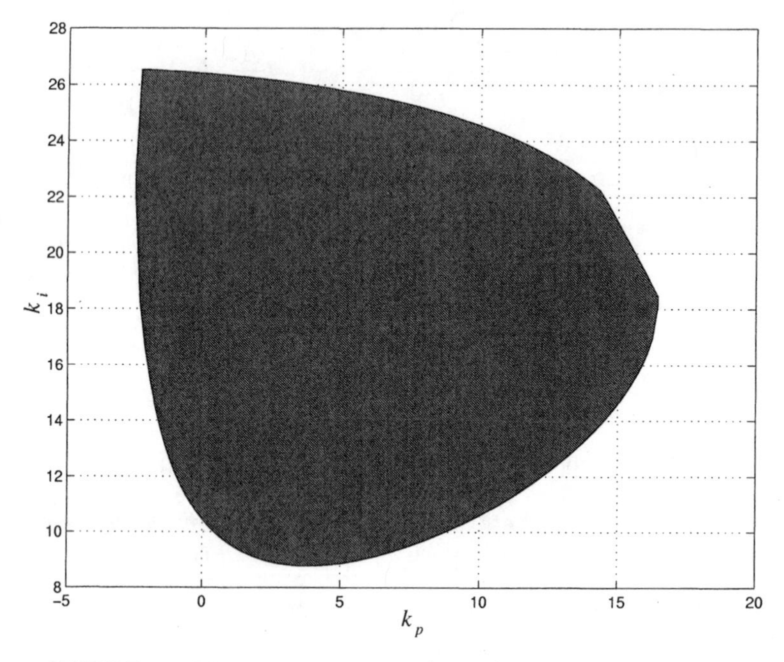

FIGURE 3.2. The stabilizing set of $(k_p,\ k_i)$ values (Example 3.4).

△

3.4 Notes and References

The results presented in this chapter were first developed by Ho, Datta, and Bhattacharyya (see [18]). A comprehensive discussion of the theory underlying the design of PI stabilizers for delay-free linear time-invariant systems is given in [10]. The root locus ideas presented in Section 3.3 are taken from Appendix A of [10].

4

PID Stabilization of Delay-Free Linear Time-Invariant Systems

In this chapter we first consider the problem of stabilizing a continuous-time system described by a rational transfer function using a PID controller. A solution based on the Generalized Hermite-Biehler Theorem of Chapter 2 is presented. The solution provided here characterizes the entire family of stabilizing controllers in terms of a family of linear programming problems. The discrete-time counterpart is then solved.

4.1 Introduction

The constant gain stabilization problem considered in Chapter 3 can be solved by several classical approaches such as the root locus technique, the Nyquist stability criterion, and the Routh-Hurwitz criterion. However, the same is not true for PID stabilization where none of these classical techniques are of much help. In the case of the Routh-Hurwitz criterion, the solution of the PID stabilization problem involves inequalities that are highly nonlinear and make this a complicated task. In this chapter we show how the Generalized Hermite-Biehler Theorem presented in Section 2.3 can be used to develop an elegant procedure for determining stabilizing PID gains k_p, k_i, and k_d for a given plant described by a rational transfer function.

The characterization of all stabilizing PID controllers involves the solution of a linear programming problem. This characterization is analogous to the YJBK parametrization of all stabilizing controllers with the difference that the YJBK parametrization cannot incorporate constraints on the controller order or structure whereas the characterization presented here includes the PID structural constraints from the very beginning.

The chapter is organized as follows. In Section 4.2, we provide a solution to the problem of characterizing all the stabilizing PID controllers for a delay-free, continuous-time, linear-time invariant system. Section 4.3 solves the same problem for the case of a discrete-time plant.

4.2 A Characterization of All Stabilizing PID Controllers

As in Sections 3.2 and 3.3, we consider the feedback control system of Fig. 3.1 where now $C(s)$ has the following form:

$$C(s) = k_p + \frac{k_i}{s} + k_d s = \frac{k_i + k_p s + k_d s^2}{s}. \tag{4.1}$$

The closed-loop characteristic polynomial becomes

$$\delta(s, k_p, k_i, k_d) = sD(s) + (k_i + k_d s^2)N(s) + k_p sN(s). \tag{4.2}$$

As mentioned in Chapter 1, the derivative term is sometimes replaced by $\frac{s}{1+T_d s}$, where T_d is a small positive value that is usually fixed. It is easy to see that this case can be handled by considering the same controller (4.1) and replacing the plant by $\frac{N(s)}{(1+T_d s)D(s)}$.

The problem of stabilization using a PID controller is to determine the values of k_p, k_i, and k_d for which the closed-loop characteristic polynomial $\delta(s, k_p, k_i, k_d)$ is Hurwitz. However, it is clear from (4.2) that all three parameters k_p, k_i, and k_d affect both the even and odd parts of $\delta(s, k_p, k_i, k_d)$. Following the procedure presented in Section 3.2, we now construct a new polynomial whose even part depends on (k_i, k_d) and whose odd part depends on k_p.

Consider the even-odd decompositions

$$\begin{aligned} N(s) &= N_e(s^2) + sN_o(s^2) \\ D(s) &= D_e(s^2) + sD_o(s^2). \end{aligned}$$

Define

$$N^*(s) = N(-s) = N_e(s^2) - sN_o(s^2).$$

Also let n, m be the degrees of $\delta(s, k_p, k_i, k_d)$ and $N(s)$, respectively. Now, multiplying $\delta(s, k_p, k_i, k_d)$ by $N^*(s)$ and examining the resulting polynomial, we obtain

$$\begin{aligned} & l(\delta(s, k_p, k_i, k_d)N^*(s)) - r(\delta(s, k_p, k_i, k_d)N^*(s)) \\ = \; & (l(\delta(s, k_p, k_i, k_d)) - r(\delta(s, k_p, k_i, k_d))) - (l(N(s)) - r(N(s))) . \end{aligned}$$

$\delta(s, k_p, k_i, k_d)$ of degree n is Hurwitz if and only if

$$l(\delta(s, k_p, k_i, k_d)) = n$$

and

$$r(\delta(s, k_p, k_i, k_d)) = 0.$$

Therefore, in view of Theorem 2.4, we have the following.

Lemma 4.1 *$\delta(s, k_p, k_i, k_d)$ is Hurwitz if and only if*

$$\sigma_i(\delta(s, k_p, k_i, k_d)N^*(s)) = n - (l(N(s)) - r(N(s))) \,. \tag{4.3}$$

Our task now is to determine those values of k_p, k_i, k_d for which (4.3) holds. It is straightforward to verify that

$$\begin{aligned}\delta(s, k_p, k_i, k_d)N^*(s) \;=\; & [s^2(N_e(s^2)D_o(s^2) - D_e(s^2)N_o(s^2)) \\ & +(k_i + k_d s^2)(N_e(s^2)N_e(s^2) - s^2 N_o(s^2)N_o(s^2))] \\ & +s[D_e(s^2)N_e(s^2) - s^2 D_o(s^2)N_o(s^2) \\ & +k_p(N_e(s^2)N_e(s^2) - s^2 N_o(s^2)N_o(s^2))].\end{aligned} \tag{4.4}$$

Substituting $s = j\omega$, we obtain

$$\delta(j\omega, k_p, k_i, k_d)N^*(j\omega) = p(\omega,\ k_i,\ k_d) + jq(\omega,\ k_p)$$

where

$$\begin{aligned}p(\omega,\ k_i,\ k_d) &= p_1(\omega) + (k_i - k_d\omega^2)p_2(\omega) \\ q(\omega,\ k_p) &= q_1(\omega) + k_p q_2(\omega) \\ p_1(\omega) &= -\omega^2(N_e(-\omega^2)D_o(-\omega^2) - D_e(-\omega^2)N_o(-\omega^2)) \\ p_2(\omega) &= N_e(-\omega^2)N_e(-\omega^2) + \omega^2 N_o(-\omega^2)N_o(-\omega^2) \\ q_1(\omega) &= \omega(D_e(-\omega^2)N_e(-\omega^2) + \omega^2 D_o(-\omega^2)N_o(-\omega^2)) \\ q_2(\omega) &= \omega(N_e(-\omega^2)N_e(-\omega^2) + \omega^2 N_o(-\omega^2)N_o(-\omega^2)) \,.\end{aligned}$$

Also, define

$$\begin{aligned}p_f(\omega, k_i, k_d) &= \frac{p(\omega, k_i, k_d)}{(1+\omega^2)^{\frac{m+n}{2}}} \\ q_f(\omega, k_p) &= \frac{q(\omega, k_p)}{(1+\omega^2)^{\frac{m+n}{2}}}.\end{aligned}$$

Note from these expressions that k_i, k_d appear affinely in $p(\omega,\ k_i,\ k_d)$ while k_p appears affinely in $q(\omega,\ k_p)$. Furthermore, for every fixed k_p, the zeros of $q(\omega, k_p)$ will not depend on k_i or k_d and so we can use the approach of

Section 3.2 to determine stabilizing values for k_i and k_d. Since there are two variables here, we are no longer able to obtain a closed form solution. Instead, a linear programming problem has to be solved for each fixed k_p. As k_p is varied, we will have a one-parameter family of linear programming problems to solve. Before formally stating the main result on PID stabilization, we first introduce some definitions. These definitions are essentially analogous to those introduced in Section 3.2 with the only difference that the present definitions *are conditioned on k_p being held at some fixed value.*

Definition 4.1 *Let the integers m, n and the function $q_f(\omega,\ k_p)$ be as already defined. For a given fixed k_p, let $0 = \omega_0 < \omega_1 < \omega_2 < \cdots < \omega_{l-1}$ be the real, non-negative, distinct finite zeros of $q_f(\omega,\ k_p)$ with odd multiplicities. Define a sequence of numbers $i_0, i_1, i_2, \ldots, i_l$ as follows:*

(i)

$$i_0 = \begin{cases} \operatorname{sgn}[p_{1_f}^{(k_n)}(0)] & \text{if } N^*(s) \text{ has a zero of multiplicity } k_n \text{ at the origin} \\ \alpha & \text{otherwise} \end{cases}$$

where $\alpha \in \{-1, 1\}$ and

$$p_{1_f}(\omega) := \frac{p_1(\omega)}{(1+\omega^2)^{\frac{(m+n)}{2}}},$$

(ii) For $t = 1, 2, \cdots, l-1$:

$$i_t = \begin{cases} 0 & \text{if } N^*(j\omega_t) = 0 \\ \alpha & \text{otherwise} \end{cases},$$

(iii)

$$i_l = \begin{cases} \alpha & \text{if } n+m \text{ is even} \\ 0 & \text{if } n+m \text{ is odd} \end{cases}.$$

With $i_0, i_1, \ldots$ defined in this way, we define the string $\mathcal{I} : N \to R$ as the following sequence of numbers:

$$\mathcal{I} := \{i_0, i_1, \ldots, i_l\}.$$

Define A_{k_p} to be the set of all possible strings $\mathcal{I}$ that can be generated to satisfy the preceding requirements.

Next we introduce the "imaginary signature" $\gamma(\mathcal{I})$ associated with any element $\mathcal{I} \in A_{k_p}$. This definition is motivated by Theorem 4.1 to follow.

Definition 4.2 *Let the integers m, n and the functions $q(\omega,\ k_p)$, $q_f(\omega, k_p)$ be as already defined. For a given fixed k_p, let $0 = \omega_0 < \omega_1 < \omega_2 < \cdots < \omega_{l-1}$ be the real, non-negative, distinct finite zeros of $q_f(\omega,\ k_p)$ with odd multiplicities. Also define $\omega_l = \infty$. For each string $\mathcal{I} = \{i_0, i_1, \ldots\}$ in A_{k_p}, let $\gamma(\mathcal{I})$ denote the "imaginary signature" associated with the string $\mathcal{I}$ defined by*

$$\begin{aligned}\gamma(\mathcal{I}) \quad &:= \quad [i_0 - 2i_1 + 2i_2 + \cdots + (-1)^{l-1}2i_{l-1} + (-1)^l i_l] \\ &\quad\quad \cdot(-1)^{l-1}\mathrm{sgn}[q(\infty,\ k_p)] \,. \end{aligned} \tag{4.5}$$

Note that referring to $\gamma(\mathcal{I})$ as the "imaginary signature" of $\mathcal{I}$ can be justified as in Section 3.2.

Definition 4.3 *The set $F^*_{k_p}$ of feasible strings for the PID stabilization problem is defined as*

$$F^*_{k_p} = \{\mathcal{I} \in A_{k_p} | \gamma(\mathcal{I}) = n - (l(N(s)) - r(N(s)))\} \,.$$

We now present the main result of this section.

Theorem 4.1 (Main Result on PID Stabilization) *The PID stabilization problem, with a fixed k_p, is solvable for a given plant with rational transfer function $G(s)$ if and only if the following conditions hold:*

(i) $F^*_{k_p}$ is not empty, i.e., at least one feasible string exists and

(ii) There exists a string $\mathcal{I} = \{i_0, i_1, \ldots\} \in F^*_{k_p}$ and values of k_i and k_d such that $\forall\, t = 0, 1, 2, \ldots$ for which $N^*(j\omega_t) \neq 0$

$$p(\omega_t, k_i, k_d) i_t \ > \ 0 \tag{4.6}$$

where $p(\omega, k_i, k_d)$ is as already defined.

Furthermore, if there exist values of k_i and k_d such that the above condition is satisfied for the feasible strings $\mathcal{I}_1, \mathcal{I}_2, \ldots, \mathcal{I}_s \in F^*_{k_p}$, then the set of stabilizing (k_i, k_d) values corresponding to the fixed k_p is the union of the (k_i, k_d) values satisfying (4.6) for $\mathcal{I}_1, \mathcal{I}_2, \ldots, \mathcal{I}_s$.

Proof. From (4.3), we know that $\delta(s, k_p, k_i, k_d)$ is Hurwitz if and only if

$$\sigma_i(\delta(s, k_p, k_i, k_d)N^*(s)) = n - (l(N(s)) - r(N(s))).$$

Thus, for a fixed k_p, it follows that $\delta(s, k_p, k_i, k_d)$ is Hurwitz if and only if there exists $\mathcal{I} \in F^*_{k_p}$ where

$$\mathcal{I} \quad = \quad \{i_0, i_1, \ldots\}$$

$$\begin{aligned}
i_0 &= \text{sgn}[p_f^{(k_n)}(0,k_i,k_d)] \\
i_j &= \text{sgn}[p_f(\omega_j,k_i,k_d)],\ j=1,2,\ldots,l-1 \\
i_l &= \begin{cases} \text{sgn}[p_f(\omega_l,k_i,k_d)] & \text{if } n+m \text{ is even} \\ 0 & \text{if } n+m \text{ is odd} \end{cases}.
\end{aligned}$$

As in the proof of Theorem 3.1, it can be easily shown that if $N^*(j\omega_t)=0$ for some $t=0,1,2,\cdots$, then the value of the corresponding entry $i_t \in \mathcal{I}$ is predetermined and is independent of k_i and k_d. The definition of A_{k_p} already accounts for such special cases. Focusing now on the other case, i.e., $N^*(j\omega_t) \neq 0$, leads us to (4.6). This completes the proof of (i) and (ii). The characterization of all stabilizing (k_i, k_d) values, corresponding to the fixed k_p, now follows as an immediate consequence. ■

Remark 4.1 *It should be noted that since the constraint set is linear, the admissible set for (4.6) is either a convex polygon or an intersection of half planes, which is again a* convex set. *Therefore, for each fixed k_p, the region in the (k_i, k_d) plane for which $\delta(s,k_p,k_i,k_d)$ is Hurwitz is either a union of convex sets or is empty.*

By using Theorem 4.1 and sweeping over all real values of k_p, we can determine the entire set of values of (k_p,k_i,k_d) for which $\delta(s,k_p,k_i,k_d)$ is Hurwitz. As in the case of PI stabilization, the range of k_p values over which the sweeping has to be carried out can be *a priori* narrowed down by using the root locus ideas presented in Section 3.3.

We now present some examples to illustrate the details of the calculations.

Example 4.1 *Consider the problem of choosing stabilizing PID gains for the plant described by the rational transfer function $G(s)=\frac{N(s)}{D(s)}$, where*

$$\begin{aligned}
N(s) &= s^3-4s^2+s+2 \\
D(s) &= s^5+8s^4+32s^3+46s^2+46s+17
\end{aligned}$$

and

$$C(s) = k_p+\frac{k_i}{s}+k_d s = \frac{k_i+k_p s+k_d s^2}{s}.$$

The closed-loop characteristic polynomial is

$$\delta(s,k_p,k_i,k_d) = sD(s)+(k_i+k_d s^2)N(s)+k_p sN(s).$$

Write

$$\begin{aligned}
D(s) &= D_e(s^2)+sD_o(s^2) \\
N(s) &= N_e(s^2)+sN_o(s^2)
\end{aligned}$$

where

$$\begin{aligned} D_e(s^2) &= 8s^4 + 46s^2 + 17 \\ D_o(s^2) &= s^4 + 32s^2 + 46 \\ N_e(s^2) &= -4s^2 + 2 \\ N_o(s^2) &= s^2 + 1. \end{aligned}$$

Now

$$\begin{aligned} N^*(s) &= N(-s) = N_e(s^2) - sN_o(s^2) \\ &= (-4s^2 + 2) - s(s^2 + 1). \end{aligned}$$

Therefore, from (4.4) we obtain

$$\begin{aligned} \delta(s, k_p, k_i, k_d)N^*(s) &= [s^2(-12s^6 - 180s^4 - 183s^2 + 75) \\ &\quad +(k_i + k_d s^2)(-s^6 + 14s^4 - 17s^2 + 4)] \\ &\quad +s[(-s^8 - 65s^6 - 246s^4 - 22s^2 + 34) \\ &\quad +k_p(-s^6 + 14s^4 - 17s^2 + 4)] \end{aligned}$$

so that

$$\begin{aligned} \delta(j\omega, k_p, k_i, k_d)N^*(j\omega) &= [p_1(\omega) + (k_i - k_d\omega^2)p_2(\omega)] \\ &\quad +j[q_1(\omega) + k_p q_2(\omega)] \end{aligned}$$

where

$$\begin{aligned} p_1(\omega) &= -12\omega^8 + 180\omega^6 - 183\omega^4 - 75\omega^2 \\ p_2(\omega) &= \omega^6 + 14\omega^4 + 17\omega^2 + 4 \\ q_1(\omega) &= -\omega^9 + 65\omega^7 - 246\omega^5 + 22\omega^3 + 34\omega \\ q_2(\omega) &= \omega^7 + 14\omega^5 + 17\omega^3 + 4\omega. \end{aligned}$$

Let us now fix k_p at a value of 1. Then,

$$\begin{aligned} q(\omega, 1) &= q_1(\omega) + q_2(\omega) \\ &= -\omega^9 + 66\omega^7 - 232\omega^5 + 39\omega^3 + 38\omega\ . \end{aligned}$$

Thus the real, non-negative, distinct finite zeros of $q_f(\omega, 1)$ with odd multiplicities are

$$\omega_0 = 0,\ \omega_1 = 0.74230,\ \omega_2 = 1.86590,\ \omega_3 = 7.89211.$$

Since $n+m = 9$ which is odd, and $N^*(s)$ has no $j\omega$-axis roots, from Definition 4.1, the set A_1 becomes

$$A_1 = \left\{ \begin{array}{ll} \{-1,-1,-1,-1,0\} & \{1,-1,-1,-1,0\} \\ \{-1,-1,-1,1,0\} & \{1,-1,-1,1,0\} \\ \{-1,-1,1,-1,0\} & \{1,-1,1,-1,0\} \\ \{-1,-1,1,1,0\} & \{1,-1,1,1,0\} \\ \{-1,1,-1,-1,0\} & \{1,1,-1,-1,0\} \\ \{-1,1,-1,1,0\} & \{1,1,-1,1,0\} \\ \{-1,1,1,-1,0\} & \{1,1,1,-1,0\} \\ \{-1,1,1,1,0\} & \{1,1,1,1,0\}. \end{array} \right\}.$$

Since

$$l(N(s)) - r(N(s)) = -1$$

and

$$(-1)^{l-1}\mathrm{sgn}[q(\infty,1)] = 1,$$

it follows using Definition 4.3 that every string

$$\mathcal{I} = \{i_0, i_1, i_2, i_3, i_4\} \in F_1^*$$

must satisfy

$$i_0 - 2i_1 + 2i_2 - 2i_3 + i_4 = 7.$$

Hence

$$F_1^* = \{1,\ -1,\ 1,\ -1,\ 0\}.$$

Thus it follows from Theorem 4.1 that the stabilizing (k_i, k_d) values corresponding to $k_p = 1$ must satisfy the string of inequalities:

$$\left\{ \begin{array}{l} p_1(\omega_0) + (k_i - k_d\omega_0^2)p_2(\omega_0) > 0 \\ p_1(\omega_1) + (k_i - k_d\omega_1^2)p_2(\omega_1) < 0 \\ p_1(\omega_2) + (k_i - k_d\omega_2^2)p_2(\omega_2) > 0 \\ p_1(\omega_3) + (k_i - k_d\omega_3^2)p_2(\omega_3) < 0\,. \end{array} \right.$$

Substituting for ω_0, ω_1, ω_2, and ω_3 in the above expressions, we obtain

$$\left\{ \begin{array}{l} k_i > 0 \\ k_i - 0.55101k_d < 3.81670 \\ k_i - 3.48158k_d > -12.19183 \\ k_i - 62.28540k_d < 464.03862\,. \end{array} \right. \tag{4.7}$$

The admissible set of values of $(k_i,\ k_d)$ for which (4.7) holds can be solved by linear programming and is shown in Fig. 4.1.

Similarly, for k_p fixed at 5, we have

$$\begin{aligned} q(\omega,5) &= q_1(\omega) + 5q_2(\omega) \\ &= -\omega^9 + 70\omega^7 - 176\omega^5 + 107\omega^3 + 54\omega\,. \end{aligned}$$

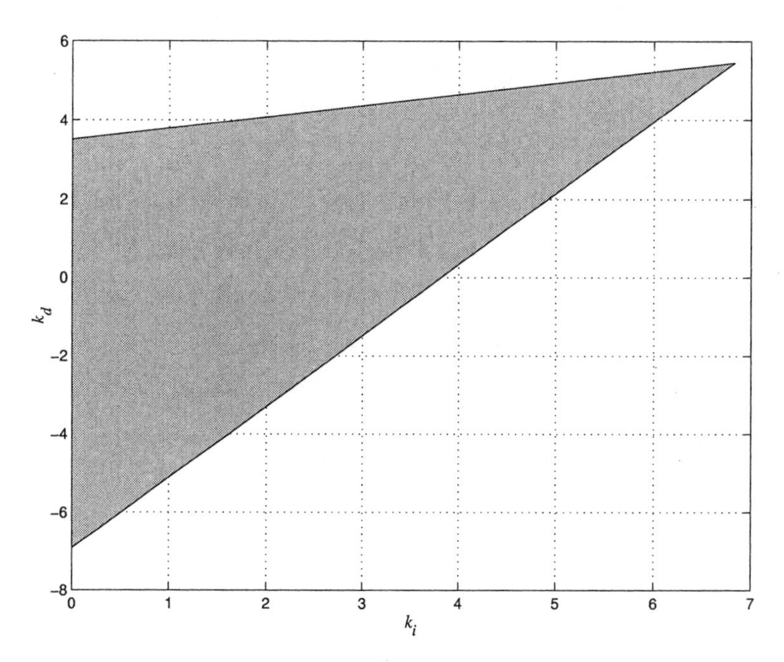

FIGURE 4.1. The stabilizing set of (k_i, k_d) values when $k_p = 1$ (Example 4.1).

Then the real, non-negative, distinct finite zeros of $q_f(\omega, 5)$ with odd multiplicities are

$$\omega_0 = 0, \ \omega_1 = 8.21054.$$

Since $n + m = 9$ which is odd, and $N^*(s)$ has no $j\omega$-axis roots, from Definition 4.1, the set A_5 becomes

$$A_5 = \left\{ \begin{array}{ll} \{-1,-1,0\} & \{1,1,0\} \\ \{-1,1,0\} & \{1,-1,0\} \end{array} \right\}.$$

Since

$$l(N(s)) - r(N(s)) = -1$$

and

$$(-1)^{l-1}\text{sgn}[q(\infty, 5)] = 1,$$

it follows using Definition 4.3 that every string $\mathcal{I} = \{i_0, i_1, i_2\} \in F_5^*$ must satisfy

$$i_0 - 2i_1 + i_2 = 7.$$

Hence

$$F_5^* = \phi$$

so that for $k_p = 5$, there are no stabilizing values for k_i and k_d.

Using the root locus ideas of Section 3.3, the range of k_p values over which the sweeping needs to be carried out was narrowed down to

$$k_p \in (-8.5, 4.23337).$$

By sweeping over different k_p values in this interval and following the procedure illustrated above, we can generate the set of (k_p, k_i, k_d) values for which $\delta(s, k_p, k_i, k_d)$ is Hurwitz. This set is sketched in Fig. 4.2. △

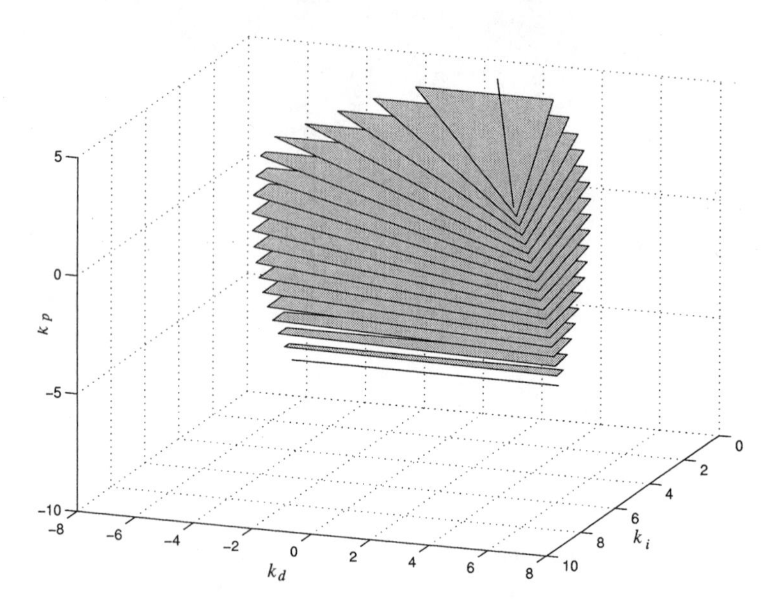

FIGURE 4.2. The stabilizing set of (k_p, k_i, k_d) values (Example 4.1).

Example 4.2 *Consider now the following plant:*

$$\begin{aligned} N(s) &= s^3 + 4s^2 + 2s + 9 \\ D(s) &= s^4 + 4s^3 + 5s^2 + 8s + 16\,. \end{aligned}$$

As in the previous example, we consider the problem of determining the stabilizing PID gains for this plant. The closed-loop characteristic polynomial is

$$\delta(s, k_p, k_i, k_d) = sD(s) + (k_i + k_d s^2)N(s) + k_p s N(s).$$

The even-odd decompositions for the polynomials $N(s)$ and $D(s)$ are given by

$$\begin{aligned} D(s) &= D_e(s^2) + sD_o(s^2) \\ N(s) &= N_e(s^2) + sN_o(s^2) \end{aligned}$$

where

$$\begin{aligned} N_e(s^2) &= 4s^2 + 9 \\ N_o(s^2) &= s^2 + 2 \\ D_e(s^2) &= s^4 + 5s^2 + 16 \\ D_o(s^2) &= 4s^2 + 8\,. \end{aligned}$$

Now

$$\begin{aligned} N^*(s) &= N(-s) = N_e(s^2) - sN_o(s^2) \\ &= (4s^2 + 9) - s(s^2 + 2). \end{aligned}$$

Therefore, from (4.4) we obtain

$$\begin{aligned} \delta(s, k_p, k_i, k_d)N^*(s) &= [s^2(-s^6 + 9s^4 + 42s^2 + 40) + (k_i + k_d s^2) \\ &\quad (-s^6 + 12s^4 + 68s^2 + 81)] + s[(13s^4 + 93s^2 \\ &\quad +144) + k_p(-s^6 + 12s^4 + 68s^2 + 81)] \end{aligned}$$

so that

$$\begin{aligned} \delta(j\omega, k_p, k_i, k_d)N^*(j\omega) &= [p_1(\omega) + (k_i - k_d\omega^2)p_2(\omega)] \\ &\quad +j[q_1(\omega) + k_p q_2(\omega)] \end{aligned}$$

where

$$\begin{aligned} p_1(\omega) &= -\omega^8 - 9\omega^6 + 42\omega^4 - 40\omega^2 \\ p_2(\omega) &= \omega^6 + 12\omega^4 - 68\omega^2 + 81 \\ q_1(\omega) &= 13\omega^5 - 93\omega^3 + 144\omega \\ q_2(\omega) &= \omega^7 + 12\omega^5 - 68\omega^3 + 81\omega. \end{aligned}$$

Using the root locus ideas of Section 3.3, the range of k_p values over which the sweeping needs to be carried out was narrowed down to

$$k_p \in (-20.6272, -1.7778) \cup (-0.3311, 6.1639).$$

By sweeping over different k_p values in these two intervals and using Theorem 4.1 repeatedly, we obtained two disconnected sets of (k_p, k_i, k_d) values for which $\delta(s, k_p, k_i, k_d)$ is Hurwitz. These sets are sketched in Fig. 4.3. $\triangle$

4.3 PID Stabilization of Discrete-Time Plants

In this section we present a procedure for determining the set of all PID gains that can stabilize a given discrete-time plant of arbitrary order. This procedure is based on the continuous-time results of the last section and the bilinear transformation. It is remarkable to note that the linear programming nature of the continuous-time solution can be preserved under the bilinear transformation with a suitable reparametrization.

In the analysis of discrete-time systems, the problem of determining the Schur stability of a given polynomial can be converted to the problem of determining the Hurwitz stability of another polynomial using the so-called bilinear transformation. Among the several bilinear transformations

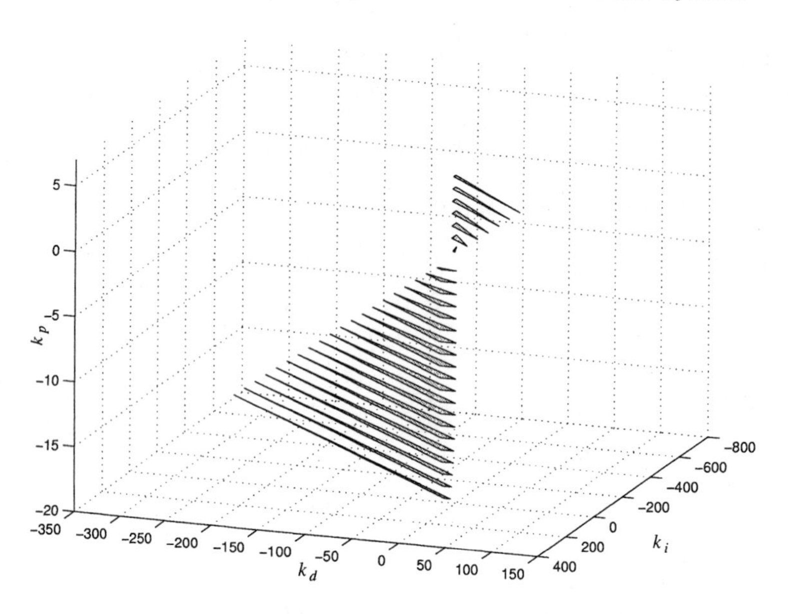

FIGURE 4.3. The stabilizing set of (k_p, k_i, k_d) values (Example 4.2).

available in the literature, we concentrate on the bilinear transformation $\mathcal{W}$ defined as follows. Given any polynomial $x(z)$

$$\mathcal{W}\{x(z)\} = x\left(\frac{w+1}{w-1}\right) = y(w)$$

where $y(w)$ is a rational function of w.

As the following lemma shows, the bilinear transformation $\mathcal{W}$ maps the roots of $x(z)$ located inside (on or outside) the unit circle to the zeros of $y(w)$ in $\mathcal{C}^-$ (on the imaginary axis or in $\mathcal{C}^+$). Additionally, a root of $x(z)$ at $z = 1$ is mapped to a zero of $y(w)$ at $w = \infty$. Thus, the Schur stability of a polynomial $x(z)$ is equivalent to the Hurwitz stability of $y(w)$, provided the numerator and denominator of $y(w)$ are of the same degree. Moreover, provided $x(z)$ has no roots at $z = 1$, the root distribution of $x(z)$ with respect to the unit circle is identical to the root distribution of the numerator of $y(w)$ with respect to the imaginary axis.

Consider a polynomial $\delta_z(z)$ of degree n given by

$$\delta_z(z) = a_0 + a_1 z + \cdots + a_{n-1} z^{n-1} + a_n z^n \,.$$

Then, $\mathcal{W}\{\delta_z(z)\}$ is given by

$$\mathcal{W}\{\delta_z(z)\} = \frac{\delta(w)}{(w-1)^n} \tag{4.8}$$

where $\delta(w) = b_o + b_1 w + \cdots + b_{m-1} w^{m-1} + b_m w^m$ is a polynomial in w of degree $m \leq n$.

Lemma 4.2 *Let n_i, n_o, n_b be the numbers of roots of $\delta_z(z)$ located inside, outside, and on the unit circle, respectively. Furthermore, let $l(\delta)$, $r(\delta)$, and $i(\delta)$ denote the numbers of roots of $\delta(w)$ in $\mathcal{C}^-$, $\mathcal{C}^+$, and on the imaginary axis, respectively. Then, we have*

1. *$n - m =$ the number of root of $\delta_z(z)$ at $z = 1$.*
2. *$n_i = l(\delta)$, $n_o = r(\delta)$.*
3. *$n_b = i(\delta) + (n - m)$.*

Proof. We start by rewriting the polynomial $\delta_z(z)$ as follows:

$$\delta_z(z) = k \prod_{i=1}^{n} (z - z_i)$$

where z_i, $i = 1, 2, \ldots, n$ are the roots of $\delta_z(z)$. Clearly

$$\mathcal{W}\{\delta_z(z)\} = k \prod_{i=1}^{n} \mathcal{W}\{z - z_i\}\,.$$

Let us now concentrate on the factor $\mathcal{W}(z - z_i)$. First let us assume that $z_i \neq 1$. Then from the definition of $\mathcal{W}$ we have

$$\begin{aligned}
\mathcal{W}\{z - z_i\} &= \frac{w+1}{w-1} - z_i \\
&= \frac{(1 - z_i)(w - \frac{z_i+1}{z_i-1})}{w-1} \\
&= c_i \frac{w - w_i}{w-1}
\end{aligned}$$

where $c_i = 1 - z_i$, and $w_i = \frac{z_i+1}{z_i-1}$. If we assume $z_i = x_i + jy_i$, then

$$\begin{aligned}
w_i &= \frac{x_i + 1 + jy_i}{x_i - 1 + jy_i} \\
&= \frac{x_i^2 + y_i^2 - 1}{(x_i - 1)^2 + y_i^2} - j\frac{2y_i}{(x_i - 1)^2 + y_i^2}\,.
\end{aligned} \tag{4.9}$$

We now consider the following three cases:

Case 1: z_i is inside the unit circle. Then, $x_i^2 + y_i^2 < 1$, so that from (4.9) it follows that $\text{Re}[w_i] < 0$.

Case 2: z_i is outside the unit circle. In this case $x_i^2 + y_i^2 > 1$, so that from (4.9) it follows that $\text{Re}[w_i] > 0$.

Case 3: z_i is located on the unit circle. In this case $x_i^2 + y_i^2 = 1$, so that from (4.9) it follows that $\text{Re}[w_i] = 0$ and w_i lies on the imaginary axis.

We now consider the case $z_i = 1$ where direct computation yields

$$\mathcal{W}\{z - z_i\} = \frac{2}{w-1}\,.$$

Thus, in this case the numerator of $\mathcal{W}\{z - z_i\}$ has degree one less than its denominator. The proof of the lemma is obtained by applying the above observations to each of the factors $(z - z_i)$. ■

Lemma 4.2 can be used to study the closed-loop stability of a discrete-time system. For example, let us consider the standard unity feedback configuration in Fig. 3.1 where now the system is of the discrete-time type:

$$\begin{aligned} G_z(z) &= \frac{N_z(z)}{D_z(z)} \\ C_z(z) &= \frac{B_z(z)}{A_z(z)} . \end{aligned}$$

Then the characteristic equation of the closed-loop system is given by

$$\delta_z(z) = A_z(z)D_z(z) + B_z(z)N_z(z) . \tag{4.10}$$

Suppose that the polynomials $A_z(z)$, $B_z(z)$, $D_z(z)$, $N_z(z)$ have degrees n_c, m_c, n, m, respectively. Let us also assume that $G_z(z)$ and $C_z(z)$ are proper so that

$$m_c \leq n_c , \; m \leq n . \tag{4.11}$$

Applying the bilinear transformation to $G_z(z)$ and $C_z(z)$, we obtain

$$\begin{aligned} G(w) &= \mathcal{W}\{G_z(z)\} = \frac{N(w)}{D(w)} \\ C(w) &= \mathcal{W}\{C_z(z)\} = \frac{B(w)}{A(w)} \end{aligned}$$

where $G(w)$ and $C(w)$ represent the new plant and controller in the w domain.

Similarly, applying the bilinear transformation to $\delta_z(z)$ and taking into account (4.11), we have

$$\begin{aligned} \mathcal{W}\{\delta_z(z)\} &= \mathcal{W}\{A_z(z)\}\mathcal{W}\{D_z(z)\} + \mathcal{W}\{B_z(z)\}\mathcal{W}\{N_z(z)\} \\ &= \frac{A_o(w)}{(w-1)^{n_c}} \cdot \frac{D_o(w)}{(w-1)^{n}} + \frac{B_o(w)}{(w-1)^{m_c}} \cdot \frac{N_o(w)}{(w-1)^{m}} \\ &= \frac{A_o(w) \cdot D_o(w)}{(w-1)^{n+n_c}} \\ &\quad + \frac{B_o(w)(w-1)^{n_c-m_c} \cdot N_o(w)(w-1)^{n-m}}{(w-1)^{n+n_c}} . \end{aligned} \tag{4.12}$$

The following relationships are easily verified:

$$\begin{aligned} A(w) &= A_o(w) \\ B(w) &= B_o(w)(w-1)^{n_c-m_c} \\ D(w) &= D_o(w) \\ N(w) &= N_o(w)(w-1)^{n-m} . \end{aligned} \tag{4.13}$$

Hence, the numerator of (4.12) can be expressed as

$$\delta(w) = A(w)D(w) + B(w)N(w) . \tag{4.14}$$

This allows us to state the following result.

Lemma 4.3 *Suppose $\delta_z(z)$ in (4.10) has no roots at $z = 1$. Then the $[G_z(z), C_z(z)]$ closed-loop system in the z domain is Schur stable if and only if the $[G(w), C(w)]$ closed-loop system in the w domain is Hurwitz stable.*

It is important to point out that in Lemma 4.3, it is assumed that $\delta_z(z)$ has no roots at $z = 1$. Without this assumption, it is not possible to infer the Schur stability of $\delta_z(z)$ by simply ascertaining the Hurwitz stability of the polynomial $\delta(w)$ in (4.14).

We will now make use of this lemma to characterize the set of stabilizing P, PI, and PID gains for a given discrete-time plant. Consider the discrete-time P, PI, and PID controllers described by

$$\begin{aligned}
\text{P:} \quad C_z(z) &= k_p , \\
\text{PI:} \quad C_z(z) &= k_p + k_i \frac{1}{1 - z^{-1}} = \frac{(k_p + k_i)z - k_p}{z - 1} , \\
\text{PID:} \quad C_z(z) &= k_p + k_i \frac{1}{1 - z^{-1}} + k_d \frac{1 - 2z^{-1} + z^{-2}}{1 - z^{-1}} \\
&= \frac{(k_p + k_i + k_d)z^2 - (k_p + 2k_d)z + k_d}{z^2 - z} .
\end{aligned}$$

Also consider the w-domain counterparts obtained by substituting $z = \frac{w+1}{w-1}$

$$\begin{aligned}
\text{P:} \quad \frac{B(w)}{A(w)} &= \frac{k_p}{1} , \\
\text{PI:} \quad \frac{B(w)}{A(w)} &= \frac{k_i w + 2k_p + k_i}{2} , \\
\text{PID:} \quad \frac{B(w)}{A(w)} &= \frac{k_i w^2 + 2(k_p + k_i)w + 2k_p + k_i + 4k_d}{2(w + 1)} .
\end{aligned}$$

According to (4.14), the corresponding w-domain closed-loop characteristic polynomials are

$$\begin{aligned}
\text{P:} \quad \delta(w) &= D(w) + k_p N(w) , \\
\text{PI:} \quad \delta(w) &= 2D(w) + (k_i w + 2k_p + k_i)N(w) , \\
\text{PID:} \quad \delta(w) &= 2(w + 1)D(w) + [k_i w^2 + 2(k_p + k_i)w + 2k_p \\
&\quad + k_i + 4k_d]N(w) .
\end{aligned}$$

In view of Lemma 4.3, it follows that as long as $\delta_z(z)$ has no roots at $z = 1$, the Hurwitz stability of each of the above w-domain polynomials will guarantee the Schur stability of the corresponding closed-loop system. The case of $\delta_z(z)$ having a root at $z = 1$ arises when a PI or a PID controller is being used and the plant has a zero at $z = 1$. However, in such cases, there is an unstable pole-zero cancellation, and so the discrete-time closed-loop system is anyway internally unstable, regardless of the controller parameter values. Thus these cases can be handled by concluding instability directly without having to go through any bilinear transformation. For all other cases, we proceed as follows to find the controller parameter values that make $\delta(w)$ Hurwitz stable.

As in the previous section, we multiply (4.14) by the factor $N(-w)$ to obtain

$$\delta^*(w) = N(-w)\delta(w) .$$

If we now consider the even-odd decompositions

$$\begin{aligned} N(w) &= N_e(w^2) + wN_o(w^2) \\ D(w) &= D_e(w^2) + wD_o(w^2) , \end{aligned}$$

we can obtain the following expressions for $\delta^*(w)$:

(1) For a P controller

$$\begin{aligned} \delta^*(w) &= p(w^2, k_p) + wq(w^2) \\ &= [k_p p_1(w^2) + p_2(w^2)] + wq(w^2) \end{aligned}$$

where

$$\begin{aligned} p_1(w^2) &= N_e^2(w^2) - w^2 N_o^2(w^2) \\ p_2(w^2) &= D_e(w^2)N_e(w^2) - w^2 D_o(w^2)N_o(w^2) \\ q(w^2) &= D_o(w^2)N_e(w^2) - D_e(w^2)N_o(w^2) . \end{aligned}$$

(2) For a PI controller

$$\begin{aligned} \delta^*(w) &= p(w^2, k_p, k_i) + wq(w^2, k_i) \\ &= [k_p p_1(w^2) + k_i p_2(w^2) + p_3(w^2)] + w[k_i q_1(w^2) + q_2(w^2)] \end{aligned}$$

where

$$\begin{aligned} p_1(w^2) &= 2(N_e^2(w^2) - w^2 N_o^2(w^2)) \\ p_2(w^2) &= N_e^2(w^2) - w^2 N_o^2(w^2) \\ p_3(w^2) &= 2(D_e(w^2)N_e(w^2) - w^2 D_o(w^2)N_o(w^2)) \\ q_1(w^2) &= N_e^2(w^2) - w^2 N_o^2(w^2) \\ q_2(w^2) &= 2(D_o(w^2)N_e(w^2) - D_e(w^2)N_o(w^2)) . \end{aligned}$$

(3) For a PID controller

$$\begin{aligned}\delta^*(w) &= p(w^2, k_p, k_i, k_d) + wq(w^2, k_p, k_i) \\ &= [k_p p_1(w^2) + k_i p_2(w^2) + k_d p_3(w^2) + p_4(w^2)] \\ &\quad + w[k_p q_1(w^2) + k_i q_2(w^2) + q_3(w^2)]\end{aligned}$$

where

$$\begin{aligned}p_1(w^2) &= 2(N_e^2(w^2) - w^2 N_o^2(w^2)) \\ p_2(w^2) &= (1 + w^2)(N_e^2(w^2) - w^2 N_o^2(w^2)) \\ p_3(w^2) &= 4(N_e^2(w^2) - w^2 N_o^2(w^2)) \\ p_4(w^2) &= 2(N_e(w^2)D_e(w^2) + w^2 N_e(w^2)D_o(w^2) \\ &\quad - w^2 N_o(w^2)D_e(w^2) - w^2 N_o(w^2)D_o(w^2)) \\ q_1(w^2) &= 2(N_e^2(w^2) - w^2 N_o^2(w^2)) \\ q_2(w^2) &= 2(N_e^2(w^2) - w^2 N_o^2(w^2)) \\ q_3(w^2) &= 2(N_e(w^2)D_e(w^2) + N_e(w^2)D_o(w^2) - N_o(w^2)D_e(w^2) \\ &\quad - w^2 N_o(w^2)D_o(w^2)) .\end{aligned}$$

Note that for a P controller, k_p appears only in the even part of $\delta^*(w)$ and so we can use the approach of Section 3.2 to obtain the set of all k_p values that make $\delta(w)$ Hurwitz stable. Similarly, we note that for a PI controller, k_i appears in both the even and odd parts of $\delta^*(w)$ whereas k_p appears only in the even part. Thus all that we need to do is sweep over all real values of k_i and find the stabilizing set of k_p values at each stage following the procedure outlined in Section 3.3.

In the case of the PID controller, the situation gets a little more involved since now two parameters (k_p and k_i) appear in both the even and odd parts of $\delta^*(w)$. However, we can simplify this by noting that $q_1(w^2) = q_2(w^2)$. Thus we can combine k_p and k_i together by using the substitution

$$k_i = k_s - k_p . \tag{4.15}$$

With this substitution, we have

$$\begin{aligned}\delta^*(w) &= \bar{p}(w^2, k_p, k_s, k_d) + w\bar{q}(w^2, k_s) \\ &= [k_p \bar{p_1}(w^2) + k_s \bar{p_2}(w^2) + k_d \bar{p_3}(w^2) + \bar{p_4}(w^2)] \\ &\quad + w[k_s \bar{q_1}(w^2) + \bar{q_2}(w^2)]\end{aligned} \tag{4.16}$$

where

$$\begin{aligned}\bar{p_1}(w^2) &= (1 - w^2)(N_e^2(w^2) - w^2 N_o^2(w^2)) \\ \bar{p_2}(w^2) &= (1 + w^2)(N_e^2(w^2) - w^2 N_o^2(w^2)) \\ \bar{p_3}(w^2) &= 4(N_e^2(w^2) - w^2 N_o^2(w^2))\end{aligned}$$

$$\begin{aligned}
\bar{p_4}(w^2) &= 2(N_e(w^2)D_e(w^2) + w^2 N_e(w^2)D_o(w^2) - w^2 N_o(w^2)D_e(w^2) \\
&\quad -w^2 N_o(w^2)D_o(w^2)) \\
\bar{q_1}(w^2) &= 2(N_e^2(w^2) - w^2 N_o^2(w^2)) \\
\bar{q_2}(w^2) &= 2(N_e(w^2)D_e(w^2) + N_e(w^2)D_o(w^2) - N_o(w^2)D_e(w^2) \\
&\quad -w^2 N_o(w^2)D_o(w^2)) .
\end{aligned}$$

From (4.16), it is clear that we can now proceed as in Section 4.2: fix k_s and then use linear programming to solve for the stabilizing values of k_p and k_d. Once the stabilizing values of (k_p, k_d, k_s) have been obtained, the stabilizing values of (k_p, k_i, k_d) can be obtained using the following transformation:

$$\begin{bmatrix} k_p \\ k_i \\ k_d \end{bmatrix} = \begin{bmatrix} 1 & 0 & 0 \\ -1 & 0 & 1 \\ 0 & 1 & 0 \end{bmatrix} \cdot \begin{bmatrix} k_p \\ k_d \\ k_s \end{bmatrix} . \tag{4.17}$$

Example 4.3 *Consider a discrete-time plant given by the transfer function* $G_z(z) = \frac{N_z(z)}{D_z(z)}$, *where*

$$\begin{aligned}
N_z(z) &= z + 1 \\
D_z(z) &= z^2 - 0.8z + 0.12 .
\end{aligned}$$

Our objective is to determine the set of discrete-time PID controllers that stabilize the plant $G_z(z)$. *Using the bilinear transformation, we obtain the* w*-domain plant* $G(w) = \frac{N(w)}{D(w)}$, *where*

$$\begin{aligned}
N(w) &= 2w(w-1) \\
D(w) &= 0.32w^2 + 1.76w + 1.92 .
\end{aligned}$$

If we now combine the controller parameters k_p *and* k_i *as in (4.15), the closed-loop characteristic polynomial is given by*

$$\begin{aligned}
\delta^*(w) &= [k_p\bar{p_1}(w^2) + k_s\bar{p_2}(w^2) + k_d\bar{p_3}(w^2) + \bar{p_4}(w^2)] \\
&\quad +w[k_s\bar{q_1}(w^2) + \bar{q_2}(w^2)]
\end{aligned}$$

where

$$\begin{aligned}
&\bar{p_1}(w^2) = -4w^6 + 8w^4 - 4w^2 \quad && \bar{q_1}(w^2) = 8w^4 - 8w^2 \\
&\bar{p_2}(w^2) = 4w^6 - 4w^2 && \bar{q_2}(w^2) = 1.28w^4 + 23.04w^2 + 7.68 \\
&\bar{p_3}(w^2) = 16w^4 - 16w^2 && \\
&\bar{p_4}(w^2) = 9.6w^4 + 22.4w^2 . &&
\end{aligned}$$

As in the continuous-time case, we can determine the range of k_s *values to be swept over by examining the odd part of* $\delta^*(w)$. *The range of* k_s *values so determined is*

$$0 < k_s < 0.8808$$

where $k_s = k_p + k_i$. *For each value in this range, we can determine the set of stabilizing values in the* k_p-k_d *space. Then, we can use the transformation (4.17) to obtain the stabilizing values of* (k_p, k_i, k_d) *for this problem. Figure 4.4 shows the resulting stabilizing region.* △

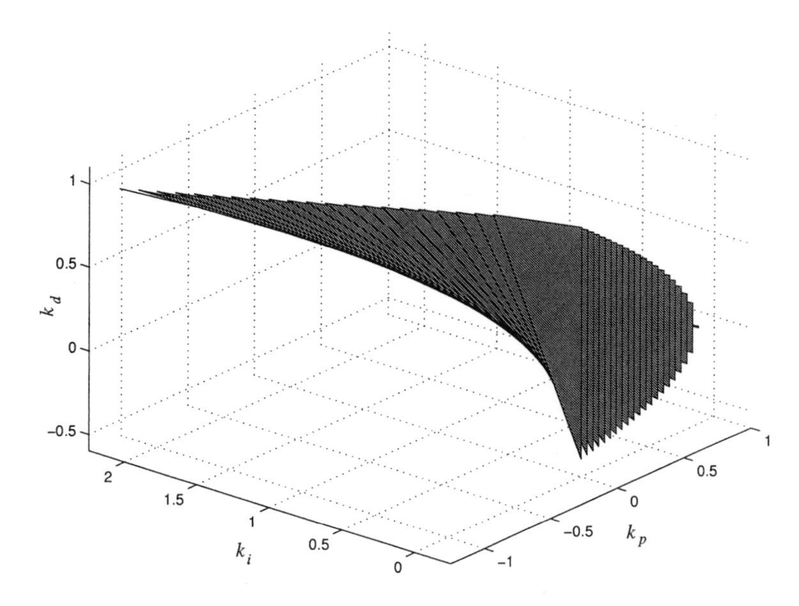

FIGURE 4.4. The stabilizing set of (k_p, k_i, k_d) values (Example 4.3).

4.4 Notes and References

The characterization of all stabilizing PID controllers for a given delay-free, linear-time invariant plant was developed by Ho, Datta, and Bhattacharyya [19]. For an extensive description of this characterization and its applications to optimal and robust design of PID controllers, the reader is referred to [10]. The results in Section 4.3 on PID stabilization of discrete-time plants were developed by Xu, Datta, and Bhattacharyya [50]. A detailed treatment of the bilinear transformation can be found in [33].

5

Preliminary Results for Analyzing Systems with Time Delay

In this chapter we present an important generalization of the Hermite-Biehler Theorem due to Pontryagin for characteristic equations or systems that contain time-delay terms. Some other results for analyzing systems with time delay are also introduced and will be the basis for the work presented in the following chapters. Proofs of all results are omitted and the reader is referred to the extensive literature on the subject.

5.1 Introduction

In the previous chapter we have seen how the classical Hermite-Biehler Theorem can be generalized and used to solve the problem of finding the set of all the stabilizing PID controllers for a linear time-invariant system described by a rational transfer function. However, when the system under study involves time delays, the Hermite-Biehler Theorem for polynomials cannot be used.

Linear time-invariant systems with delays give rise to characteristic functions known as quasi-polynomials. Pontryagin was one of the first researchers to study these quasi-polynomials. He derived necessary and sufficient conditions for the roots of a given quasi-polynomial to have negative real parts. Furthermore, he used such conditions to study the stability of certain classes of quasi-polynomials. These and some other preliminary results are described in this chapter and will be used throughout the book to study the stability of systems with time delays. Even though these results cannot be used to easily check stability, they can form the basis of useful strategies for solving some fixed-order and fixed-structure stabilization problems for systems with time delays.

The chapter is organized as follows. Section 5.2 introduces the concept of characteristic equations for delay systems. In Section 5.3 we discuss the limitations of using the Padé approximation to approximate a pure time delay by a rational transfer function. This is done mainly to motivate the need for working with quasi-polynomials while dealing with systems containing time delays. In Section 5.4 we present an extension of the Hermite-Biehler Theorem for quasi-polynomials due to Pontryagin along with some other useful results. Section 5.5 shows how to apply the previous results to the analysis of systems with time delay. Finally, Section 5.6 provides the reader with some alternative tools for studying the stability of quasi-polynomials.

5.2 Characteristic Equations for Delay Systems

Delays are present in a system when a signal or physical variable originating in one part of a system becomes available in another part after a lapse of time. For example, the change of flow rate in a pipeline becomes known downstream after a lapse of time determined by the length of the pipe and velocity of flow. Delays can also happen due to the time associated with the transmission of information to remote locations and in digital control systems due to the time involved in computing control signals from measured data.

The block in Fig. 5.1 can represent time delay:

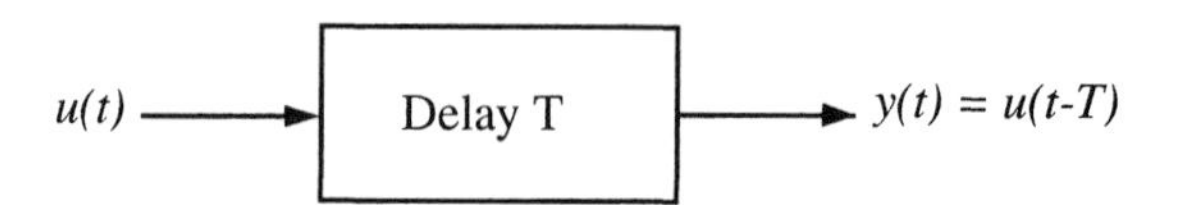

FIGURE 5.1. Delay representation.

In a dynamic feedback system where delay is present, the system equation may take the form

$$\dot{y}(t) + ay(t-T) = u(t) \ . \tag{5.1}$$

The block diagram representation of (5.1) is depicted in Fig. 5.2.

If there is a delay in the input, the system equation may take the form

$$\dot{y}(t) + ay(t) = u(t-T) \tag{5.2}$$

with the block diagram depicted in Fig. 5.3 or, if the delay is within the loop,

$$\dot{y}(t) = -ay(t-T) + u(t-T) \tag{5.3}$$

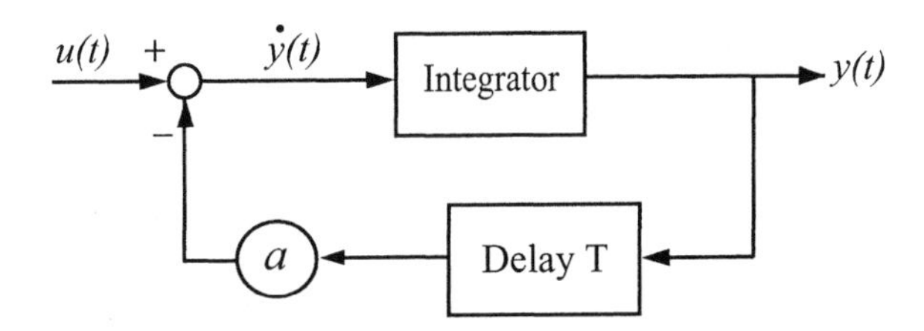

FIGURE 5.2. A feedback system with delay.

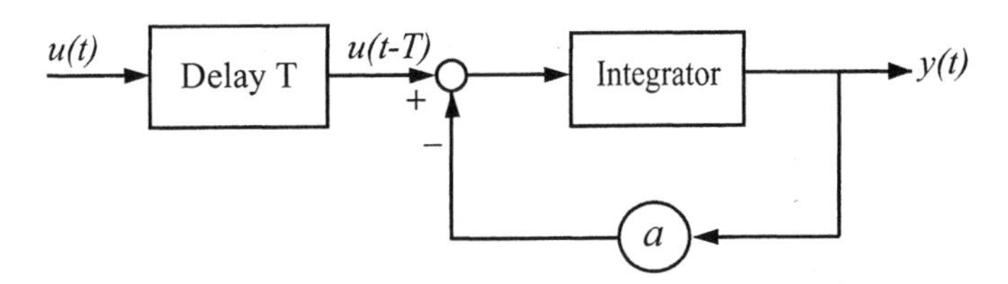

FIGURE 5.3. Input delay.

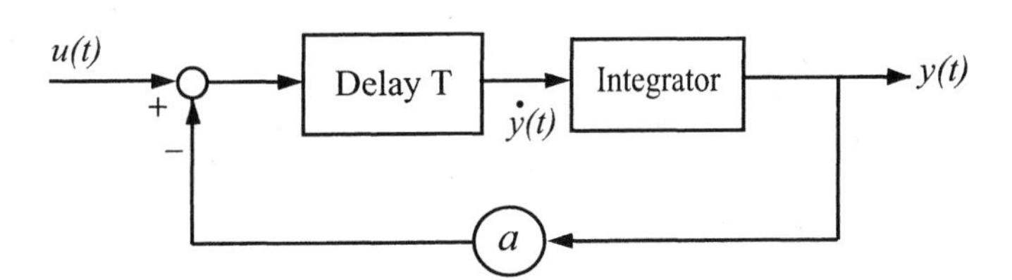

FIGURE 5.4. Delay within the loop.

with the block diagram depicted in Fig. 5.4.

A higher-order sysem with multiple delays might be represented by the equation

$$\ddot{y}(t) + a_1\dot{y}(t - T_1) + a_0 y(t - T_0) = u(t) \tag{5.4}$$

with the corresponding block diagram depicted in Fig. 5.5. The system (5.4) can be represented in state variable form by introducing

$$y(t) = x_1(t)\ ,\ \ \dot{y}(t) = x_2(t)$$

and writing

$$\begin{aligned} \begin{pmatrix} \dot{x}_1(t) \\ \dot{x}_2(t) \end{pmatrix} &= \begin{pmatrix} 0 & 1 \\ 0 & 0 \end{pmatrix}\begin{pmatrix} x_1(t) \\ x_2(t) \end{pmatrix} + \begin{pmatrix} 0 & 0 \\ -a_0 & 0 \end{pmatrix}\begin{pmatrix} x_1(t - T_0) \\ x_2(t - T_0) \end{pmatrix} \\ &+ \begin{pmatrix} 0 & 0 \\ 0 & -a_1 \end{pmatrix}\begin{pmatrix} x_1(t - T_1) \\ x_2(t - T_1) \end{pmatrix} + \begin{pmatrix} 0 \\ 1 \end{pmatrix} u(t)\ . \end{aligned} \tag{5.5}$$

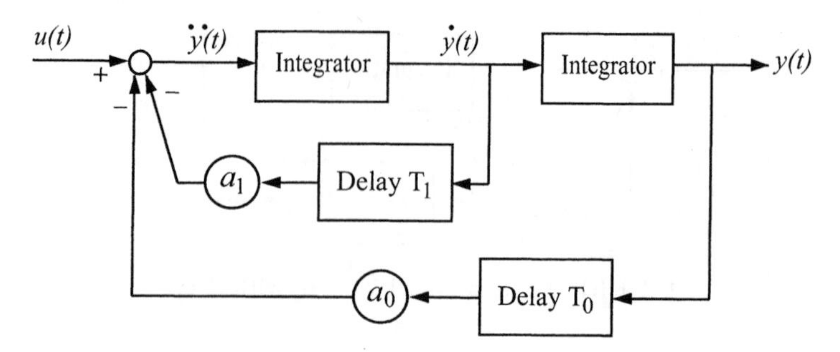

FIGURE 5.5. Multiple delays.

More generally, a linear time-invariant system with l distinct delays $T_1, \ldots,$ T_l may be represented in state-space form as

$$\dot{x}(t) = A_0 x(t) + \sum_{i=1}^{l} A_i x(t - T_i) + Bu(t) . \tag{5.6}$$

To discuss the stability of systems such as (5.1)—(5.6), it is usual to examine the solutions $y(t)$ with $u(t) \equiv 0$ and study the behavior of $y(t)$ as $t \to \infty$. For this purpose consider, for example, the system (5.4) with $u(t) \equiv 0$, and let $y(t) = e^{st}$ be a proposed solution of

$$\ddot{y}(t) + a_1 \dot{y}(t - T_1) + a_0 y(t - T_0) \equiv 0 . \tag{5.7}$$

Then we have

$$\left(s^2 + a_1 e^{-sT_1} s + a_0 e^{-sT_0}\right) e^{st} \equiv 0$$

so that s must satisfy

$$s^2 + a_1 s e^{-sT_1} + a_0 e^{-sT_0} = 0 . \tag{5.8}$$

Equation (5.8) is the characteristic equation of (5.4) or (5.7), and the location of its roots (or zeros) determine the stability of the system represented by (5.4). In particular, if any roots lie in the closed right half plane, the system is unstable as the solution grows without bound.

The characteristic equation associated with (5.6) can be shown to be

$$\begin{aligned} \delta(s) &:= \det\left(sI - A_0 - \sum_{i=1}^{l} e^{-sT_i} A_i\right) \\ &= P_0(s) + \sum_{k=1}^{m} P_k(s) e^{-L_k s} \end{aligned} \tag{5.9}$$

where L_k, $k = 1, 2, \ldots, m$, are sums of the T_i and

$$P_0(s) = s^n + \sum_{i=0}^{n-1} a_i s^i \tag{5.10}$$

$$P_k(s) = \sum_{i=0}^{n-1} (b_k)_i s^i . \tag{5.11}$$

We note that in (5.9) there is no delay term associated with the highest-order derivative. Such systems are referred to as *retarded delay systems.* When the highest-derivative term contains delays we may have an equation such as

$$\ddot{y}(t - T_2) + \alpha_1 \dot{y}(t - T_1) + \alpha_0 y(t - T_0) = u(t) \tag{5.12}$$

with characteristic equation

$$e^{-sT_2} s^2 + \alpha_1 e^{-sT_1} s + \alpha_0 e^{-sT_0} = 0 . \tag{5.13}$$

Such systems (with delays in the highest-derivative terms) are called *neutral delay systems.* In both retarded and neutral delay systems, stability is equivalent to the condition that all the roots of the characteristic equation lie in the open left half plane (LHP).

There are important differences between the nature of the roots of characteristic equations for retarded and neutral delay systems. In a retarded system there can only be a finite number of right half plane (RHP) roots, a condition that does not hold for all neutral systems. The stability of retarded systems is equivalent to the absence of closed RHP roots. For neutral systems certain root chains can approach the imaginary axis from the LHP and thus destroy stability. This can be avoided by insisting that all roots lie strictly to the left of a line Re$[s]= -\alpha$, for $\alpha > 0$. The fact that retarded systems have a finite number of RHP roots means that one can count the number of roots crossing into the RHP through the stability boundary and keep track of the number of RHP roots as some parameters vary. This has significant implications for stability analysis.

Finally, we mention that in equations such as (5.8), (5.9), or (5.13), the delays may be integer multiples of a common positive number τ. In such cases, the delays are said to be commensurate and the characteristic equation takes the form

$$\delta(s) = a_0(s) + a_1(s)e^{-\tau s} + a_2(s)e^{-2\tau s} + \cdots + a_k(s)e^{-k\tau s}$$

where $a_i(s)$, $i = 0, 1, \ldots, k$, are polynomials. Thus $\delta(s)$ is a polynomial in the two variables s and $v := e^{-s\tau}$. There are extensive results on such quasi-polynomials by Pontryagin and others, which will be used in this book.

The reader is referred to the expert book by Gu, Kharitonov, and Chen [15] for a detailed and more complete discussion of stability issues in time-delay systems.

In the present book we will deal only with the case of a single delay in the feedback loop, representing delay in control action or delayed measurements. Even for this simple case, the stability problem from the synthesis point of view is complex and challenging, as we shall see.

5.3 Limitations of the Padé Approximation

The Padé approximation is often used to approximate a pure time delay by a rational transfer function. A logical approach is to use this approximation for the stability analysis and design of controllers for time-delay systems; since this approximation reduces a time-delay system to one with a *rational* transfer function, for P, PI, and PID controllers, the results presented in Chapters 3 and 4 can be employed.

In this section we show via examples that PI and PID controllers that stabilize a system obtained by such an approximation of the time delay may actually be destabilizing for the true system. This will constitute one of the motivations for developing a new theory for the study of PID controllers for time-delay systems. There is usually some qualitative agreement for small values of the time delay, but the behavior may be very different for large values of the time delay. Furthermore, these examples will also show that the qualitative behavior improves with increasing order of the Padé approximant, but at the expense of greater algebraic complexity.

To show the limitations of the Padé approximation, we consider a simple first-order model with time delay described by the following transfer function:

$$G(s) = \frac{k}{1+sT} e^{-sL} \,. \tag{5.14}$$

Here k represents the steady-state gain, T is the time constant, and L is the time delay of the system. As in Section 3.2, we consider the feedback control system of Fig. 3.1 where the controller transfer function $C(s)$ is chosen to be a PID, i.e.,

$$C(s) = k_p + \frac{k_i}{s} + k_d s = \frac{k_i + k_p s + k_d s^2}{s} \tag{5.15}$$

and the plant $G(s)$ being stabilized is described by (5.14). For a given delay-free plant of arbitrary order, the results in Section 4.2 allow us to characterize all stabilizing PID controllers. Clearly these results cannot be directly applied to the PID stabilization of the plant model (5.14) since it is not a *rational* transfer function. This difficulty can, however, be overcome by approximating the time-delay term by a properly chosen Padé approximation. The Padé approximation for the term e^{-sL} is given by

$$e^{-sL} \cong \frac{N_r(sL)}{D_r(sL)}$$

where

$$\begin{aligned} N_r(sL) &= \sum_{k=0}^{r} \frac{(2r-k)!}{k!(r-k)!}(-sL)^k \\ D_r(sL) &= \sum_{k=0}^{r} \frac{(2r-k)!}{k!(r-k)!}(sL)^k \end{aligned}$$

and r represents the order of the approximation. For example, the third-order Padé approximation ($r = 3$) is given by

$$\frac{N_3(sL)}{D_3(sL)} = \frac{-L^3s^3 + 12L^2s^2 - 60Ls + 120}{L^3s^3 + 12L^2s^2 + 60Ls + 120}.$$

Since this approximation is of finite order, the results of Section 4.2 can be used to characterize all stabilizing PID controllers for the resulting rational transfer function model. Thus we are now in a position to compare the stabilizing sets of PID parameters obtained using (5.14) and its Padé approximants of different orders. This is carried out in the next two subsections.

5.3.1 Using a First-Order Padé Approximation

The first-order Padé approximation of the time-delay term is

$$e^{-sL} \cong \frac{2 - Ls}{2 + Ls}.$$

Using this approximation in (5.14), we obtain the following rational transfer function $G_m(s)$:

$$G_m(s) = \frac{k}{(Ts+1)} \frac{(-Ls+2)}{(Ls+2)}.$$

With the PID controller given by (5.15), the closed-loop characteristic polynomial becomes

$$\delta(s, k_p, k_i, k_d) = s(Ts+1)(Ls+2) + (k_i + k_p s + k_d s^2)(k)(-Ls+2)$$

which can be rewritten as

$$\delta(s, k_p, k_i, k_d) = (Ts^2 + s)(Ls+2) + (k_d' s^2 + k_i')(-Ls+2) + k_p' s(-Ls+2)$$

where $k_d' = kk_d$, $k_i' = kk_i$, $k_p' = kk_p$.

Now by using the results of Section 4.2 we can obtain an analytical characterization of all stabilizing PID controllers for $G_m(s)$. The stabilizing (k_p, k_i, k_d) values must satisfy the following inequalities:

$$\left\{ \begin{array}{rcl} k_i & > & 0 \\ k_i - [\frac{4(1+kk_p)}{L(4T+L-kk_pL)}]k_d & < & \frac{2(1+kk_p)(2T+L-kk_pL)}{kL(4T+L-kk_pL)} \\ k_d & < & \frac{T}{k} \end{array} \right. \tag{5.16}$$

and

$$-\frac{1}{k} < k_p < \frac{1}{k}\left(1 + \frac{4T}{L}\right). \tag{5.17}$$

Note that the set of inequalities (5.16) has a special structure. For a fixed k_p, (5.16) becomes a set of linear inequalities in terms of k_i, k_d and can

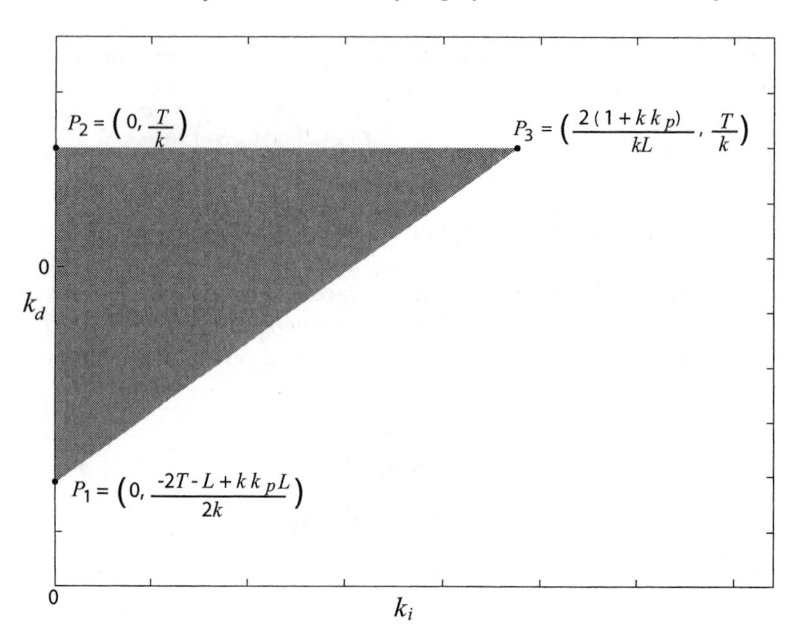

FIGURE 5.6. The stabilizing set of $(k_i,\, k_d)$ values for a fixed k_p.

be solved by linear programming. It is easy to show that for a fixed k_p satisfying (5.17), the admissible set of (k_i, k_d) values for which (5.16) holds, is given by the triangle shown in Fig. 5.6.

The coordinates of the vertices P_1, P_2, and P_3 of this triangular stabilizing region are

$$\left\{ \begin{array}{lcl} P_1 & = & (0, \frac{-2T-L+kk_pL}{2k}) \\ P_2 & = & (0, \frac{T}{k}) \\ P_3 & = & (\frac{2(1+kk_p)}{kL}, \frac{T}{k}). \end{array} \right. \tag{5.18}$$

Now the question of interest is whether the first-order Padé approximant accurately captures the actual set of stabilizing PID parameters for the original time-delay system. As the following example illustrates, the set in Fig. 5.6 can contain controller parameter values that lead to an unstable closed-loop system. Thus the first-order Padé approximation can prove inadequate for determining the set of stabilizing PID parameters for a time-delay system.

Example 5.1 *Consider the plant*

$$G(s) = \frac{1.6667}{1+2.9036s} e^{-0.2475s}.$$

The first-order Padé approximation yields

$$G_m(s) = \frac{1.6667}{(1+2.9036s)} \frac{(-0.1238s+1)}{(0.1238s+1)}.$$

Using (5.17) we obtained $k_p \in (-0.6, 28.7555)$ *as the necessary condition to be satisfied by any stabilizing* k_p *value. We fixed* k_p *at* 8.4467, *which is the value suggested by the Ziegler-Nichols step response method. Using (5.18) we obtained the set of* (k_i, k_d) *values that stabilize* $G_m(s)$ *for this fixed value of* k_p. *This set is sketched in Fig. 5.7.*

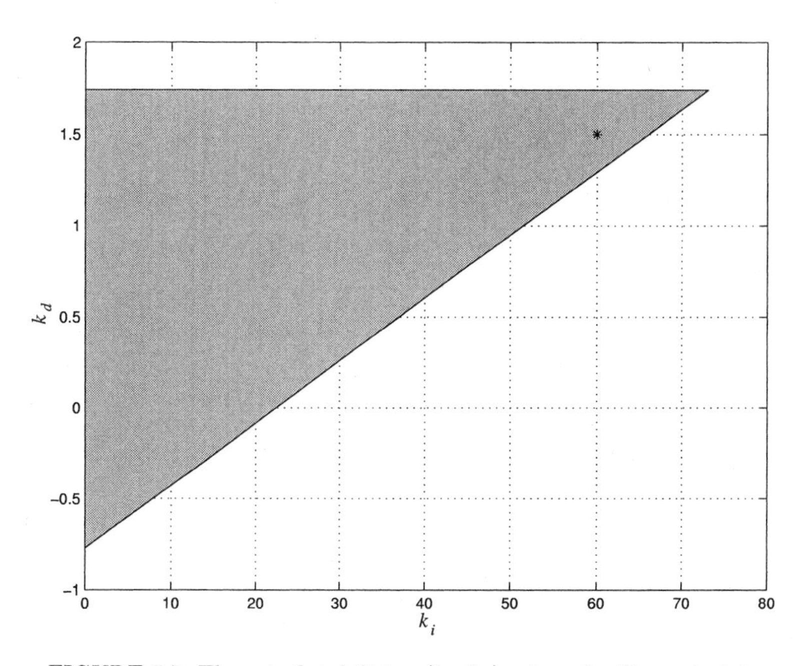

FIGURE 5.7. The set of stabilizing (k_i, k_d) values for Example 5.1.

We now set the PID controller parameters to $k_p = 8.4467$, $k_i = 60$, $k_d = 1.5$, *denoted by* * *in Fig. 5.7. Notice that this point is contained inside the region depicted in Fig. 5.7. However, as Fig. 5.8 shows, this set of values leads to an unstable closed-loop system with the true delay. Here, the input* $r(t)$ *to the system is a unit step applied at* $t = 5$ *seconds, and* $y(t)$ *represents the output of the system. It is clear from this that some stabilizing* (k_i, k_d) *values obtained on the basis of a Padé approximation can lead to a closed-loop system that in reality is unstable.* $\triangle$

5.3.2 Using Higher-Order Padé Approximations

When higher-order Padé approximations are used, it is no longer possible to obtain analytical expressions for the stabilizing PID gain values. However, a computational characterization is still possible by making use of the results of Section 4.2.

We now define as *approximate stabilizing sets*, the sets generated by approximating the time delay of the system with a Padé approximation

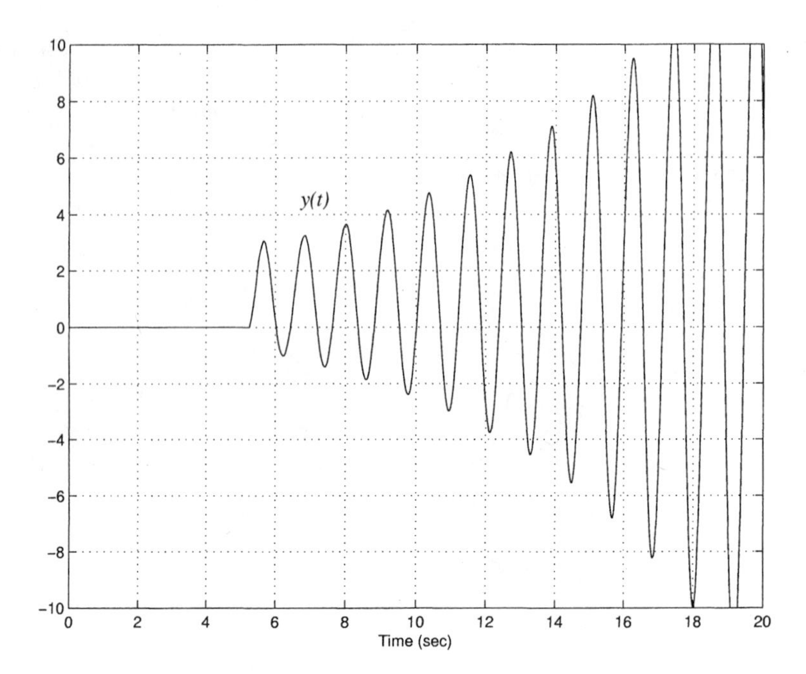

FIGURE 5.8. Step response of the system in Example 5.1.

and then using the results in Section 4.2. As the following examples show, using a second-order Padé approximation still fails to adequately capture the set of stabilizing PID gain values, i.e., it contains controller parameter values that lead to an unstable closed-loop system. However, as higher-order Padé approximations are used, the situation seems to improve since the approximate sets tend to converge toward a set that appears to be the *true stabilizing* set.

Example 5.2 *Consider again the plant used in Example 5.1. We now approximate the time-delay term using the second-, third-, and fifth-order Padé approximations. For each case we obtain the rational transfer function approximation* $G_m^i(s)$, *where* $i = 2, 3, 5$, *indicates the order of the approximation. For instance,*

$$G_m^3(s) = \frac{-1.6667s^3 + 80.8097s^2 - 1632.52s + 13192.1}{2.9036s^4 + 141.781s^3 + 2892.54s^2 + 23961.7s + 7915.09}.$$

As in Example 5.1, the controller parameter k_p *is set to* 8.4467. *Applying the results of Section 4.2 to the transfer functions* $G_m^i(s)$, *we obtain the sets of all stabilizing* (k_i, k_d) *values for* $k_p = 8.4467$. *The set corresponding to* $G_m^2(s)$ *is sketched in Fig. 5.9 with solid lines. The sets corresponding to* $G_m^3(s)$ *and* $G_m^5(s)$ *are both sketched with dashed lines and are essentially overlapping with each other.*

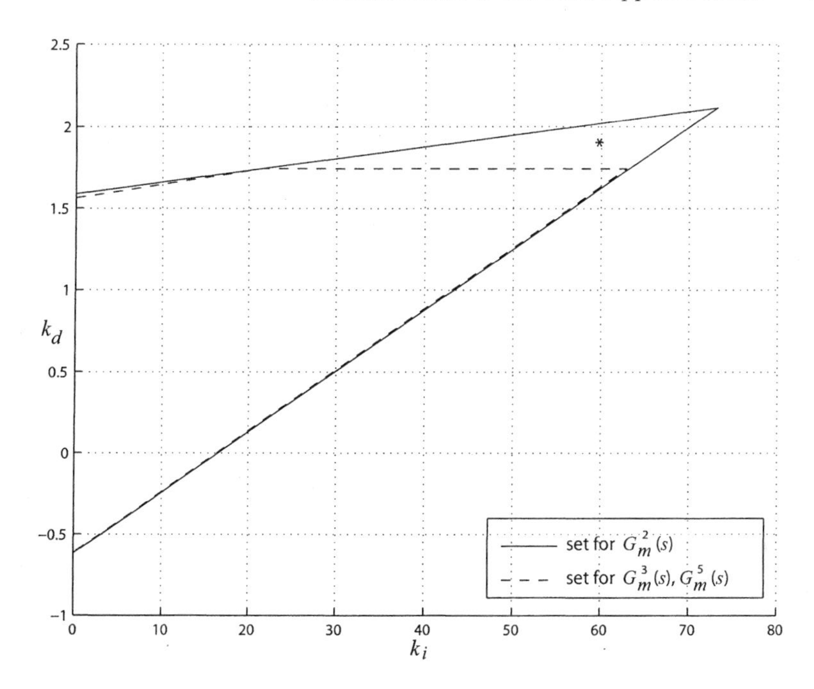

FIGURE 5.9. The set of stabilizing (k_i, k_d) values for Example 5.2.

*As in Example 5.1 we can take values inside the region corresponding to $G_m^2(s)$ but outside $G_m^3(s)$ (such as the one denoted by *) and show that the corresponding closed-loop system is unstable. Thus we conclude that while the second-order Padé approximation fails to adequately capture the actual stabilizing set in the (k_i, k_d)-plane, the third- and fifth-order Padé approximations apparently do a better job.* △

We now consider a system with a larger time delay.

Example 5.3 *Consider the following first-order plant with deadtime:*

$$G(s) = \frac{1}{1+s} e^{-10s} .$$

Let us approximate the time-delay term using the first-, second-, third-, fifth-, seventh-, and ninth-order Padé approximations. As in the previous example, we obtain the rational transfer function approximations $G_m^i(s)$, $i = 1, 2, 3, 5, 7, 9$. Next, we set the controller parameter k_p to 0.5 and compute the set of stabilizing (k_i,k_d) values using the results in Section 4.2. Figure 5.10 shows the stabilizing controller sets C_m^1, C_m^2 obtained for $G_m^1(s)$ and $G_m^2(s)$ in solid and dashed lines, respectively.

Figure 5.11 shows the stabilizing sets C_m^3, C_m^5 obtained for $G_m^3(s)$ and $G_m^5(s)$ in solid and dashed lines, respectively. The sets corresponding to $G_m^7(s)$ and $G_m^9(s)$ are similar and are represented with a dash-dotted line.

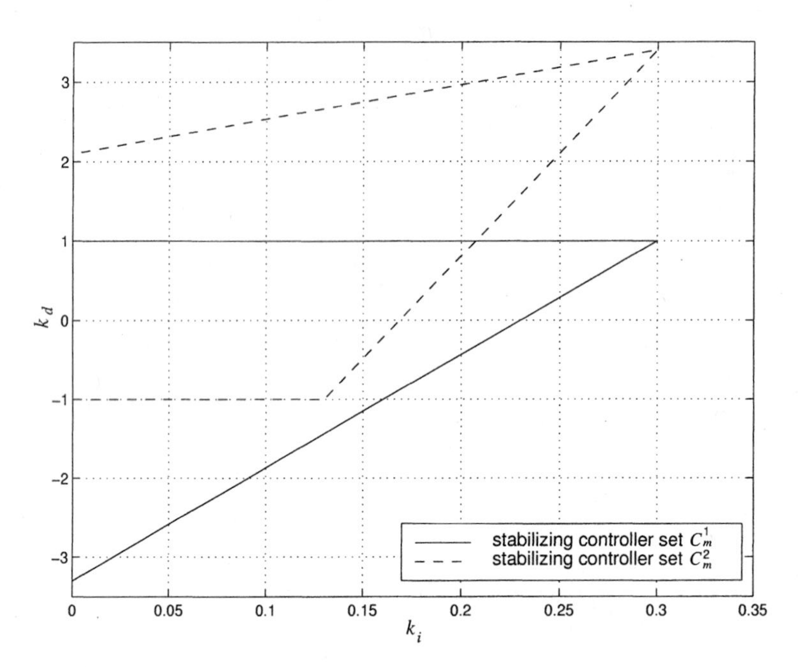

FIGURE 5.10. The set of stabilizing (k_i, k_d) values for Example 5.3.

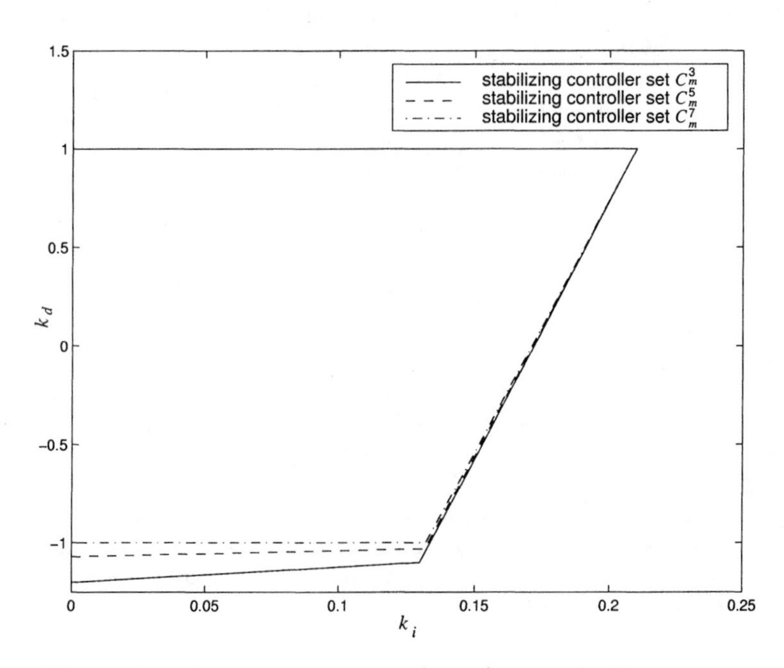

FIGURE 5.11. The set of stabilizing (k_i, k_d) values for Example 5.3.

For higher-order Padé approximations of the time delay, the set of (k_i, k_d) values seems to converge toward a possible true set. As in previous examples, we can show that for lower-order approximations we obtain sets that contain controller gain values that lead to an unstable behavior of the closed-loop system. △

From the previous examples we can make the following observations:

1. For small values of the time delay, the approximate sets easily converge to the possible true sets. However, the convergence becomes more difficult as the value of the time delay increases.

2. The convergence of the approximate set to a possible true set improves with increased order of the Padé approximation.

We conclude that the Padé approximation is far from being a satisfactory tool for ensuring the stability of the resulting control design. The main problem lies in the fact that it is not *a priori* clear as to what order of the approximation will yield a stabilizing set of controller parameters accurately approximating the true set. The previous examples also showed that by increasing the order of the approximation, the approximate set can be made to closely approach the possible true set but at the cost of a greater algebraic complexity.

5.4 The Hermite-Biehler Theorem for Quasi-Polynomials

The stabilization of delay-free systems is relatively easy to study because the number of roots of their characteristic equations is finite. However, when time delays are introduced, this ease of analysis disappears: the number of roots is no longer finite, making the establishment of stability quite a difficult task. To complicate matters, it can be shown that the Hermite-Biehler Theorem for Hurwitz polynomials presented in Chapter 2 does not carry over to arbitrary functions $F(s)$ of the complex variable s. Counterexamples can be found that show that the interlacing property introduced in Section 2.2 (see Fig. 2.2) is neither necessary nor sufficient to guarantee stability [5].

We will now study functions of the form $f(s, e^s)$, where $f(s,t)$ is a polynomial in two variables and is called a quasi-polynomial. Before presenting the results, we introduce some preliminary definitions. Let $f(s,t)$ be a polynomial in two variables with real or complex coefficients defined as follows:

$$f(s,t) = \sum_{h=0}^{M} \sum_{k=0}^{N} a_{hk} s^h t^k .$$

Definition 5.1 *$f(s,t)$ is said to have a principal term if there exists a nonzero coefficient a_{hk} where both indices have maximal values. Without loss of generality, we will denote the principal term as $a_{MN}s^M t^N$. This means that for each other term $a_{hk}s^h t^k$, for $a_{hk} \neq 0$, we have either $M > h$, $N > k$; or $M = h$, $N > k$; or $M > h$, $N = k$.*

For example, $f(s,t) = 3s + t^2$ does not have a principal term but the polynomial $f(s,t) = s^2 + t + 2s^2 t$ does. We now state the first result of Pontryagin.

Theorem 5.1 *If the polynomial $f(s,t)$ does not have a principal term, then the function $F(s) = f(s, e^s)$ has an infinite number of zeros with arbitrarily large positive real parts.*

If $f(s,t)$ does have a principal term, the main result of Pontryagin is to show that the Hermite-Biehler Theorem extends to the class of functions $F(s) = f(s, e^s)$. Before presenting this generalization of the Hermite-Biehler Theorem, we first need to study the zeros of functions of the form $g(s, \cos(s), \sin(s))$.

Let $g(s,u,v)$ be a polynomial with real coefficients in the variables s, u, and v, which we represent in the form

$$g(s,u,v) = \sum_{h=0}^{M} \sum_{k=0}^{N} s^h \phi_h^{(k)}(u,v) \,. \tag{5.19}$$

Here $\phi_h^{(k)}(u,v)$ is a polynomial of degree k, homogeneous in u and v. We assume that $\phi_h^{(k)}(u,v)$ is not divisible by u^2+v^2, i.e.,

$$\phi_h^{(k)}(1, \pm j) \neq 0 \,. \tag{5.20}$$

The principal term in the polynomial in (5.19) is the term $s^M \phi_M^{(N)}(u,v)$, for which h and k simultaneously attain maximum values, that is, for all h and k, we have either $M > h$, $N > k$; or $M = h$, $N > k$; or $M > h$, $N = k$. Furthermore, let $\phi^{*(N)}(u,v)$ denote the coefficient of s^M in (5.19), i.e.,

$$\phi^{*(N)}(u,v) = \sum_{k=0}^{N} \phi_M^{(k)}(u,v) \,.$$

We now consider the transcendental function

$$G(s) = g(s, \cos(s), \sin(s))$$

which assumes real values for real values of the argument. We will give necessary and sufficient conditions for the function $G(s)$ to have only real zeros in terms of its behavior in the real domain. To this end, let

$$\Phi^{*(N)}(s) := \phi^{*(N)}(\cos(s), \sin(s))$$

be a function that is clearly periodic with period 2π.

Theorem 5.2 *Let $g(s,u,v)$ be a polynomial with principal term given by $s^M \phi_M^{(N)}(u,v)$. If η is such that $\Phi^{*(N)}(\eta + j\omega)$ does not take the value zero for real ω, then starting from some sufficiently large value of l, the function $G(s)$ will have exactly $4lN+M$ zeros in the strip $-2l\pi+\eta \le \mathrm{Re}[s] \le 2l\pi+\eta$. Thus for the function $G(s)$ to have only real roots, it is necessary and sufficient that in the interval*

$$-2l\pi + \eta \le \mathrm{Re}[s] \le 2l\pi + \eta \ ,$$

it have exactly $4lN + M$ real roots starting with some sufficiently large l.

Let us now return to the function $F(s) = f(s, e^s)$ in the presence of a principal term. Recall that the function $f(s,t)$ was given by

$$f(s,t) = \sum_{h=0}^{M} \sum_{k=0}^{N} a_{hk} s^h t^k \tag{5.21}$$

and the principal term was $a_{MN} s^M t^N$. We rewrite this function as follows:

$$f(s,t) = s^M X^{*(N)}(t) + \sum_{h=0}^{M-1} \sum_{k=0}^{N} a_{hk} s^h t^k \tag{5.22}$$

where $X^{*(N)}(t)$ is the coefficient of s^M. Hence $X^{*(N)}(t)$ is given by

$$X^{*(N)}(t) = \sum_{k=0}^{N} a_{Mk} t^k \ .$$

We now introduce the following definition for interlacing.

Definition 5.2 *Let $F(s) = f(s, e^s)$, where $f(s,t)$ is a polynomial with a principal term, and write*

$$F(j\omega) = F_r(\omega) + jF_i(\omega)$$

where $F_r(\omega)$ and $F_i(\omega)$ represent, respectively, the real and imaginary parts of $F(j\omega)$. Let $\omega_{r1}, \omega_{r2}, \omega_{r3}, \ldots,$ denote the real roots of $F_r(\omega)$, and let $\omega_{i1}, \omega_{i2}, \omega_{i3}, \ldots,$ denote the real roots of $F_i(\omega)$, both arranged in ascending order of magnitude. Then we say that the roots of $F_r(\omega)$ and $F_i(\omega)$ interlace if they satisfy the following property:

$$\omega_{r1} < \omega_{i1} < \omega_{r2} < \omega_{i2} < \cdots \ .$$

In this definition we have

$$F_r(\omega) = g_r(\omega, \cos(\omega), \sin(\omega)) \ , \ \ F_i(\omega) = g_i(\omega, \cos(\omega), \sin(\omega))$$

where $g_r(\omega, u, v)$ and $g_i(\omega, u, v)$ are polynomials.

After these preliminaries, we present the generalization of the Hermite-Biehler Theorem to the quasi-polynomial $F(s) = f(s, e^s)$.

Theorem 5.3 *Let $F(s) = f(s, e^s)$, where $f(s,t)$ is a polynomial with a principal term, and write*

$$F(j\omega) = F_r(\omega) + jF_i(\omega)$$

where $F_r(\omega)$ and $F_i(\omega)$ represent, respectively, the real and imaginary parts of $F(j\omega)$. If all the roots of $F(s)$ lie in the open LHP, then the roots of $F_r(\omega)$ and $F_i(\omega)$ are real, simple, interlacing, and

$$F_i^{'}(\omega)F_r(\omega) - F_i(\omega)F_r^{'}(\omega) > 0 \tag{5.23}$$

for each ω in $(-\infty, \infty)$, where $F_r^{'}(\omega)$ and $F_i^{'}(\omega)$ denote the first derivative with respect to ω of $F_r(\omega)$ and $F_i(\omega)$, respectively.

Moreover, in order that all the roots of $F(s)$ lie in the open LHP, it is sufficient that one of the following conditions be satisfied:

1. *All the roots of $F_r(\omega)$ and $F_i(\omega)$ are real, simple, and interlacing and the inequality (5.23) is satisfied for at least one value of ω;*

2. *All the roots of $F_r(\omega)$ are real and for each root $\omega = \omega_r$, condition (5.23) is satisfied, i.e., $F_i(\omega_r)F_r^{'}(\omega_r) < 0$;*

3. *All the roots of $F_i(\omega)$ are real and for each root $\omega = \omega_i$, condition (5.23) is satisfied, i.e., $F_i^{'}(\omega_i)F_r(\omega_i) > 0$.*

We need to point out here that condition (5.23) is analogous to the monotonic phase increase property already introduced in Chapter 2. Moreover, we see that this property has to hold for each ω in $(-\infty, \infty)$.

To conclude this section, we present the following theorem. It gives additional information concerning the existence of roots of the function $F(s)$ in the open right half of the complex plane.

Theorem 5.4 *Let $F(s) = f(s, e^s)$, where $f(s,t)$ is the polynomial defined in (5.22). If the function $X^{*(N)}(e^s)$ has roots in the open RHP, then the function $F(s)$ has an unbounded set of zeros in the open RHP. If all the zeros of the function $X^{*(N)}(e^s)$ lie in the open LHP, then the function $F(s)$ can only have a bounded set of zeros in the open RHP.*

5.5 Applications to Control Theory

Many problems in process control engineering involve time delays. As discussed in Section 5.2, these time delays lead to dynamic models with characteristic equations of the form

$$\delta(s) = d(s) + e^{-sL_1}n_1(s) + e^{-sL_2}n_2(s) + \cdots + e^{-sL_m}n_m(s) \tag{5.24}$$

where $d(s)$, $n_i(s)$ for $i = 1, 2, ..., m$, are polynomials with real coefficients, and

(A1) $\deg[d(s)] = q$ and $\deg[n_i(s)] \leq q$ for $i = 1, 2, ..., m$;

(A2) $0 < L_1 < L_2 < \cdots < L_m$;

(A3) $L_i = \alpha_i L_1$, $i = 2, ..., m$, and α_i are non-negative integers.

Based on Pontryagin's results, a suitable extension of Theorem 5.3 can be developed to study the stability of this class of quasi-polynomials. Instead of (5.24) we can consider the quasi-polynomial

$$\begin{aligned} \delta^*(s) &= e^{sL_m}\delta(s) \\ \Rightarrow \delta^*(s) &= e^{sL_m}d(s) + e^{s(L_m - L_1)}n_1(s) + e^{s(L_m - L_2)}n_2(s) \\ &\quad + \cdots + n_m(s) \ . \end{aligned} \tag{5.25}$$

Notice that the new quasi-polynomial $\delta^*(s)$ becomes of the form $f(s, e^s)$ since, in view of (A3), the system exhibits *commensurate* delays, i.e., delays that are related by integers.

Since e^{sL_m} does not have any finite zeros in the complex plane, the zeros of $\delta(s)$ are identical to those of $\delta^*(s)$. Furthermore, in view of (A1) and (A2), the quasi-polynomial $\delta^*(s)$ has a principal term since the coefficient of the term containing the highest powers of s and e^s is nonzero.

As mentioned before, the quasi-polynomial $\delta^*(s)$ has an infinite number of roots. However, any bounded region of the complex plane contains only a finite number of its roots. Roots that are far from the origin can be assigned to a finite number of asymptotic chains. The geometry of these chains has been carefully studied in the past and they determine the following classes of quasi-polynomials:

1. *Retarded-type quasi-polynomials* (or delay-type quasi-polynomials): this first class consists of quasi-polynomials whose asymptotic chains go *deep* into the open LHP.

2. *Neutral-type quasi-polynomials*: this second class consists of quasi-polynomials that along with delay-type chains contain at least one asymptotic chain of roots in a vertical strip of the complex plane.

3. *Forestall-type quasi-polynomials*: this last class consists of quasi-polynomials with at least one asymptotic chain that goes *deep* into the open RHP.

It turns out that any quasi-polynomial of either delay or neutral type has the principal term, and vice versa: every quasi-polynomial with the principal term belongs to one of these two classes. It then follows that our quasi-polynomial $\delta^*(s)$ in (5.25) is either of the *delay* or of the *neutral* type.

Now for quasi-polynomials of either delay or neutral type, stability is defined as follows.

Definition 5.3 *A delay-type quasi-polynomial is said to be stable if and only if all its roots have negative real parts.*

Definition 5.4 *A neutral-type quasi-polynomial is said to be stable if there exists a positive number σ such that the real parts of all its roots are less than $-\sigma$.*

The reason why we have a *stronger* condition for the stability of neutral type quasi-polynomials is that we need to exclude, for this case, the possibility of an asymptotic chain of roots converging to the imaginary axis. However, notice that we can always make a shift of the independent variable $s \rightarrow \bar{s} = s + \sigma$, so that the negativity of the real parts of the roots of the quasi-polynomial $\bar{\delta}^*(\bar{s}) = \delta^*(\bar{s} - \sigma)$ with respect to $\bar{s}$ would imply the stability of the original time-invariant system with delays.

With these definitions of stability in hand, the stability of the system with characteristic equation (5.24) is equivalent to the condition that all the zeros of $\delta^*(s)$ be in the open LHP. We will say equivalently that $\delta^*(s)$ is Hurwitz or is stable. The following theorem, which is an immediate consequence of Theorem 5.3, gives necessary and sufficient conditions for the stability of $\delta^*(s)$.

Theorem 5.5 *Let $\delta^*(s)$ be given by (5.25), and write*

$$\delta^*(j\omega) = \delta_r(\omega) + j\delta_i(\omega) \ ,$$

where $\delta_r(\omega)$ and $\delta_i(\omega)$ represent, respectively, the real and imaginary parts of $\delta^*(j\omega)$. Under conditions (A1) and (A2), $\delta^*(s)$ is stable if and only if

1. *$\delta_r(\omega)$ and $\delta_i(\omega)$ have only simple, real roots and these interlace.*
2. *$\delta_i^{'}(\omega_o)\delta_r(\omega_o) - \delta_i(\omega_o)\delta_r^{'}(\omega_o) > 0$, for some ω_o in $(-\infty, \infty)$*

where $\delta_r^{'}(\omega)$ and $\delta_i^{'}(\omega)$ denote the first derivative with respect to ω of $\delta_r(\omega)$ and $\delta_i(\omega)$, respectively.

In the rest of this book, we will make use of this theorem to provide solutions to the P, PI, and PID stabilization problems for systems with time delay. Notice that the second condition establishes that the phase increase property needs to be checked at a single real frequency ω_o, provided Condition 1 is true.

Remark 5.1 *In Theorem 5.5 above, the requirement in Condition 1 that $\delta_r(\omega)$ and $\delta_i(\omega)$ have only simple,* real *roots is not a superfluous one. This is borne out by the following example.*

Example 5.4 *Consider a first-order system given by*

$$G(s) = \frac{1}{2s+1}$$

and a PI controller arranged in the closed-loop configuration shown in Fig. 3.1. Recall that the PI controller has the following transfer function:

$$C(s) = k_p + \frac{k_i}{s} = \frac{k_p s + k_i}{s} .$$

If the parameters of the PI controller are $k_p = 1.8$ *and* $k_i = 0.2$*, then the characteristic equation of the closed-loop system is*

$$\delta(s) = 2s^2 + 2.8s + 0.2$$

and it is stable.

Let us now consider the same first-order model but with a time delay of 10 *seconds, i.e,*

$$G(s) = \frac{1}{2s+1} e^{-10s} .$$

As in the delay-free case, we consider the same PI controller parameters: $k_p = 1.8$ *and* $k_i = 0.2$*. With these values, the characteristic equation of the closed-loop system is given by*

$$\delta(s) = 2s^2 + s + (1.8s + 0.2)e^{-10s} .$$

For analyzing the stability we consider

$$\delta^*(s) = e^{10s}\delta(s) = (2s^2 + s)e^{10s} + 1.8s + 0.2 .$$

Thus the real and imaginary parts of $\delta^*(j\omega)$ *are given by*

$$\begin{aligned} \delta_r(\omega) &= 0.2 - \omega \sin(10\omega) - 2\omega^2 \cos(10\omega) \\ \delta_i(\omega) &= \omega[1.8 + \cos(10\omega) - 2\omega \sin(10\omega)] . \end{aligned}$$

Using these expressions we can check if the quasi-polynomial $\delta^*(s)$ *satisfies the interlacing property. Figure 5.12 shows the plot of the real and imaginary parts of* $\delta^*(j\omega)$*. It is clear from this plot that the roots of the real and imaginary parts interlace for all* $\omega > 0$*. Notice that the interlacing condition needs to be checked only up to a finite frequency. This follows from the fact that the phasor of* $\left.\frac{1.8s+0.2}{2s^2+s}\right|_{s=j\omega}$ *tends to zero as* ω *tends to* $+\infty$*. This ensures that the quasi-polynomial* $\delta^*(s)$ *has the monotonic phase property for a sufficiently large* ω*. Therefore, the interlacing condition needs to be verified only for a low frequency range.*

Since interlacing holds for all ω*, we might be tempted to think that it is unnecessary to check if the roots of* $\delta_r(\omega)$ *and* $\delta_i(\omega)$ *are all real.*

Next we check the monotonic phase property at $\omega_o = 0$*. At this frequency we have* $\delta_i(\omega_o) = 0$*. Thus, in Theorem 5.5 we have*

$$\delta_i'(\omega_o)\delta_r(\omega_o) = (2.8)(0.2) > 0$$

which indicates that the monotonic phase property holds. However, as Fig. 5.13 shows, the system is unstable. In this figure, we have sketched the system response to a unit step input occurring at $t = 5$ *seconds.* △

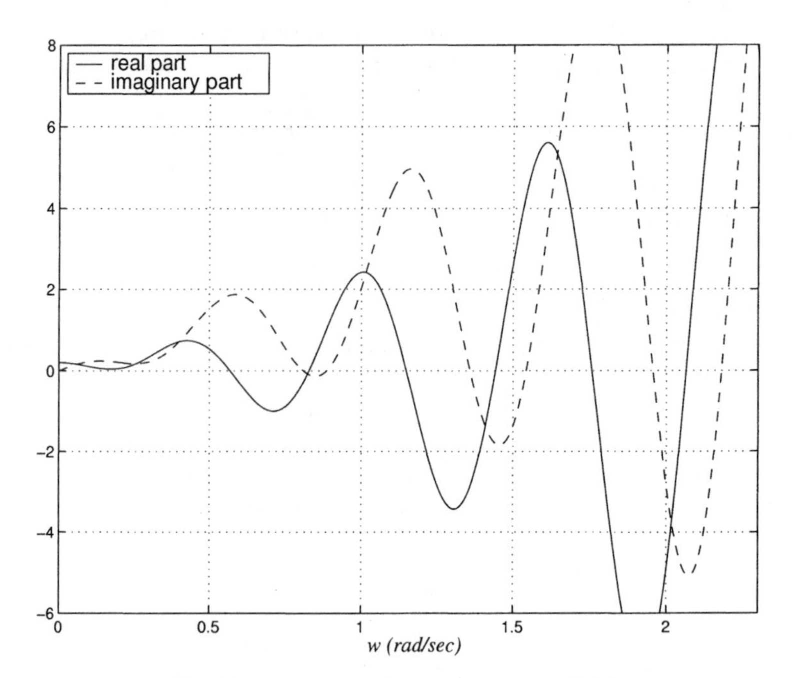

FIGURE 5.12. Plot of the real and imaginary parts of $\delta^*(s)$ for Example 5.4.

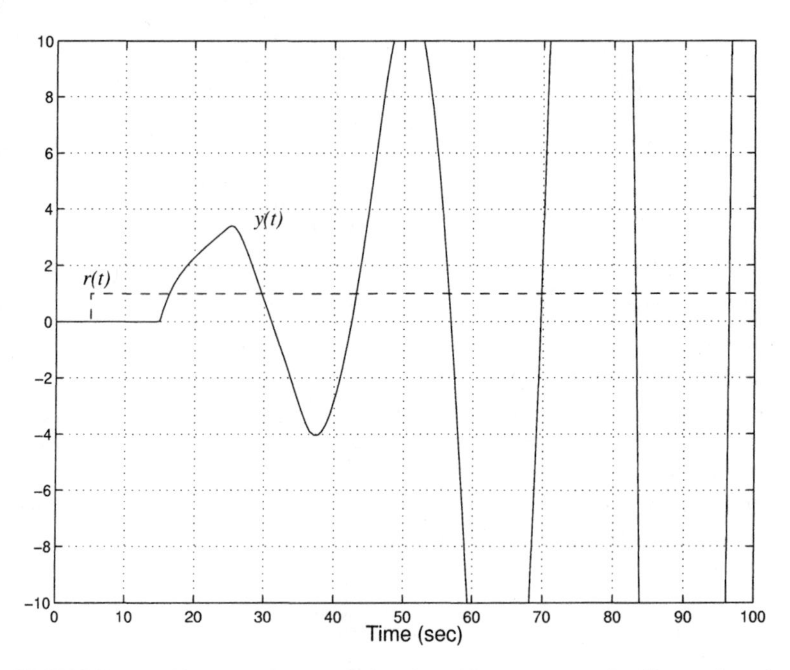

FIGURE 5.13. Time response of the closed-loop system for Example 5.4.

The previous example illustrates the case of a time-delay system that satisfies the interlacing and monotonic phase properties but fails to be stable. The reason for this behavior lies in the nature of the roots (zeros) of $\delta_r(\omega)$ and $\delta_i(\omega)$: they are not all real. Thus, a crucial step in applying Theorem 5.5 to check stability is to ensure that $\delta_r(\omega)$ and $\delta_i(\omega)$ have only *real* roots. Such a property can be checked by using the following key result, also due to Pontryagin.

Theorem 5.6 *Let M and N denote the highest powers of s and e^s, respectively, in $\delta^*(s)$. Let η be an appropriate constant such that the coefficients of terms of highest degree in $\delta_r(\omega)$ and $\delta_i(\omega)$ do not vanish at $\omega = \eta$. Then for the equations $\delta_r(\omega) = 0$ or $\delta_i(\omega) = 0$ to have only real roots, it is necessary and sufficient that in each of the intervals*

$$-2l\pi + \eta \leq \omega \leq 2l\pi + \eta \ , \ l = l_o, l_o + 1, l_o + 2, \ldots$$

$\delta_r(\omega)$ or $\delta_i(\omega)$ have exactly $4lN + M$ real roots for a sufficiently large l_o.

We will now show how Theorem 5.6 can be used to determine the nature of the roots of $\delta_r(\omega)$ or $\delta_i(\omega)$ in Example 5.4.

First let us make the following change of variables: $\hat{s} = 10s$. Then the expression for $\delta^*(s)$ can be rewritten as

$$\hat{\delta}^*(\hat{s}) = (0.02\hat{s}^2 + 0.1\hat{s})e^{\hat{s}} + 0.18\hat{s} + 0.2 \ .$$

We see that for the new quasi-polynomial in $\hat{s}$, $M = 2$ and $N = 1$. Also, the real and imaginary parts of $\hat{\delta}^*(j\hat{\omega})$ are given by

$$\begin{aligned} \hat{\delta}_r(\hat{\omega}) &= 0.2 - 0.1\hat{\omega}\sin(\hat{\omega}) - 0.02\hat{\omega}^2\cos(\hat{\omega}) \\ \hat{\delta}_i(\hat{\omega}) &= \hat{\omega}[0.18 + 0.1\cos(\hat{\omega}) - 0.02\hat{\omega}\sin(\hat{\omega})] \ . \end{aligned}$$

We now focus on the imaginary part of $\hat{\delta}^*(j\hat{\omega})$. From the previous expressions we can compute the roots of $\hat{\delta}_i(\hat{\omega}) = 0$, i.e.,

$$\hat{\omega}[0.18 + 0.1\cos(\hat{\omega}) - 0.02\hat{\omega}\sin(\hat{\omega})] = 0 \ .$$

Then

$$\begin{aligned} \hat{\omega} = 0 \ , \ \text{or} \\ 0.18 + 0.1\cos(\hat{\omega}) - 0.02\hat{\omega}\sin(\hat{\omega}) &= 0 \ . \end{aligned} \tag{5.26}$$

From this we see that one root of the imaginary part is $\hat{\omega}_o = 0$. The positive real roots of (5.26) are

$$\begin{array}{ll} \hat{\omega}_1 = 8.0812 & \hat{\omega}_2 = 8.8519 \\ \hat{\omega}_3 = 13.5896 & \hat{\omega}_4 = 15.4332 \\ \hat{\omega}_5 = 19.5618 & \hat{\omega}_6 = 21.8025 \\ \vdots & \vdots \end{array}$$

Next we choose $\eta = \frac{\pi}{4}$ to satisfy the requirement imposed by Theorem 5.6 that $\sin(\eta) \neq 0$. Figure 5.14 shows the root distribution of $\hat{\delta}_i(\hat{\omega})$ and will enable us to apply Theorem 5.6 to this example.

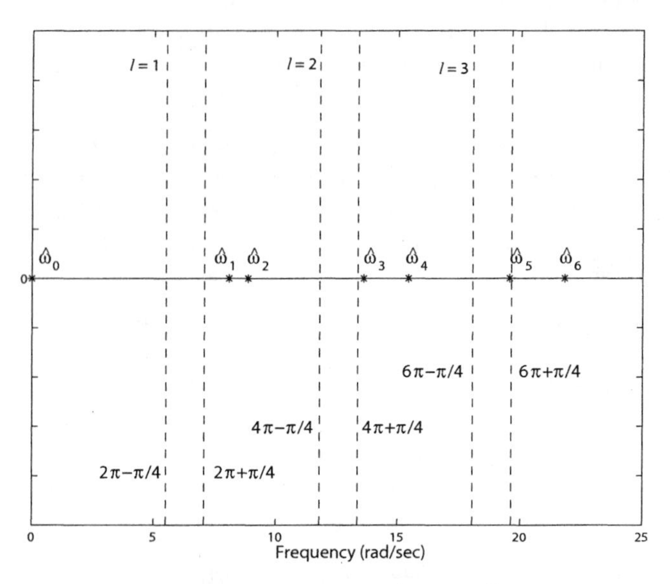

FIGURE 5.14. Root distribution of $\hat{\delta}_i(\hat{\omega})$.

From this figure we see that $\hat{\delta}_i(\hat{\omega})$ has only one real root in the interval $[0, 2\pi - \frac{\pi}{4}] = [0, \frac{7\pi}{4}]$: the root at the origin. Since $\hat{\delta}_i(\hat{\omega})$ is an odd function of $\hat{\omega}$, it follows that in the interval $[-\frac{7\pi}{4}, \frac{7\pi}{4}]$, $\hat{\delta}_i(\hat{\omega})$ will have only one real root. Also observe from the same figure that $\hat{\delta}_i(\hat{\omega})$ has no real roots in the interval $[\frac{7\pi}{4}, \frac{9\pi}{4}]$. Thus $\hat{\delta}_i(\hat{\omega})$ has only one real root in the interval

$$\left[-2\pi + \frac{\pi}{4}, 2\pi + \frac{\pi}{4}\right],$$

which does not sum up to $4N + M = 6$ for $l_o = 1$.

Let us now make $l_o = 2$ so now the requirement on the number of real roots is $8N + M = 10$. From Fig. 5.14 we see that in the interval

$$\left[-4\pi + \frac{\pi}{4}, 4\pi + \frac{\pi}{4}\right]$$

the function $\hat{\delta}_i(\hat{\omega})$ has only five real roots. Following the same procedure for $l = 3, 4, \ldots$, we see that the number of real roots of $\hat{\delta}_i(\hat{\omega})$ in the interval

$$\left[-2l\pi + \frac{\pi}{4}, 2l\pi + \frac{\pi}{4}\right]$$

is always less than $4lN + M = 4l + 2$. Hence, from Theorem 5.6 we conclude that the roots of $\hat{\delta}_i(\hat{\omega})$ are not all real.

5.6 Stability of Time-Delay Systems with a Single Delay

In Section 5.4 we presented a general stability criterion for systems with time delay. This criterion gives necessary and sufficient conditions under which the roots of the characteristic equation of the closed-loop system have negative real parts. In this section, we consider an alternative analysis for the case of time-delay systems with a single delay, i.e., a system with the following characteristic equation:

$$\delta(s) = d(s) + n(s)e^{-sL} \tag{5.27}$$

where $d(s)$ and $n(s)$ are polynomials with real coefficients, $\deg[d(s)] = q$, $\deg[n(s)] = p$, $q \geq p$, and $L > 0$ is the time delay of the system. Moreover, we assume that any common factors of $d(s)$ and $n(s)$ have been removed.

The condition for stability is that all the roots of the characteristic equation (5.27) lie in the open left half of the complex s plane. Thus, in our case, the basic problem of stability is that of determining the range (or ranges) of values of L for which this occurs. One way to answer this stability question is to develop a systematic procedure to analyze the behavior of the roots of (5.27) as L increases from 0 to ∞.

In this section, we will introduce one such procedure due to Walton and Marshall [49]. This procedure consists of three basic steps. The first step is to examine the stability of (5.27) for $L = 0$ and determine the number of roots, if any, of $\delta(s) = 0$ not lying in the open LHP. The second step considers the case of an infinitesimally small positive L. For this value there will be an infinite number of new roots and it is necessary to find out where in the complex plane these roots have arisen. The third and final step is to find positive values of L, if any, at which there are roots of $\delta(s) = 0$ lying on the imaginary axis and then to determine whether these roots merely touch the axis or whether they cross from one half plane to the other with increasing L. Roots crossing from left to right are considered *destabilizing* and those crossing from right to left are considered *stabilizing*.

We will now use this procedure to study the movement of the roots of $\delta(s) = 0$ with an increasing $L > 0$. In particular, we will determine for which values of L these roots do not all lie in the open LHP, i.e., the regions of instability. To explicitly show the dependence of the characteristic equation on the time delay L, we rewrite the characteristic equation as

$$\delta(s, L) = d(s) + n(s)e^{-Ls} = 0 \,. \tag{5.28}$$

Step 1. We start by examining the stability at $L = 0$, i.e., we have

$$\delta(s, 0) = d(s) + n(s) = 0 \,.$$

This is a delay-free problem to which any of the classical methods may be applied. If the system is found to be unstable it will then be necessary to determine how many zeros lie in the open RHP or on the imaginary axis.

Step 2. In this step we increment L from 0 to an infinitesimally small and positive number. In this situation the number of roots changes from being finite to infinite and we need to determine where in the complex plane these new roots arise. Notice that for an infinitesimally small L, the new roots must come in *at infinity*; otherwise, the expression e^{-Ls} would be approximately equal to unity and there would not be any new roots. Consequently, if $p < q$, (5.28) can be satisfied for large s if and only if e^{-Ls} is large, i.e., $\mathrm{Re}(s) < 0$. Thus in this case the new roots all lie in the open LHP. The case where $p = q$ involves more details and will be presented later in this section.
Step 3. In this step we have to consider potential crossing points on the imaginary axis. By taking complex conjugates of the quantities involved in the definition of $\delta(s, L)$, it follows that if $\delta(s, L) = 0$ has a root at $s = j\omega$, then it also has a root at $s = -j\omega$. This implies that the roots cross or touch the imaginary axis in conjugate pairs and therefore it suffices to consider positive values of ω. A special case is $s = 0$ and this will be analyzed later in this section. Substituting $s = \pm j\omega$ in (5.28) we obtain

$$\begin{aligned} d(j\omega) + n(j\omega)e^{-jL\omega} &= 0 \\ d(-j\omega) + n(-j\omega)e^{jL\omega} &= 0 . \end{aligned} \tag{5.29}$$

Elimination of the exponential terms yields

$$d(j\omega)d(-j\omega) - n(j\omega)n(-j\omega) = 0 . \tag{5.30}$$

The expression on the left-hand side of this equation is a polynomial in ω^2 and, for convenience, we denote it by

$$W(\omega^2) = d(j\omega)d(-j\omega) - n(j\omega)n(-j\omega) . \tag{5.31}$$

It should be clear that only the real, non-negative zeros of $W(\omega^2)$ are of interest since only these can lead to potential crossing points $s = \pm j\omega$. As a consequence of this if there are no positive roots of $W(\omega^2) = 0$ (notice that this is a function of ω^2) then there are no values of L for which $\delta(j\omega, L) = 0$. This leads us to the following important remark.

Remark 5.2 *If $p < q$ and $W(\omega^2)$ has no positive real roots, then there is no change in stability, i.e., if the system is stable at $L = 0$, then it will be stable for all $L \geq 0$, whereas if it is unstable for $L = 0$, then it will be unstable for all $L \geq 0$.*

If there is a real value of ω such that

$$d(j\omega) + n(j\omega)e^{-jL\omega} = 0$$

then

$$e^{-jL\omega} = -\frac{d(j\omega)}{n(j\omega)} . \tag{5.32}$$

Notice that $n(j\omega) \neq 0$, otherwise, from (5.30), $d(j\omega)$ would also be zero, which is not possible since we assumed that any common factors of $d(s)$ and $n(s)$ had been removed.

Equation (5.32) will yield real positive values of L if and only if

$$\left| -\frac{d(j\omega)}{n(j\omega)} \right| = 1$$

which is indeed true from (5.30). Thus for any real $\omega \neq 0$ satisfying $W(\omega^2) = 0$, there exist real positive L such that $\delta(j\omega, L) = 0$ and these are given by

$$\cos(L\omega) = \mathrm{Re}\left[-\frac{d(j\omega)}{n(j\omega)} \right] , \ \sin(L\omega) = \mathrm{Im}\left[\frac{d(j\omega)}{n(j\omega)} \right] . \tag{5.33}$$

From these expressions it follows that if L_o denotes the smallest value of L (for a particular value of ω) satisfying this, then

$$L = L_o + \frac{2\pi k}{\omega} , \ k = 0, 1, 2, \ldots$$

are also solutions. Hence, for each ω satisfying $W(\omega^2) = 0$, there is an infinite number of values of L at which the roots cross the imaginary axis.

We now consider the special case $s = 0$. In this case, instead of (5.29) and (5.30) we have only one equation

$$\begin{aligned} d(0) + n(0) &= 0 \\ \Rightarrow d(0) + e^{-L \cdot 0} n(0) &= 0 \end{aligned} \tag{5.34}$$

for all finite L. Thus the system is unstable for all values of L and for our analysis this solution can be ignored if (5.34) is satisfied.

Once we have found a value of L at which there is a root of the characteristic equation (5.28) on the imaginary axis, we need to determine its behavior for slightly smaller and slightly larger values of L. This means that we need to find out if the root crosses the imaginary axis and in which direction or if it merely touches the imaginary axis. We can achieve this by considering the root s of the characteristic equation $\delta(s, L) = 0$ as an explicit function of L, i.e., $s = f(L)$, and analyzing the expression $\mathrm{Re}(\frac{ds}{dL})$ evaluated at the root $s = j\omega$. Then, we have

- If $\mathrm{Re}(\frac{ds}{dL}) > 0$, then the root crosses the imaginary axis from left to right, i.e., it is destabilizing.
- If $\mathrm{Re}(\frac{ds}{dL}) < 0$, then the root crosses the imaginary axis from right to left, i.e., it is stabilizing.
- If $\mathrm{Re}(\frac{ds}{dL}) = 0$, then it is necessary to consider higher-order derivatives.

If we differentiate equation (5.28) with respect to L we obtain

$$\frac{ds}{dL} = \frac{sn(s)e^{-Ls}}{d'(s) + n'(s)e^{-Ls} - n(s)Le^{-Ls}}$$

where $n'(s)$ and $d'(s)$ denote the first derivative with respect to s of $n(s)$ and $d(s)$, respectively. From (5.28) we can rewrite this expression as

$$\frac{ds}{dL} = -s\left[\frac{d'(s)}{d(s)} - \frac{n'(s)}{n(s)} + L\right]^{-1}.$$

We now evaluate this expression at $s = j\omega$ and find the sign of the real part:

$$\begin{aligned} S \triangleq \operatorname{sgn}\left[\operatorname{Re}\left(\frac{ds}{dL}\right)\right] &= -\operatorname{sgn}\left[\operatorname{Re}\left(j\omega\left(\frac{d'(j\omega)}{d(j\omega)} - \frac{n'(j\omega)}{n(j\omega)} + L\right)^{-1}\right)\right] \\ &= -\operatorname{sgn}\left[\operatorname{Re}\left(\frac{1}{j\omega}\left(\frac{d'(j\omega)}{d(j\omega)} - \frac{n'(j\omega)}{n(j\omega)} + L\right)\right)\right] \end{aligned}$$

since $\operatorname{sgn}[\operatorname{Re}(a+jb)] = \operatorname{sgn}[\operatorname{Re}(\frac{1}{a+jb})]$. Then,

$$S = \operatorname{sgn}\left[\operatorname{Re}\left(\frac{1}{j\omega}\left(\frac{n'(j\omega)}{n(j\omega)} - \frac{d'(j\omega)}{d(j\omega)}\right)\right)\right].$$

If we consider

$$\frac{n'(j\omega)}{n(j\omega)} - \frac{d'(j\omega)}{d(j\omega)} = a(\omega) + jb(\omega)$$

then

$$\begin{aligned} \operatorname{Re}\left(\frac{1}{j\omega}(a(\omega) + jb(\omega))\right) &= \frac{b(\omega)}{\omega} \\ &= \operatorname{Im}\left(\frac{1}{\omega}(a(\omega) + jb(\omega))\right). \end{aligned}$$

Thus,

$$S = \operatorname{sgn}\left[\operatorname{Im}\left(\frac{1}{\omega}\left(\frac{n'(j\omega)}{n(j\omega)} - \frac{d'(j\omega)}{d(j\omega)}\right)\right)\right]$$

which is independent of L. This implies that even though there is an infinite number of values of L associated with each value of ω that make $\delta(j\omega, L) = 0$, the behavior of the roots at these points will always be the same. Hence, we may classify solutions of $W(\omega^2)$ as

- stabilizing if $S = -1$ or

- destabilizing if $S = +1$.

Since at $s = j\omega$ we have $W(\omega^2) = 0$ then from (5.31) we have $d(j\omega)d(-j\omega) = n(j\omega)n(-j\omega)$. Thus,

$$\begin{aligned} S &= \operatorname{sgn}\left[\operatorname{Im}\left(\frac{1}{\omega}\left(\frac{n'(j\omega)n(-j\omega)}{d(j\omega)d(-j\omega)} - \frac{d'(j\omega)}{d(j\omega)}\right)\right)\right] \\ &= \operatorname{sgn}\left[\operatorname{Im}\left(\frac{1}{\omega}\left(\frac{n'(j\omega)n(-j\omega) - d'(j\omega)d(-j\omega)}{d(j\omega)d(-j\omega)}\right)\right)\right] \\ &= \operatorname{sgn}\left[\operatorname{Im}\left(\frac{1}{\omega}\left(n'(j\omega)n(-j\omega) - d'(j\omega)d(-j\omega)\right)\right)\right] \end{aligned}$$

since $d(j\omega)d(-j\omega) = |d(j\omega)|^2 > 0$. Now, using the property $\operatorname{Im}(z) = \frac{z-\bar{z}}{2j}$, for any complex number z, we have

$$\begin{aligned} S = \operatorname{sgn}\Big[\frac{1}{2j\omega}\Big(& n'(j\omega)n(-j\omega) - n(j\omega)n'(-j\omega) - \\ & d'(j\omega)d(-j\omega) + d(j\omega)d'(-j\omega)\Big)\Big] \end{aligned}$$

which finally leads us to

$$S = \operatorname{sgn}[W'(\omega^2)] \tag{5.35}$$

in which the prime denotes differentiation with respect to ω^2. Hence, we can use (5.35) to determine whether a root is destabilizing or stabilizing.

Step 2: Special Case. We mentioned earlier that the case $q = p$ in Step 2 involved more details. In this case it is possible for all the roots of the characteristic equation to lie in the open LHP but for the system to be unstable. For example, consider the following system:

$$s + 1 + se^{-Ls} = 0\ .$$

Detailed calculations show that the new roots for infinitesimally small $L > 0$ are just in the LHP but the system is not stable in the sense of Definition 5.4. Therefore, we will need to further analyze the case where $q = p$.

Toward this end, suppose that $s = \sigma + j\omega$ is a new root of $\delta(s, L) = 0$ for infinitesimally small L. As mentioned earlier, since L is infinitesimally small, the new root must come in *at infinity*. Since $q = p$, (5.28) can be satisfied for large $s = \sigma + j\omega$ if and only if $e^{-(\sigma + j\omega)L}$ is a real number. This happens if $e^{-j\omega L}$ equates to unity, or $\cos(\omega L) - j\sin(\omega L) = 1$. This implies that $\omega L = 2l\pi$, $l = 0, 1, ...$, and since L is infinitesimally small, we thus conclude that $|\omega| >> |\sigma|$. Hence, we now have

$$e^{-L\sigma} = \left|\frac{d(s)}{n(s)}\right|_{s=\sigma+j\omega} \approx \left|\frac{d(j\omega)}{n(j\omega)}\right|\ . \tag{5.36}$$

From this expression we conclude that $\sigma > 0$ if and only if $|d(j\omega)| < |n(j\omega)|$ or, equivalently, $W(\omega^2) < 0$ for large ω. Thus we conclude that the system is unstable for $q = p$ if $W(\omega^2) < 0$ for large ω.

For the case of stability in the sense of Definition 5.4 we require that the new roots lie to the left of the line $\mathrm{Re}(s) = \alpha$ for some $\alpha < 0$. From (5.36), this occurs if

$$\lim_{\omega \to \infty} \left| \frac{d(j\omega)}{n(j\omega)} \right| = \left| \frac{a_q}{c_p} \right| > 1$$

where a_q and c_p denote the coefficient of the highest powers in s of polynomials $d(s)$ and $n(s)$, respectively. Moreover, the new roots will lie in the LHP if $W(\omega^2) > 0$ for large ω. Thus, we conclude that the system is stable in the sense of Definition 5.4 for $q = p$ if $W(\omega^2) > 0$ for large ω and this occurs if and only if $|a_q| > |c_p|$.

We can now summarize the previous discussion as follows.

- **Step 1.** Examine the stability at $L = 0$.
- **Step 2.** Consider an infinitesimally small positive L. If $q > p$, all the new roots will lie in the open LHP and this step can be omitted. If $q = p$, the location of the new roots is determined by the sign of $W(\omega^2)$ for large ω.
- **Step 3.** Determine the positive roots of $W(\omega^2) = 0$, the corresponding positive values of L, and the nature of these roots. If there are no repeated roots, then the stabilizing and destabilizing roots alternate. For example, if the largest root is destabilizing, we can then label the roots in descending order as destabilizing, stabilizing, and so on. The same procedure can be used for the corresponding values of L in order to determine for what values of L all the roots of $\delta(s, L) = 0$ lie in the open LHP.

Next we present several examples that will clarify the concepts introduced in this section.

Example 5.5 *Let* $\delta(s, L) = s + 2e^{-Ls}$. *Then,*
(1) $\delta(s, 0) = s + 2$, *so the system is stable for* $L = 0$.
(2) Since $q = 1 > p = 0$, *we skip this step.*
(3) $d(s) = s$ *and* $n(s) = 2$, *so* $W(\omega^2) = \omega^2 - 4$. *Thus,* $W'(\omega^2) = 1$ *which is positive. We conclude that there is only one positive root of* $W(\omega^2)$ *at 4. Since* $S = \mathrm{sgn}[W'(\omega^2)] = 1$, *then this root is destabilizing. The corresponding values of* L *are given by (5.33):*

$$\cos(L\omega) = \mathrm{Re}\left[-\frac{j\omega}{2}\right] = 0 \, , \; \sin(L\omega) = \mathrm{Im}\left[\frac{j\omega}{2}\right] = 1 \, .$$

Solving for L *we obtain*

$$L = (4k+1)\frac{\pi}{4} \, , \; k = 0, 1, 2, \ldots \, .$$

This means that at $L = \frac{\pi}{4}$, two roots of $\delta(s, L) = 0$ cross from left to right of the imaginary axis. Then, at $L = \frac{5\pi}{4}$, two more cross from left to right of the imaginary axis and so on. We conclude that the only region of stability is $0 \leq L < \frac{\pi}{4}$. △

Example 5.6 *Consider $\delta(s, L) = (s + 1) + (s + 3)e^{-Ls}$. Then,*
(1) $\delta(s, 0) = 2s + 4$, so the system is stable for $L = 0$.
(2) Since $q = p = 1$, we need to consider the behavior of $W(\omega^2)$ for large ω^2. We have

$$W(\omega^2) = (j\omega + 1)(-j\omega + 1) - (j\omega + 3)(-j\omega + 3) = -8 \,.$$

Thus, since $W(\omega^2) < 0$ for large ω^2, an infinite number of new roots occur in the RHP and the system is unstable for all $L > 0$. Notice however that the system is stable *for $L = 0$.* △

Example 5.7 *Let $\delta(s, L) = s^2 + s + 4 + 2e^{-Ls}$. Then,*
(1) $\delta(s, 0) = s^2 + s + 6$, so the system is stable for $L = 0$.
(2) Since $q = 2 > p = 0$, we skip this step.
(3) $d(s) = s^2 + s + 4$ and $n(s) = 2$, so $W(\omega^2)$ is given by

$$\begin{aligned} W(\omega^2) &= \omega^4 - 7\omega^2 + 12 \\ \Rightarrow W'(\omega^2) &= 2\omega^2 - 7 \,. \end{aligned}$$

The positive roots of $W(\omega^2)$ are $\omega_1^2 = 4$ and $\omega_2^2 = 3$. The corresponding values of L satisfy (5.33), i.e.,

$$\cos(L\omega) = 0.5\omega^2 - 2 \,, \ \sin(L\omega) = 0.5\omega \,.$$

$\omega_1^2 = 4$ is the larger of the two roots and $S = \operatorname{sgn}[W'(4)] = 1$, so this root is destabilizing. The corresponding values of L are given by

$$\cos(L\omega_1) = 0 \,, \ \sin(L\omega_1) = 1$$

and hence $L_1 = (k + 1/4)\pi$, $k = 0, 1, 2, \ldots$.

On the other hand, $\omega_2^2 = 3$ is the smallest of the two roots of $W(\omega^2)$ and $S = \operatorname{sgn}[W'(3)] = -1$, so this root is stabilizing. The corresponding values of L are given by

$$\sqrt{3}L = \frac{2\pi}{3} + 2k\pi \ \Rightarrow \ L_2 = \frac{2\pi/3 + 2k\pi}{\sqrt{3}} \,.$$

Thus at $L_1 = 0.25\pi, 1.25\pi, 2.25\pi, \ldots$, a pair of roots crosses from the LHP into the RHP, whereas at $L = 0.3849\pi, 1.5396\pi, 2.6943\pi, \ldots$, a pair of roots crosses from the RHP into the LHP. We can summarize these root crossings as follows:

1. *At $L = 0.25\pi$, two roots move from the LHP to the RHP and then back to the LHP at $L = 0.3849\pi$.*

2. *At $L = 1.25\pi$, a second pair of roots crosses into the RHP and then crosses back at $L = 1.5396\pi$.*

3. *At $L = 2.25\pi$, a third pair crosses into the right and then crosses back at $L = 2.6943\pi$.*

This succession of stable and unstable regions must eventually cease since $\frac{2\pi}{\omega_2}$, the interval between successive stabilizing values, is greater than $\frac{2\pi}{\omega_1}$, that of destabilizing values. When it does, permanent instability will occur. This occurs at $L = 6.25\pi$ where a pair of roots crosses from left to right, but then at $L = 7.25\pi$ another pair follows the same pattern. Thus the roots have come to accumulate in the RHP. Moreover, since there can never be two consecutive stabilizing crossings (since $\frac{2\pi}{\omega_1} < \frac{2\pi}{\omega_2}$), no more stability intervals are possible for $L > 6.25\pi$ and instability occurs. We conclude that there is stability for

$$\begin{aligned} L &= (0, 0.25\pi) \bigcup (0.3849\pi, 1.25\pi) \bigcup (1.5396\pi, 2.25\pi) \bigcup \\ &\quad (2.6943\pi, 3.25\pi) \bigcup (3.8490\pi, 4.25\pi) \bigcup (5.0037\pi, 5.25\pi) \bigcup \\ &\quad (6.1584\pi, 6.25\pi) . \end{aligned}$$

These are the so-called stability windows *of the time-delay system.* △

This example shows that even low-order systems with time delay can exhibit a complicated behavior. As we can see, the addition of delay to a system may produce a stabilizing effect which may contradict *intuition*. Moreover, the presence of these *stability windows* constitutes another reason why a Padé approximation of the time delay may not be adequate for analyzing the stability of time-delay systems.

5.7 Notes and References

The results presented in Section 5.3 are taken from [40]. The use of the Padé approximation and its effect on stability and cost functional evaluation of time-delay systems is treated in detail in the book by Marshall, Gorecki, Korytowski, and Walton [29]. Early discussions on the pitfalls of replacing the exponential term by a rational transfer function can be found in the work of Choksy [8] and Marshall [30]. For a complete description of Pontryagin's results, the reader is referred to his original paper [37]. Extensions of these results for a certain class of quasi-polynomials can be found in the book by Bellman and Cooke [4] and in the book by Bhattacharyya, Chapellat, and Keel [5] and the references therein. A more detailed treatment of the geometry of the chains introduced in Section 5.4 can be found in [4]. Applications of the Pontryagin's results introduced in this chapter can be found in [16, 25, 30, 46]. The procedure presented in Section 5.6 for a single delay was developed by Walton and Marshall in 1987 and for more details the reader is referred to [29] and [49].

An important and complete account of recent advances in the study of time-delay systems can be found in the book by Gu, Kharitonov, and Chen [15]. This work presents in detail important methods and tools developed for the study of time-delay systems. The methods are organized into three categories: (a) frequency domain tools, (b) time domain methods, and (c) input-output stability formulation. In all cases, the robust stability of linear time-invariant delay systems is also discussed in a coherent and systematic manner.

6

Stabilization of Time-Delay Systems using a Constant Gain Feedback Controller

In this chapter we present a solution to the problem of stabilizing a first- or second-order plant with time delay using a proportional (P) controller. This solution is built upon the results presented in Chapter 5 for quasi-polynomials and computes the complete set of stabilizing gains. Examples are included to illustrate the application of these results.

6.1 Introduction

In industrial control applications the plant is often modeled as a first- or second-order system with time delay and the controller is either of the P, PI, or PID type. The solution presented here makes use of the results presented in Chapter 5. The solution to the proportional control case is developed first because it serves as a stepping stone for tackling the more complicated cases of stabilization using a PI or a PID controller. These cases will be analyzed in later chapters. The proportional control stabilization problem for first-order systems with time delay can be solved using other techniques such as the Nyquist criterion and its variations. The approach presented here, however, allows a clear understanding of the relationship between the time delay exhibited by a system and its stabilization using a constant gain controller.

In Section 6.2 we present and solve the stabilization problem of a first-order system with time delay using a constant gain controller. In Section 6.3 we present and solve the stabilization problem of a second-order system with time delay using a constant gain controller. The cases of an open-loop stable and an open-loop unstable plant are both considered.

6.2 First-Order Systems with Time Delay

Systems with step responses like the one shown in Fig. 6.1 are commonly modeled as first-order processes with a time delay which can be mathematically described by the transfer function

$$G(s) = \frac{k}{1+Ts}e^{-Ls} . \tag{6.1}$$

Here k represents the steady-state gain of the plant, L the time delay, and T the time constant.

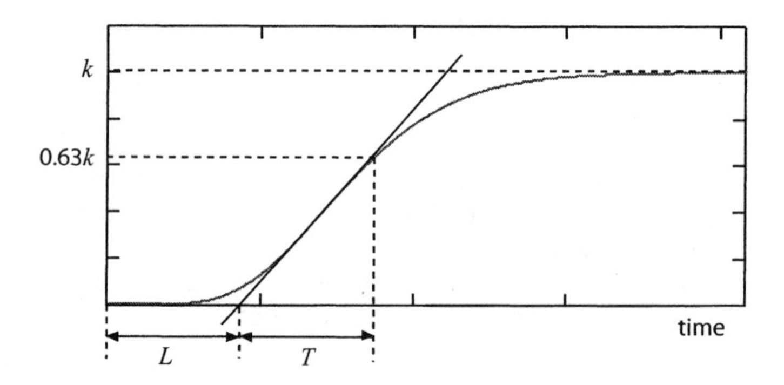

FIGURE 6.1. Open-loop step response.

Consider now the feedback control system shown in Fig. 6.2 where r is the command signal, y is the output of the plant, $G(s)$ given by (6.1) is the plant to be controlled, and $C(s)$ is the controller. In this section we will consider the case of a constant gain controller

$$C(s) = k_c .$$

Our objective is to determine the values of the parameter k_c for which the closed-loop system is stable.

When the time delay of the plant model is zero, that is, $L = 0$, the closed-loop characteristic equation of the system is given by

$$\delta(s) = kk_c + 1 + Ts .$$

This polynomial has a single root at $s = -\frac{1+k \cdot k_c}{T}$. Thus, for instance, if we assume that the steady-state gain k is positive and $T > 0$ so that the plant

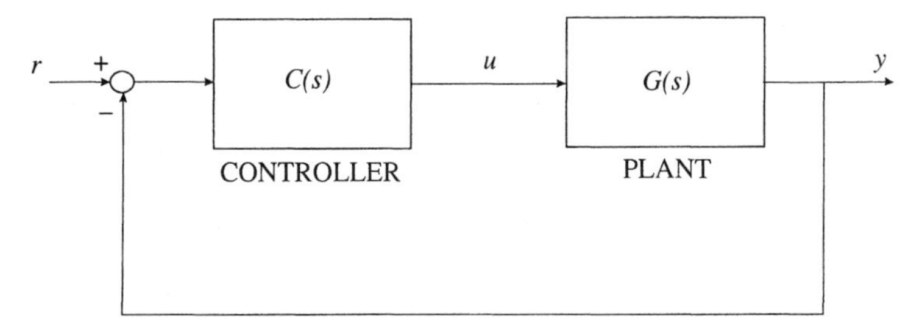

FIGURE 6.2. Feedback control system.

is open-loop stable, then to ensure the stability of the closed-loop system with $L = 0$, we must have

$$k_c > -\frac{1}{k} .$$

If, on the other hand, the plant is open-loop unstable, i.e., $T < 0$, then to ensure the stability of the closed-loop system with $L = 0$, we must have

$$k_c < -\frac{1}{k} . \tag{6.2}$$

Now let us consider the case where the time delay of the model is different from zero and try to determine the set of all stabilizing gains. The closed-loop characteristic equation of the system is given by

$$\delta(s) = kk_c e^{-Ls} + 1 + Ts .$$

In order to study the stability of the closed-loop system, we need to determine if all the roots of the above expression lie in the open LHP. Due to the presence of the exponential term e^{-Ls}, the number of roots of the expression $\delta(s)$ is infinite, which makes such a stability check extremely difficult. Instead, we can invoke Theorem 5.5 to determine the set of stabilizing gains k_c as follows.

First we consider the quasi-polynomial $\delta^*(s)$ defined by

$$\delta^*(s) = e^{Ls}\delta(s) = kk_c + (1 + Ts)e^{Ls} .$$

Substituting $s = j\omega$, we have

$$\delta^*(j\omega) = \delta_r(\omega) + j\delta_i(\omega)$$

where

$$\begin{aligned} \delta_r(\omega) &= \cos(L\omega) - T\omega\sin(L\omega) + kk_c \\ \delta_i(\omega) &= \sin(L\omega) + T\omega\cos(L\omega) . \end{aligned}$$

We now consider two different cases.

6.2.1 Open-Loop Stable Plant

In this subsection we give a closed-form solution to the constant gain stabilization problem for the case of an open-loop stable plant. This means that the time constant T of the plant satisfies $T > 0$. Moreover, we assume that $k > 0$ and $L > 0$.

Theorem 6.1 *Under the above assumptions on k and L, the set of all stabilizing gains k_c for a given open-loop stable plant with transfer function $G(s)$ as in (6.1) is given by*

$$-\frac{1}{k} < k_c < \frac{T}{kL}\sqrt{z_1^2 + \frac{L^2}{T^2}} \tag{6.3}$$

where z_1 is the solution of the equation

$$\tan(z) = -\frac{T}{L}z$$

in the interval $(\frac{\pi}{2}, \pi)$.

Proof. With the change of variables $z = L\omega$ the real and imaginary parts of $\delta^*(j\omega)$ can be rewritten as

$$\delta_r(z) = \cos(z) - \frac{T}{L}z\sin(z) + kk_c \tag{6.4}$$

$$\delta_i(z) = \sin(z) + \frac{T}{L}z\cos(z)\,. \tag{6.5}$$

By Theorem 5.5, we need to check two conditions to ensure the stability of the quasi-polynomial $\delta^*(s)$.

Step 1. We first check condition 2 of Theorem 5.5:

$$E(\omega_o) = \delta_i'(\omega_o)\delta_r(\omega_o) - \delta_i(\omega_o)\delta_r'(\omega_o) > 0$$

for some ω_o in $(-\infty, \infty)$. Let us take $\omega_o = 0$, so $z_o = 0$. Thus $\delta_i(z_o) = 0$ and $\delta_r(z_o) = 1 + kk_c$. We also have

$$\begin{aligned} \delta_i'(z) &= \left[\left(1 + \frac{T}{L}\right)\cos(z) - \frac{T}{L}z\sin(z)\right] \\ \Rightarrow E(z_o) &= \left(1 + \frac{T}{L}\right)(1 + kk_c)\,. \end{aligned}$$

By our initial assumption $T > 0$ and $L > 0$. Thus, for $k_c > -\frac{1}{k}$ we have $E(z_o) > 0$.

Step 2. We now check condition 1 of Theorem 5.5: the interlacing of the roots of $\delta_r(z)$ and $\delta_i(z)$. From (6.5) we can compute the roots of the imaginary part, i.e., $\delta_i(z) = 0$. This gives us the following equation:

$$\sin(z) + \frac{T}{L}z\cos(z) = 0$$

$$\Leftrightarrow \tan(z) = -\frac{T}{L}z\,. \tag{6.6}$$

An analytical solution of (6.6) is difficult to find. However, we can plot the two terms involved in this equation, i.e., $\tan(z)$ and $-\frac{T}{L}z$, to study the behavior of the roots of $\delta_i(z)$. Figure 6.3 shows this plot. Clearly the non-negative real roots of the imaginary part are

$$z_o = 0,\ z_1 \epsilon \left(\frac{\pi}{2}, \pi\right),\ z_2 \epsilon \left(\frac{3\pi}{2}, 2\pi\right),\ z_3 \epsilon \left(\frac{5\pi}{2}, 3\pi\right),$$

and so on, where the roots z_i, for $i = 1, 2, 3, \ldots$ satisfy (6.6).

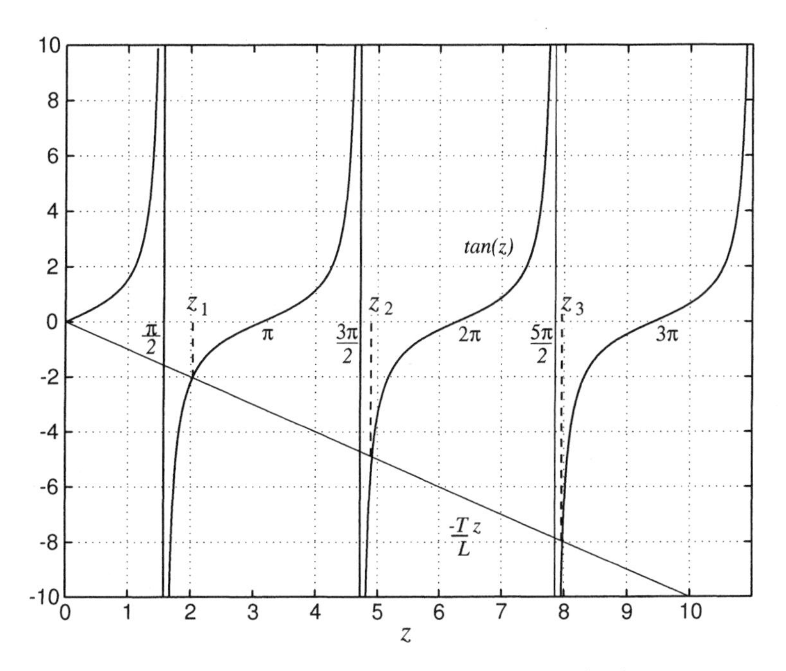

FIGURE 6.3. Plot of the terms involved in (6.6).

Let us now use Theorem 5.6 to check if $\delta_i(z)$ has only real roots. Substituting $s_1 = Ls$ in the expression for $\delta^*(s)$, we see that for the new quasi-polynomial in s_1, $M = 1$ and $N = 1$. Next we choose $\eta = \frac{\pi}{4}$ to satisfy the requirement that $\cos(\eta) \neq 0$. From Fig. 6.3 it can be shown that in the interval $[0, 2\pi - \frac{\pi}{4}] = [0, \frac{7\pi}{4}]$, $\delta_i(z) = 0$ has three real roots including a root at the origin. Since $\delta_i(z)$ is an odd function it follows that in the interval $[-\frac{7\pi}{4}, \frac{7\pi}{4}]$, $\delta_i(z)$ will have five real roots. Also observe that $\delta_i(z)$ does not have any real roots in $(\frac{7\pi}{4}, \frac{9\pi}{4}]$. Thus $\delta_i(z)$ has $4N + M = 5$ real roots in the interval $[-2\pi + \frac{\pi}{4}, 2\pi + \frac{\pi}{4}]$. Moreover, $\delta_i(z)$ has two real roots in each of the intervals $[2l\pi + \frac{\pi}{4}, 2(l+1)\pi + \frac{\pi}{4}]$ and $[-2(l+1)\pi + \frac{\pi}{4}, -2l\pi + \frac{\pi}{4}]$ for $l = 1, 2, \ldots$. Hence it follows that $\delta_i(z)$ has exactly $4lN + M$ real roots in

$[-2l\pi + \frac{\pi}{4}, 2l\pi + \frac{\pi}{4}]$ for $l = 1, 2, ...$, which by Theorem 5.6 implies that $\delta_i(z)$ has only real roots.

We now evaluate $\delta_r(z)$ at the roots of the imaginary part $\delta_i(z)$. For $z_o = 0$, using (6.4) we obtain

$$\begin{aligned} \delta_r(z_o) &= \cos(0) - \frac{T}{L} 0 \cdot \sin(0) + kk_c \\ \Rightarrow \delta_r(z_o) &= 1 + kk_c \, . \end{aligned} \tag{6.7}$$

For z_1, using (6.4) we obtain

$$\begin{aligned} \delta_r(z_1) &= \cos(z_1) - \frac{T}{L} z_1 \sin(z_1) + kk_c \\ &= \cos(z_1) + \tan(z_1)\sin(z_1) + kk_c \ \ \text{[using (6.6)]} \\ &= \frac{1}{\cos(z_1)} + kk_c \, . \end{aligned}$$

From Fig. 6.3, since $z_1 \epsilon (\frac{\pi}{2}, \pi)$ we obtain

$$\begin{aligned} \cos(z_1) &= -\frac{L}{\sqrt{T^2 z_1^2 + L^2}} \\ \Rightarrow \delta_r(z_1) &= -\frac{T}{L}\sqrt{z_1^2 + \frac{L^2}{T^2}} + kk_c \, . \end{aligned} \tag{6.8}$$

A similar analysis for $z_2 \epsilon (\frac{3\pi}{2}, 2\pi)$, $z_3 \epsilon (\frac{5\pi}{2}, 3\pi)$, and so on, gives us

$$\delta_r(z_2) = \frac{T}{L}\sqrt{z_2^2 + \frac{L^2}{T^2}} + kk_c \tag{6.9}$$

$$\delta_r(z_3) = -\frac{T}{L}\sqrt{z_3^2 + \frac{L^2}{T^2}} + kk_c \tag{6.10}$$

$$\vdots$$

From Step 1 we have that $k_c > -\frac{1}{k}$. Thus from (6.7) we see that $\delta_r(z_o) > 0$. Then, interlacing the roots of $\delta_r(z)$ and $\delta_i(z)$ is equivalent to $\delta_r(z_1) < 0$, $\delta_r(z_2) > 0$, $\delta_r(z_3) < 0$, and so on. Using this fact and (6.7)–(6.10) we obtain

$$\begin{aligned} \delta_r(z_o) > 0 \ &\Rightarrow \ k_c > -\frac{1}{k} =: M_0 \\ \delta_r(z_1) < 0 \ &\Rightarrow \ k_c < \frac{T}{kL}\sqrt{z_1^2 + \frac{L^2}{T^2}} =: M_1 \\ \delta_r(z_2) > 0 \ &\Rightarrow \ k_c > -\frac{T}{kL}\sqrt{z_2^2 + \frac{L^2}{T^2}} =: M_2 \\ \delta_r(z_3) < 0 \ &\Rightarrow \ k_c < \frac{T}{kL}\sqrt{z_3^2 + \frac{L^2}{T^2}} =: M_3 \\ &\vdots \end{aligned}$$

Since $z_1 < z_2 < z_3 < \cdots$, we conclude that $|M_1| < |M_2| < |M_3| < \cdots$. Intersecting the bounds previously found for k_c, we conclude that for the interlacing property to hold, we must have

$$-\frac{1}{k} < k_c < \frac{T}{kL}\sqrt{z_1^2 + \frac{L^2}{T^2}} \,.$$

For values of k_c in this range, the interlacing property and the fact that the roots of $\delta_i(z)$ are all real can be used in Theorem 5.6 to guarantee that $\delta_r(z)$ also has only real roots. To see this, note that since $\delta_i(z)$ has five real roots in the interval $[-2\pi + \frac{\pi}{4}, 2\pi + \frac{\pi}{4}]$, then by the interlacing property we conclude that $\delta_r(z)$ has at least four real roots in the same interval. The number of roots of $\delta_r(z)$ in the strip $-2\pi + \frac{\pi}{4} \leq z \leq 2\pi + \frac{\pi}{4}$ is exactly five. This additional root cannot be complex since it can be shown that complex roots appear in complex conjugate pairs for any quasi-polynomial with real coefficients. Thus, this additional root must be real and we conclude that $\delta_r(z)$ has $4N+M = 5$ real roots in the interval $[-2\pi+\frac{\pi}{4}, 2\pi+\frac{\pi}{4}]$. Hence, by Theorem 5.6, $\delta_r(z)$ has only real roots. In this manner, all the conditions of Theorem 5.5 are satisfied and this completes the proof. ∎

By explicitly evaluating the derivative with respect to L, it can be shown that the upper bound for k_c given in Theorem 6.1 is a monotonically decreasing function of the time delay L of the system. Thus, if the delay of the system is reduced from L_1 to L_2, with k_c fixed, the system remains stable since now a larger range of k_c can be tolerated. This is formalized in the following lemma.

Lemma 6.1 *The upper bound for k_c given in (6.3) is a monotonically decreasing function of the time delay L of the system.*

Proof. Let us define k_u to be the upper bound for k_c given in (6.3), i.e.,

$$k_u \triangleq \frac{T}{kL}\sqrt{z_1^2 + \frac{L^2}{T^2}} \,.$$

We can rewrite k_u as follows:

$$k_u = \frac{1}{k}\sqrt{\frac{T^2}{L^2}z_1^2 + 1} \,.$$

Now we find the derivative of k_u with respect to L as follows:

$$\begin{aligned}\frac{dk_u}{dL} &= \frac{1}{2k\sqrt{\frac{T^2}{L^2}z_1^2+1}}\left[\frac{-2T^2}{L^3}z_1^2 + 2\frac{T^2}{L^2}z_1\frac{dz_1}{dL}\right] \\ &= \frac{T^2 z_1}{kL^3\sqrt{\frac{T^2}{L^2}z_1^2+1}}\left[-z_1 + L\frac{dz_1}{dL}\right]\end{aligned}$$

where $\frac{dz_1}{dL}$ represents the derivative of the parameter z_1 (as defined in Theorem 6.1) with respect to L. Since $k > 0$, $L > 0$, and $z_1 > 0$, the first factor in the previous expression is always positive. We now need to find the sign of the expression $-z_1 + L\frac{dz_1}{dL}$. This will determine the sign of $\frac{dk_u}{dL}$.

Since z_1 is the solution of the equation $\tan(z) = -\frac{T}{L}z$, we have

$$\tan(z_1) = -\frac{T}{L}z_1 \,. \tag{6.11}$$

By differentiating this expression with respect to L we obtain

$$\begin{aligned}
\sec^2(z_1)\frac{dz_1}{dL} &= \frac{T}{L^2}z_1 - \frac{T}{L}\frac{dz_1}{dL} \\
\Rightarrow \frac{dz_1}{dL}\left[1 + \tan^2(z_1) + \frac{T}{L}\right] &= \frac{T}{L^2}z_1 \\
\Rightarrow \frac{dz_1}{dL}\left[1 + \frac{T^2}{L^2}z_1^2 + \frac{T}{L}\right] &= \frac{T}{L^2}z_1 \text{ [using (6.11)]} \\
\Rightarrow \frac{dz_1}{dL} &= \frac{Tz_1}{L^2 + T^2z_1^2 + TL} \,.
\end{aligned}$$

Thus we have

$$\begin{aligned}
-z_1 + L\frac{dz_1}{dL} &= -z_1 + \frac{TLz_1}{L^2 + T^2z_1^2 + TL} \\
&= -\left[1 - \frac{1}{\frac{L}{T} + \frac{T}{L}z_1^2 + 1}\right] z_1 \,.
\end{aligned}$$

Since $L > 0$, $T > 0$, and $z_1 > 0$, then

$$\frac{1}{\frac{L}{T} + \frac{T}{L}z_1^2 + 1} < 1$$

and we conclude that $-z_1 + L\frac{dz_1}{dL} < 0$. Thus we have

$$\frac{dk_u}{dL} < 0 \,,$$

which shows that k_u is a monotonically decreasing function of L. This completes the proof of the lemma. ■

Remark 6.1 *From Lemma 6.1 we can see that any gain k_c lying in the range (6.3) stabilizes all systems with a time delay less than or equal to L.*

6.2.2 Open-Loop Unstable Plant

In this subsection we present a theorem that gives a closed-form solution to the constant gain stabilization problem for the case of an open-loop unstable plant. This means that now $T < 0$. Of course, as before, we assume that $k > 0$ and $L > 0$.

Theorem 6.2 *Under the above assumptions on k and L, a necessary condition for a gain k_c to simultaneously stabilize the delay-free plant and the plant with delay is $|\frac{T}{L}| > 1$. If this necessary condition is satisfied, then the set of all stabilizing gains k_c for a given open-loop unstable plant with transfer function $G(s)$ as in (6.1) is given by*

$$\frac{T}{kL}\sqrt{z_1^2 + \frac{L^2}{T^2}} < k_c < -\frac{1}{k} \tag{6.12}$$

where z_1 is the solution of the equation

$$\tan(z) = -\frac{T}{L}z$$

in the interval $(0, \frac{\pi}{2})$.

Proof. The proof follows along the same lines as that of Theorem 6.1 and will be briefly sketched here. Again, the idea of the proof is to verify conditions 1 and 2 of Theorem 5.5.

Step 1. First we check condition 2 of Theorem 5.5:

$$E(\omega_o) = \delta_i^{'}(\omega_o)\delta_r(\omega_o) - \delta_i(\omega_o)\delta_r^{'}(\omega_o) > 0$$

for some ω_o in $(-\infty,\infty)$. Let us take $\omega_o = 0$, so $z_o = 0$. Thus $\delta_i(z_o) = 0$ and $\delta_r(z_o) = 1 + kk_c$. We also have

$$\begin{aligned} \delta_i^{'}(z) &= \left[\left(1+\frac{T}{L}\right)\cos(z) - \frac{T}{L}z\sin(z)\right] \\ \Rightarrow E(z_o) &= \left(1+\frac{T}{L}\right)(1+kk_c)\,. \end{aligned}$$

From (6.2), it is clear that from the closed-loop stability of the delay-free system, we have $1 + kk_c < 0$. Hence to have $E(z_o) > 0$, we must have $1 + \frac{T}{L} < 0$ or $\frac{T}{L} < -1$,

$$\Rightarrow \left|\frac{T}{L}\right| > 1\,.$$

Step 2. We now check condition 1 of Theorem 5.5: the interlacing of the roots of $\delta_r(z)$ and $\delta_i(z)$. The roots of $\delta_i(z)$ satisfy the following equation:

$$\tan(z) = -\frac{T}{L}z\,. \tag{6.13}$$

As before we can plot the two terms involved in this equation, i.e., $\tan(z)$ and $-\frac{T}{L}z$, to study the behavior of the roots of $\delta_i(z)$. Figure 6.4 shows this plot. Clearly the non-negative real roots of the imaginary part are

$$z_o = 0,\ z_1\epsilon\left(0,\frac{\pi}{2}\right),\ z_2\epsilon\left(\pi,\frac{3\pi}{2}\right),\ z_3\epsilon\left(2\pi,\frac{5\pi}{2}\right),$$

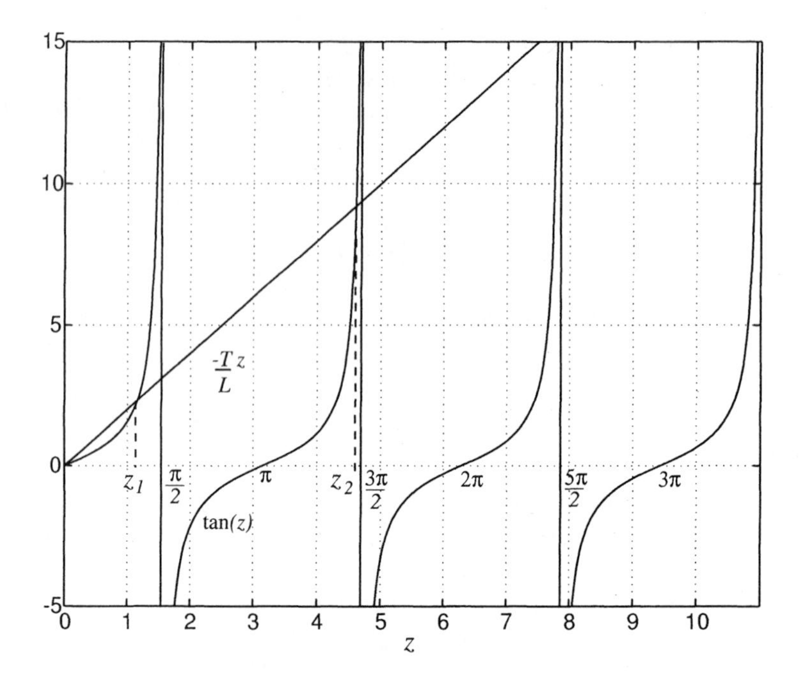

FIGURE 6.4. Plot of the terms involved in (6.13).

and so on, where the roots z_i, for $i = 1, 2, 3, ...$ satisfy (6.13). Arguing as in the proof of Theorem 6.1, we can show that $\delta_i(z)$ has only real roots.
We now evaluate $\delta_r(z)$ at the roots of the imaginary part $\delta_i(z)$. For $z_o = 0$, using (6.4) we obtain

$$\delta_r(z_o) = 1 + kk_c \,. \tag{6.14}$$

For z_1, using (6.4) and the fact that this root satisfies (6.13) we obtain

$$\delta_r(z_1) = \frac{1}{\cos(z_1)} + kk_c \,.$$

Recalling from Fig. 6.4 that $z_1 \epsilon (0, \frac{\pi}{2})$ we obtain

$$\begin{aligned} \cos(z_1) &= \frac{L}{\sqrt{T^2 z_1^2 + L^2}} \\ \Rightarrow \delta_r(z_1) &= -\frac{T}{L}\sqrt{z_1^2 + \frac{L^2}{T^2}} + kk_c \,. \end{aligned} \tag{6.15}$$

A similar analysis for $z_2 \epsilon (\pi, \frac{3\pi}{2})$, $z_3 \epsilon (2\pi, \frac{5\pi}{2})$, etc., gives us the following:

$$\delta_r(z_2) = \frac{T}{L}\sqrt{z_2^2 + \frac{L^2}{T^2}} + kk_c \tag{6.16}$$

$$\delta_r(z_3) = -\frac{T}{L}\sqrt{z_3^2 + \frac{L^2}{T^2}} + kk_c \tag{6.17}$$
$$\vdots$$

To ensure the interlacing of the roots of $\delta_r(z)$ and $\delta_i(z)$ we need $\delta_r(z_o) < 0$ (which comes from condition (6.2) for the closed-loop stability of the delay-free system), $\delta_r(z_1) > 0$, $\delta_r(z_2) < 0$, $\delta_r(z_3) > 0$, and so on. Using this fact and (6.14)–(6.17) we obtain

$$\begin{aligned}
\delta_r(z_o) < 0 \quad &\Rightarrow \quad k_c < -\frac{1}{k} =: M_0 \\
\delta_r(z_1) > 0 \quad &\Rightarrow \quad k_c > \frac{T}{kL}\sqrt{z_1^2 + \frac{L^2}{T^2}} =: M_1 \\
\delta_r(z_2) < 0 \quad &\Rightarrow \quad k_c < -\frac{T}{kL}\sqrt{z_2^2 + \frac{L^2}{T^2}} =: M_2 \\
\delta_r(z_3) > 0 \quad &\Rightarrow \quad k_c > \frac{T}{kL}\sqrt{z_3^2 + \frac{L^2}{T^2}} =: M_3 \\
&\vdots
\end{aligned}$$

Since $z_1 < z_2 < z_3 < \cdots$, we conclude that $|M_1| < |M_2| < |M_3| < \cdots$. Thus intersecting the bounds previously found for k_c we see that the interlacing property holds provided that

$$\frac{T}{kL}\sqrt{z_1^2 + \frac{L^2}{T^2}} < k_c < -\frac{1}{k}\,.$$

As in the proof of Theorem 6.1, we can show that for values of k_c in the above set $\delta_r(z)$ has only real roots so that all conditions in Theorem 5.5 are satisfied. This completes the proof. ∎

Remark 6.2 *The above results can also be obtained from the Nyquist criterion. Our analysis here of the Hermite-Biehler Theorem is done as a preparation for treating the PI and PID cases for which it is no longer obvious how the Nyquist criterion can be used.*

We now present some examples to illustrate the application of the results presented in this section.

Example 6.1 *Consider the constant gain stabilization problem, where the system is described by the first-order model with time delay (6.1). The plant parameters are $k = 1$, $L = 1.8$ seconds, and $T = 3$ seconds. Since the plant is open-loop stable we will use Theorem 6.1 to obtain the set of stabilizing gains. As the first step, we compute z_1, the solution of*

$$\tan(z) = -1.6667z$$

in the interval $(\frac{\pi}{2}, \pi)$*. This is given by* $z_1 = 1.8798$*. Then, from (6.3) the range of stabilizing constant gain values is*

$$-1 < k_c < 3.2887 .$$

Next we check if the roots of the real and imaginary parts of $\delta^*(j\omega)$ *interlace for a particular value of the controller parameter* k_c*. We now set* k_c *to* 1 *and the characteristic quasi-polynomial* $\delta^*(s)$ *of the system is given by*

$$\delta^*(s) = 1 + (3s+1)e^{1.8s} .$$

Substituting $s = j\omega$ *we obtain*

$$\begin{aligned} \delta^*(j\omega) &= [1 + \cos(1.8\omega) - 3\omega\sin(1.8\omega)] \\ &\quad + j[\sin(1.8\omega) + 3\omega\cos(1.8\omega)] . \end{aligned}$$

Figure 6.5 shows the plot of the real and imaginary parts of $\delta^*(j\omega)$*. As we can see the roots of the real and imaginary parts interlace. Figure 6.6 shows the time response of the closed-loop system* $y(t)$ *to a unit step input applied at* $t = 5$ *seconds.*

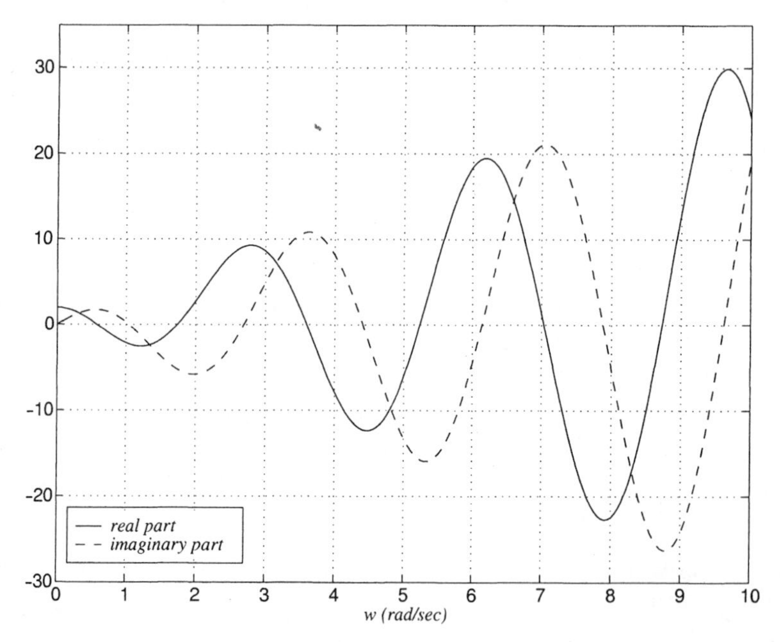

FIGURE 6.5. Plot of the real and imaginary parts of $\delta^*(j\omega)$ for Example 6.1.

We now use the Padé approximation introduced in Section 5.3 to determine approximate stabilizing ranges and compare them with the true stabilizing range. We approximate the time-delay term with the Padé approximation and then use the results in Section 3.2 to compute the set of

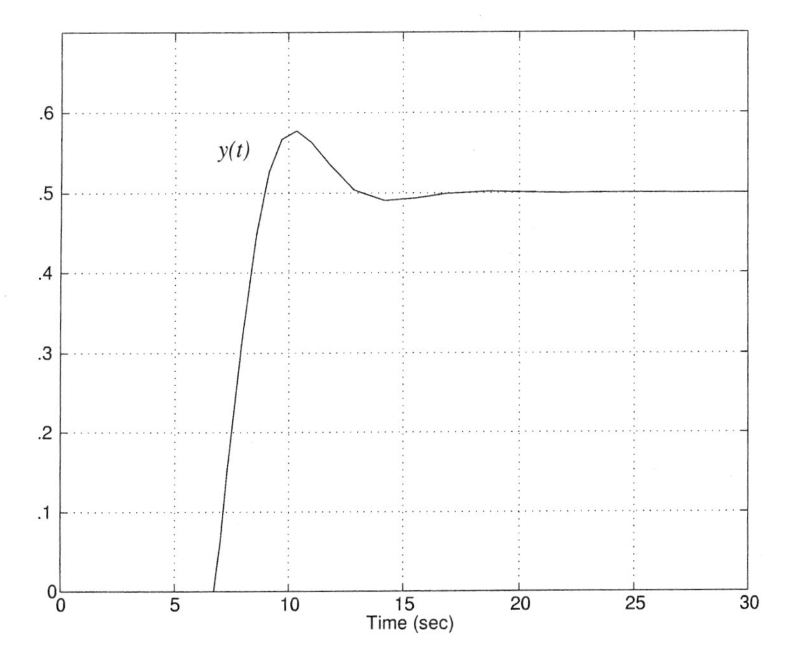

FIGURE 6.6. Time response of the closed-loop system for Example 6.1.

stabilizing feedback gains for the approximated *plant. The following results were obtained for different orders of the approximation. The reader can compare the approximate stabilizing ranges with the* true *stabilizing range that was determined earlier.*

$$\begin{array}{rl} \textit{First order:} & -1 < k_c < 4.3333 \\ \textit{Second order:} & -1 < k_c < 3.3267 \\ \textit{Third order:} & -1 < k_c < 3.2896 \\ \textit{Fifth order:} & -1 < k_c < 3.2887 \end{array}$$

Notice that Padé approximants of first, second, and third order give stabilizing gains that include destabilizing values for the true system. △

Example 6.2 *Consider again the constant gain stabilization problem for a system described by the first-order model with time delay (6.1) where now the plant parameters are* $k = 1$*,* $L = 0.5$ *seconds, and* $T = -2$*. Now the plant is open-loop unstable. Since* $\left|\frac{T}{L}\right| = 4 > 1$ *stabilization is possible. We will use Theorem 6.2 to obtain the set of stabilizing gains. First, we compute* $z_1 \epsilon (0, \frac{\pi}{2})$ *satisfying (6.13), i.e.,*

$$\tan(z) = 4z\ .$$

Solving this equation we obtain $z_1 = 1.3932$. *Thus, from (6.12) the set of stabilizing gains is given by*

$$-5.6620 < k_c < -1 \ .$$

For the controller parameter k_c *set to* -3 *the characteristic quasi-polynomial* $\delta^*(s)$ *of the system is given by*

$$\delta^*(s) = -3 + (1 - 2s)e^{0.5s} \ .$$

Substituting $s = j\omega$ *we obtain*

$$\delta^*(j\omega) = [-3 + \cos(0.5\omega) + 2\omega \sin(0.5\omega)] + j\,[\sin(0.5\omega) - 2\omega \cos(0.5\omega)] \ .$$

Figure 6.7 shows the plot of the real and imaginary parts of $\delta^*(j\omega)$. *It is clear from this plot that the roots of the real and imaginary parts interlace. Figure 6.8 shows the time response of the closed-loop system to a unit step input* r, *which verifies that the closed-loop system is indeed stable.* △

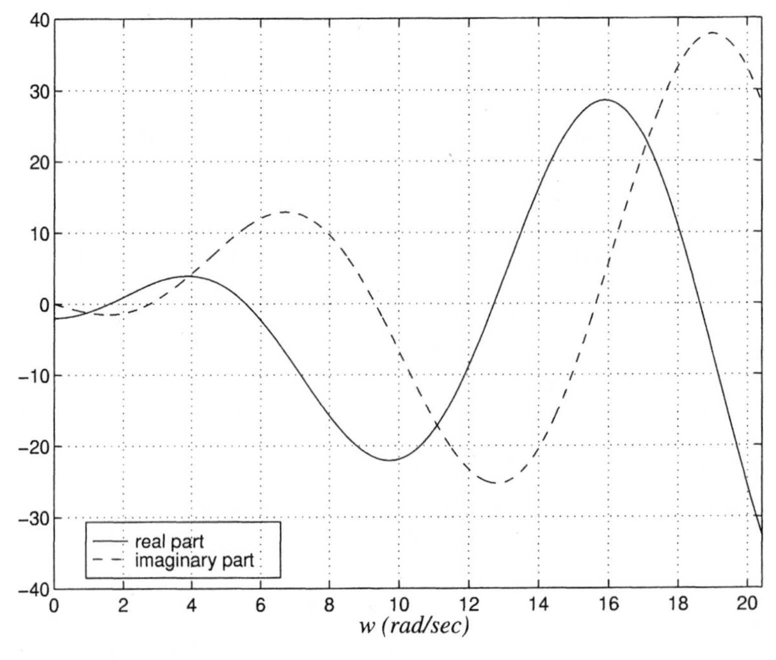

FIGURE 6.7. Plot of the real and imaginary parts of $\delta^*(j\omega)$ for Example 6.2.

6.3 Second-Order Systems with Time Delay

In this section we analyze the same stabilization problem stated in Section 6.2, but now we will consider systems with step responses like the ones

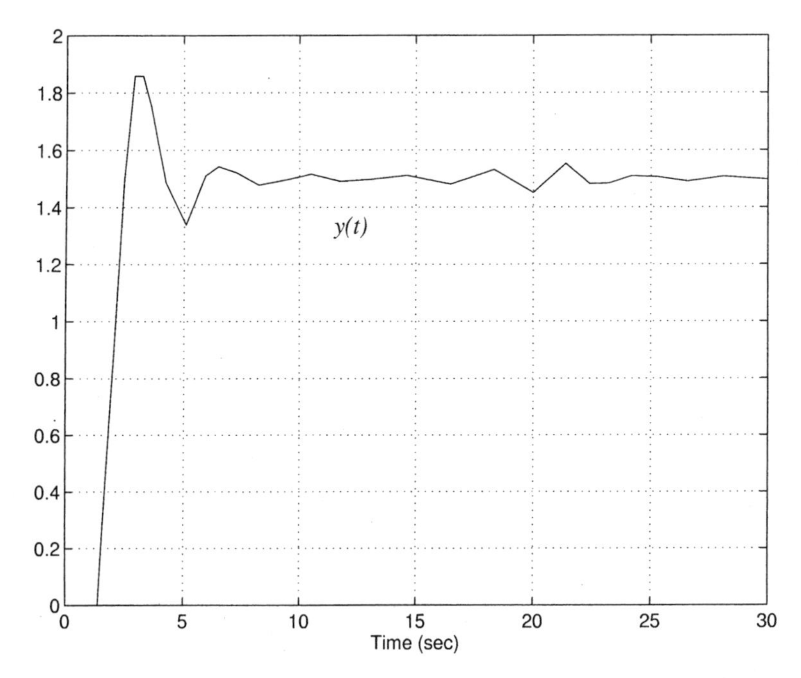

FIGURE 6.8. Time response of the closed-loop system for Example 6.2.

shown in Fig. 6.9. These systems are commonly modeled as second-order processes with a time delay which can be mathematically described by the transfer function

$$G(s) = \frac{k}{s^2 + a_1 s + a_o} e^{-Ls} . \tag{6.18}$$

Here k represents the steady-state gain of the plant, L represents the time delay, and a_1 and a_o are parameters of the plant. Again, our objective is to analytically determine the values of the parameter k_c for which the closed-loop system shown in Fig. 6.2 is stable.

When the time delay of the plant model is zero, that is, $L = 0$, the closed-loop characteristic equation of the system is given by

$$\delta(s) = s^2 + a_1 s + (a_o + kk_c) .$$

For this second-order polynomial we can determine necessary and sufficient conditions that the controller and plant parameters have to satisfy to guarantee the stability of the closed-loop system. If we assume that the steady-state gain k of the plant is positive these conditions are

$$a_1 > 0 \text{ and } k_c > -\frac{a_o}{k} . \tag{6.19}$$

Now let us consider the case where the time delay of the plant is different from zero and try to determine the set of all stabilizing gains. The closed-

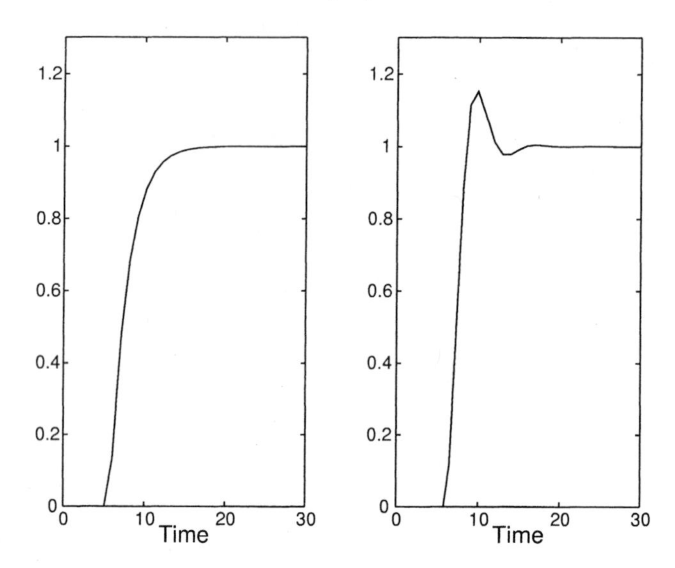

FIGURE 6.9. Open-loop step response with damped oscillations.

loop characteristic equation of the system is given by

$$\delta(s) = kk_c e^{-Ls} + s^2 + a_1 s + a_o .$$

In order to study the stability of the closed-loop system, we need to determine if all the zeros of the above expression lie in the open LHP. We can again invoke Theorem 5.5 to determine the set of all stabilizing gains k_c by proceeding as follows.

First we consider the quasi-polynomial $\delta^*(s)$ defined by

$$\delta^*(s) = e^{Ls}\delta(s) = kk_c + (s^2 + a_1 s + a_o)e^{Ls} .$$

Substituting $s = j\omega$, we have

$$\delta^*(j\omega) = \delta_r(\omega) + j\delta_i(\omega)$$

where

$$\begin{aligned} \delta_r(\omega) &= kk_c + (a_o - \omega^2)\cos(L\omega) - a_1\omega\sin(L\omega) \\ \delta_i(\omega) &= (a_o - \omega^2)\sin(L\omega) + a_1\omega\cos(L\omega) . \end{aligned}$$

For the following analysis, it is convenient to make the change of variables $z = L\omega$. Then, the real and imaginary parts of $\delta^*(j\omega)$ can be rewritten as

$$\delta_r(z) = kk_c + \left(a_o - \frac{z^2}{L^2}\right)\cos(z) - \frac{a_1}{L} z\sin(z) \tag{6.20}$$

$$\delta_i(z) = \left(a_o - \frac{z^2}{L^2}\right)\sin(z) + \frac{a_1}{L} z\cos(z) . \tag{6.21}$$

We now consider two different cases.

6.3.1 Open-Loop Stable Plant

In this subsection we give a closed-form solution to the constant gain stabilization problem for the case of an open-loop stable plant. This means that the parameters a_1 and a_o of the plant satisfy $a_1 > 0$, $a_o > 0$. Moreover, we assume that $k > 0$ and $L > 0$.

Theorem 6.3 *Under the above assumptions on k and L, the set of all stabilizing gains k_c for a given open-loop stable plant with transfer function $G(s)$ as in (6.18) is given by*

(i) If $a_1^2 \geq 2a_o$, then

$$-\frac{a_o}{k} < k_c < \frac{a_1 z_1}{kL\sin(z_1)} \tag{6.22}$$

where z_1 is the solution of the equation

$$\cot(z) = \frac{z^2 - L^2 a_o}{La_1 z}$$

in the interval $(0, \pi)$.

(ii) If $a_1^2 < 2a_o$, then

$$\max_{j=ev,ev+2} \left\{ \frac{a_1 z_j}{kL\sin(z_j)} \right\} < k_c < \min_{j=od,od+2} \left\{ \frac{a_1 z_j}{kL\sin(z_j)} \right\} \tag{6.23}$$

where z_j is the solution of the equation

$$\cot(z) = \frac{z^2 - L^2 a_o}{La_1 z}$$

in the interval $((j-1)\pi, j\pi)$; $\alpha \triangleq L\sqrt{a_o - \frac{a_1^2}{2}}$; od is an odd natural number defined as

$$od \triangleq arg \min_{j\ odd}\{\alpha - z_j\} \quad subject\ to\ \alpha - z_j \geq 0$$

and ev is an even natural number or zero defined as

$$ev \triangleq arg \min_{j\ even}\{\alpha - z_j\} \quad subject\ to\ \alpha - z_j \geq 0.$$

Proof. From Theorem 5.5, we need to check two conditions to ensure the stability of the quasi-polynomial $\delta^*(s)$.
Step 1. We first check condition 2 of Theorem 5.5:

$$E(\omega_o) = \delta_i^{'}(\omega_o)\delta_r(\omega_o) - \delta_i(\omega_o)\delta_r^{'}(\omega_o) > 0$$

for some ω_o in $(-\infty, \infty)$. Let us take $\omega_o = 0$, so $z_o = 0$. Thus $\delta_i(z_o) = 0$ and $\delta_r(z_o) = kk_c + a_o$. We also have

$$\begin{aligned} \delta_i^{'}(z) &= \left(a_o - \frac{z^2}{L^2} + \frac{a_1}{L}\right)\cos(z) - \left(\frac{2}{L^2}z + \frac{a_1}{L}z\right)\sin(z) \\ \Rightarrow E(z_o) &= \left(a_o + \frac{a_1}{L}\right)(kk_c + a_o) \,. \end{aligned}$$

By our initial assumption $a_1 > 0$ and $a_o > 0$. Thus, if we pick $k_c > -\frac{a_o}{k}$ we have that $E(z_o) > 0$.

Step 2. We now check the interlacing of the roots of $\delta_r(z)$ and $\delta_i(z)$. From (6.21) we can compute the roots of the imaginary part, i.e., $\delta_i(z) = 0$. This gives us the following equation

$$\left(a_o - \frac{z^2}{L^2}\right)\sin(z) + \frac{a_1}{L}z\cos(z) = 0 \,.$$

From this equation we clearly see that $z_o = 0$ is a root of the imaginary part. We also see that $l\pi$, $l = 1, 2, ...$, are not roots of the imaginary part. Thus for $z \neq 0$, we can rewrite the previous equation as

$$\cot(z) = \frac{z^2 - L^2 a_o}{L a_1 z} \,. \tag{6.24}$$

An analytical solution of (6.24) is difficult to find. However, we can plot the two terms involved in this equation, i.e., $\cot(z)$ and $\frac{z^2 - L^2 a_o}{L a_1 z}$, to study the nature of the real solutions. Let us denote the positive real roots of (6.24) by z_j, $j = 1, 2, ...$, arranged in increasing order of magnitude. Figure 6.10 shows the plot discussed above. Clearly the non-negative real roots of the imaginary part satisfy

$$z_1 \epsilon (0, \pi), \ z_2 \epsilon (\pi, 2\pi), \ z_3 \epsilon (2\pi, 3\pi), \ z_4 \epsilon (3\pi, 4\pi), \ \ldots \,. \tag{6.25}$$

Let us now use Theorem 5.6 to check if $\delta_i(z)$ has only real roots. Substituting $s_1 = Ls$ in the expression for $\delta^*(s)$, we see that for the new quasi-polynomial in s_1, $M = 2$ and $N = 1$. Next we choose $\eta = \frac{\pi}{4}$ to satisfy the requirement that $\sin(z)$ does not vanish at $z = \eta$. From Fig. 6.10 it can be shown that in the interval $[0, 2\pi - \frac{\pi}{4}] = [0, \frac{7\pi}{4}]$, $\delta_i(z) = 0$ has three real roots including a root at the origin. Since $\delta_i(z)$ is an odd function it follows that in the interval $[-\frac{7\pi}{4}, \frac{7\pi}{4}]$, $\delta_i(z)$ will have five real roots. Also observe that $\delta_i(z)$ has one real root in $(\frac{7\pi}{4}, \frac{9\pi}{4}]$. Thus $\delta_i(z)$ has $4N + M = 6$ real roots in the interval $[-2\pi + \frac{\pi}{4}, 2\pi + \frac{\pi}{4}]$. Moreover, $\delta_i(z)$ has two real roots in each of the intervals $[2l\pi + \frac{\pi}{4}, 2(l+1)\pi + \frac{\pi}{4}]$ and $[-2(l+1)\pi + \frac{\pi}{4}, -2l\pi + \frac{\pi}{4}]$ for $l = 1, 2, \ldots$. Hence it follows that $\delta_i(z)$ has exactly $4lN + M$ real roots in $[-2l\pi + \frac{\pi}{4}, 2l\pi + \frac{\pi}{4}]$ for $l = 1, 2, ...$, which by Theorem 5.6 implies that $\delta_i(z)$ has only real roots.

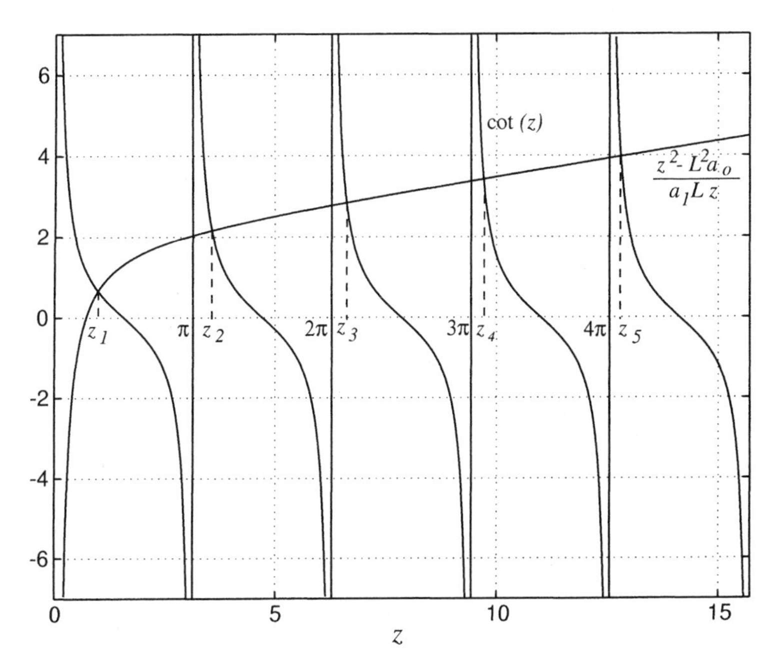

FIGURE 6.10. Plot of the terms involved in (6.24).

We now evaluate $\delta_r(z)$ at the roots of the imaginary part $\delta_i(z)$. For $z_o = 0$, using (6.20) we obtain

$$\delta_r(z_o) = kk_c + a_o \,. \tag{6.26}$$

Using (6.20) we obtain

$$\begin{aligned}
\delta_r(z_j) &= kk_c + \left(a_o - \frac{z_j^2}{L^2}\right)\cos(z_j) - \frac{a_1}{L} z_j \sin(z_j) \,, \text{ for } j = 1, 2, ..., \\
&= kk_c - \frac{a_1 z_j \cos^2(z_j)}{L \sin(z_j)} - \frac{a_1}{L} z_j \sin(z_j) \;\; \text{[using (6.24)]} \\
&= kk_c - \frac{a_1 z_j}{L \sin(z_j)} \,.
\end{aligned}$$

Thus we obtain

$$\delta_r(z_j) = k[k_c - M(z_j)] \tag{6.27}$$

where

$$M(z) = \frac{a_1 z}{kL \sin(z)} \,. \tag{6.28}$$

From Step 1 we have $k_c > -\frac{a_o}{k}$. Thus from (6.26) we see that $\delta_r(z_o) > 0$. Then, interlacing the roots of $\delta_r(z)$ and $\delta_i(z)$ is equivalent to $\delta_r(z_1) < 0$, $\delta_r(z_2) > 0$, $\delta_r(z_3) < 0$, and so on. Using this fact and (6.26)–(6.27) we obtain

$$\begin{aligned}
\delta_r(z_o) > 0 &\Rightarrow k_c > -\frac{a_o}{k} \\
\delta_r(z_1) < 0 &\Rightarrow k_c < M(z_1) =: M_1 \\
\delta_r(z_2) > 0 &\Rightarrow k_c > M(z_2) =: M_2 \\
\delta_r(z_3) < 0 &\Rightarrow k_c < M(z_3) =: M_3 \\
&\vdots
\end{aligned}$$

From (6.25) we see that z_j for odd values of j are either in the first or the second quadrant. Thus for odd values of j, $\sin(z_j) > 0$ and from (6.28), we conclude that $M(z_j) > 0$ for odd values of the parameter j. In a similar fashion, we see from (6.25) that z_j for even values of j are either in the third or the fourth quadrant. Thus for even values of j, $\sin(z_j) < 0$ and from (6.28) we conclude that $M(z_j) < 0$ for even values of the parameter j. Thus the previous set of inequalities can be rewritten as

$$k_c > -\frac{a_o}{k} \quad \text{and} \quad \max_{j=2,4,6,\ldots}\{M_j\} < k_c < \min_{j=1,3,5,\ldots}\{M_j\}\,. \tag{6.29}$$

We know that z_j, $j = 1, 2, \ldots$, satisfy (6.24). Using this, we can rewrite $M(z_j)$ defined in (6.28) as

$$\begin{aligned}
M(z_j) &= \pm\frac{1}{kL^2}\sqrt{(z_j^2 - L^2 a_o)^2 + L^2 a_1^2 z_j^2} \\
\Rightarrow M(z_j) &= \pm\frac{1}{kL^2}\sqrt{z_j^4 + L^2(a_1^2 - 2a_o)z_j^2 + L^4 a_o^2}
\end{aligned} \tag{6.30}$$

where the plus sign $(+)$ is used for odd values of j, and the minus sign $(-)$ is used for even values of j. We now consider two different cases.

Case 1: $a_1^2 \geq 2a_o$. In this case we see that $M(z_j)$ is a monotonically increasing function for odd values of j and a monotonically decreasing function for even values of j. Moreover, we see that $M(0) = -\frac{a_o}{k}$. Thus, using these observations, the bounds for k_c in (6.29) can be expressed as

$$-\frac{a_o}{k} < k_c < \frac{a_1 z_1}{kL\sin(z_1)}\,.$$

Case 2: $a_1^2 < 2a_o$. In this case, the function $|M(z_j)|$ has a minimum at $z = L\sqrt{a_o - \frac{a_1^2}{2}}$. Let us denote this minimizer as α. Then the function $|M(z_j)|$ is monotonically decreasing in the interval $[0, \alpha)$ and monotonically increasing in the interval $[\alpha, \infty)$. Let us denote by z_{od}, where *od* is an odd natural number, the z_j, j odd that minimizes $\alpha - z_j$ subject to $z_j \leq \alpha$; and z_{ev}, where *ev* is an even natural number or zero, the z_j, j even or zero that minimizes $\alpha - z_j$ subject to $z_j \leq \alpha$. Mathematically, we can express this as follows:

$$
\begin{aligned}
od &= arg \min_{j\ odd}\{\alpha - z_j\} \text{ subject to } \alpha - z_j \geq 0 \\
ev &= arg \min_{j\ even}\{\alpha - z_j\} \text{ subject to } \alpha - z_j \geq 0.
\end{aligned}
$$

The bounds for k_c in (6.29) can be expressed as

$$
\max_{j=ev,ev+2}\left\{\frac{a_1 z_j}{kL\sin(z_j)}\right\} < k_c < \min_{j=od,od+2}\left\{\frac{a_1 z_j}{kL\sin(z_j)}\right\}.
$$

Note that for values of k_c in these ranges, the interlacing property and the fact that the roots of $\delta_i(z)$ are all real can be used in Theorem 5.6 to guarantee that $\delta_r(z)$ also has only real roots. Thus all the conditions of Theorem 5.5 are satisfied and this completes the proof. ■

6.3.2 Open-Loop Unstable Plant

In this subsection we present a theorem that gives a closed-form solution to the constant gain stabilization problem for an open-loop unstable plant. From (6.19) we see that an unstable open-loop plant can be stabilized using a constant gain only if it has a *single* unstable pole. This means that an unstable but stabilizable plant must necessarily have $a_1 > 0$ and $a_o < 0$. As before, let us assume that $k > 0$ and $L > 0$.

Theorem 6.4 *Under the above assumptions on k and L, a necessary condition for a gain k_c to simultaneously stabilize the delay-free plant and the plant with delay is $|\frac{a_1}{a_o}| > L$. If this necessary condition is satisfied, then the set of all stabilizing gains k_c for a given open-loop unstable plant with transfer function $G(s)$ as in (6.18) is given by*

$$
-\frac{a_o}{k} < k_c < \frac{a_1 z_1}{kL\sin(z_1)} \tag{6.31}
$$

where z_1 is the solution of the equation

$$
\cot(z) = \frac{z^2 - L^2 a_o}{L a_1 z}
$$

in the interval $(0, \pi)$.

Proof. The proof follows along the same lines as that of Theorem 6.3 and will be briefly sketched here. Again, the idea of the proof is to verify conditions 1 and 2 of Theorem 5.5.

Step 1. First, we check condition 2 of Theorem 5.5:

$$
E(\omega_o) = \delta_i^{'}(\omega_o)\delta_r(\omega_o) - \delta_i(\omega_o)\delta_r^{'}(\omega_o) > 0
$$

for some ω_o in $(-\infty,\infty)$. Let us take $\omega_o = 0$, so $z_o = 0$. Thus $\delta_i(z_o) = 0$ and $\delta_r(z_o) = kk_c + a_o$. We also have

$$\begin{aligned} \delta_i'(z) &= \left(a_o - \frac{z^2}{L^2} + \frac{a_1}{L}\right)\cos(z) - \left(\frac{2}{L^2}z + \frac{a_1}{L}z\right)\sin(z) \\ \Rightarrow E(z_o) &= \left(a_o + \frac{a_1}{L}\right)(kk_c + a_o)\,. \end{aligned}$$

From (6.19), it is clear that from the closed-loop stability of the delay-free system, we have $(kk_c + a_o) > 0$. Hence to have $E(z_o) > 0$, we must have $a_o + \frac{a_1}{L} > 0$ or $-\frac{a_1}{a_o} > L$,

$$\Rightarrow \left|\frac{a_1}{a_o}\right| > L\,.$$

Step 2. We now check condition 1 of Theorem 5.5: the interlacing of the roots of $\delta_r(z)$ and $\delta_i(z)$. As in the previous subsection, one root of the imaginary part of $\delta^*(z)$ is $z_o = 0$ and the remaining real roots satisfy the equation

$$\cot(z) = \frac{z^2 - L^2 a_o}{L a_1 z}\,. \tag{6.32}$$

As before, we can plot the two terms involved in this equation, i.e., $\cot(z)$ and $\frac{z^2 - L^2 a_o}{L a_1 z}$ to study the behavior of the roots of $\delta_i(z)$. Figure 6.11 shows this plot when $|\frac{a_1}{a_o}| > L$. Clearly the positive real roots of the imaginary part are

$$z_1 \epsilon (0,\pi),\ z_2 \epsilon (\pi, 2\pi),\ z_3 \epsilon (2\pi, 3\pi),\ z_4 \epsilon (3\pi, 4\pi),\ \ldots\,.$$

where the roots z_i for $i = 1, 2, 3, ...$ satisfy (6.32). Arguing as in the proof of Theorem 6.3, we can show that $\delta_i(z)$ has only real roots.

We now evaluate $\delta_r(z)$ at the roots of the imaginary part $\delta_i(z)$. For $z_o = 0$, using (6.20) we obtain

$$\delta_r(z_o) = kk_c + a_o\,. \tag{6.33}$$

For z_j, $j = 1, 2, ...$, using (6.20) and (6.32) we obtain

$$\delta_r(z_j) = kk_c - \frac{a_1 z_j}{L\sin(z_j)}$$

which can be rewritten as

$$\delta_r(z_j) = k[k_c - M(z_j)] \tag{6.34}$$

where

$$M(z) = \frac{a_1 z}{kL\sin(z)}\,. \tag{6.35}$$

Now from Step 1 we have $k_c > -\frac{a_o}{k}$, which from (6.33) implies that $\delta_r(z_o) > 0$. Thus interlacing the roots of $\delta_r(z)$ and $\delta_i(z)$ is equivalent to $\delta_r(z_1) < 0$,

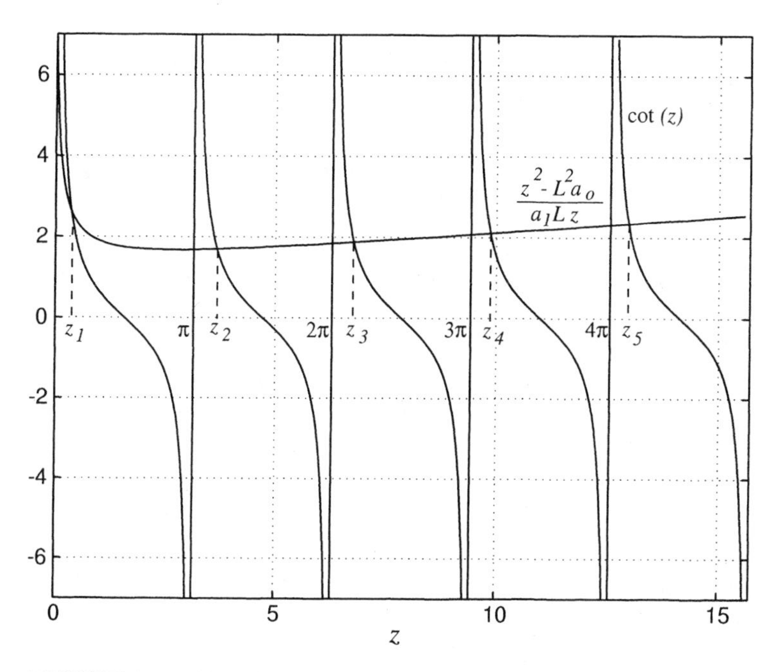

FIGURE 6.11. Plot of the terms involved in (6.32) when $|\frac{a_1}{a_o}| > L$.

$\delta_r(z_2) > 0$, $\delta_r(z_3) < 0$, and so on. Using this fact and (6.33)–(6.34) we obtain

$$\begin{aligned}
\delta_r(z_o) > 0 \quad &\Rightarrow \quad k_c > -\frac{a_o}{k} \\
\delta_r(z_1) < 0 \quad &\Rightarrow \quad k_c < M(z_1) =: M_1 \\
\delta_r(z_2) > 0 \quad &\Rightarrow \quad k_c > M(z_2) =: M_2 \\
\delta_r(z_3) < 0 \quad &\Rightarrow \quad k_c < M(z_3) =: M_3 \\
&\vdots
\end{aligned}$$

As in the proof of Theorem 6.3 we can rewrite the previous set of inequalities in compact form:

$$k_c > -\frac{a_o}{k} \quad \text{and} \quad \max_{j=2,4,6,\ldots} \{M_j\} < k_c < \min_{j=1,3,5,\ldots} \{M_j\}\,. \tag{6.36}$$

We know that z_j, $j = 1, 2, \ldots$, satisfy (6.32). Using this we can rewrite $M(z_j)$ defined in (6.35) as

$$M(z_j) = \pm\frac{1}{kL^2}\sqrt{z_j^4 + L^2(a_1^2 - 2a_o)z_j^2 + L^4 a_o^2} \tag{6.37}$$

where the plus sign (+) is used for odd values of j, and the minus sign (−) is used for even values of j.

Notice that since $a_o < 0$, we have $a_1^2 \geq 2a_o$. Thus, from (6.37) we see that $M(z_j)$ is a monotonically increasing function for odd values of j and it is a monotonically decreasing function for even values of j. Moreover, we see that $M(0) = -\frac{a_o}{k}$. Using these observations, the bounds for k_c in (6.36) reduce to

$$-\frac{a_o}{k} < k_c < \frac{a_1 z_1}{kL\sin(z_1)} .$$

Note that for values of k_c in this range, the interlacing property and the fact that the roots of $\delta_i(z)$ are all real can be used in Theorem 5.6 to guarantee that $\delta_r(z)$ also has only real roots. Thus all the conditions of Theorem 5.5 are satisfied and this completes the proof. ■

The following example illustrates how the above results can be used to solve the constant gain stabilization problem for a second-order system with time delay.

Example 6.3 *Consider the constant gain stabilization problem for a system described by the second-order model with time delay (6.18). The plant parameters are* $k = 5$, $L = 3.2$ *sec,* $a_1 = 2$, *and* $a_o = 5$. *Since the plant is open-loop stable we will use Theorem 6.3 to obtain the set of stabilizing gains. First notice that* $a_1^2 < 2a_o$. *Then, from part (ii) of Theorem 6.3, we need to compute the parameter* α. *In this case we have* $\alpha = 5.5426$. *Next we compute the parameters od and ev according to Theorem 6.3. These are given by* $od = 1$ *and* $ev = 2$. *This means that we need to compute the roots* z_j, *for* $j = 1, 2, 3, 4$ *of (6.24):*

$$\cot(z) = \frac{z^2 - 51.2}{6.4z} .$$

Solving this equation we obtain $z_1 = 2.7570$, $z_2 = 5.3080$, $z_3 = 7.6932$, *and* $z_4 = 10.3011$. *Thus, from (6.23) the set of stabilizing gains is given by*

$$\begin{aligned} \max_{j=2,4}\left\{\frac{a_1 z_j}{kL\sin(z_j)}\right\} < \quad & k_c \quad < \min_{j=1,3}\left\{\frac{a_1 z_j}{kL\sin(z_j)}\right\} \\ \Rightarrow -0.8015 < \quad & k_c \quad < 0.9186 . \end{aligned}$$

Next we check if the roots of the real and imaginary parts of $\delta^*(j\omega)$ *interlace for a particular value of the controller parameter* k_c. *We now set* k_c *to* 0.3 *and the characteristic quasi-polynomial* $\delta^*(s)$ *of the system is given by*

$$\delta^*(s) = 1.5 + (s^2 + 2s + 5)e^{3.2s} .$$

Substituting $s = j\omega$ *we obtain*

$$\begin{aligned} \delta^*(j\omega) &= \left[1.5 + (5 - \omega^2)\cos(3.2\omega) - 2\omega\sin(3.2\omega)\right] \\ &\quad + j\left[(5 - \omega^2)\sin(3.2\omega) + 2\omega\cos(3.2\omega)\right] . \end{aligned}$$

Figure 6.12 shows the plot of the real and imaginary parts of $\delta^*(j\omega)$. *As we can see the roots of the real and imaginary parts interlace. Figure 6.13*

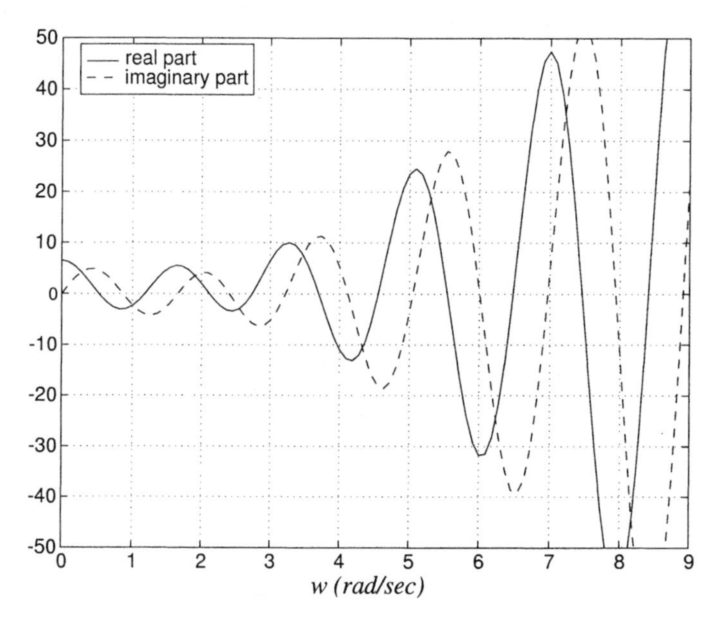

FIGURE 6.12. Plot of the real and imaginary parts of $\delta^*(j\omega)$ for Example 6.3.

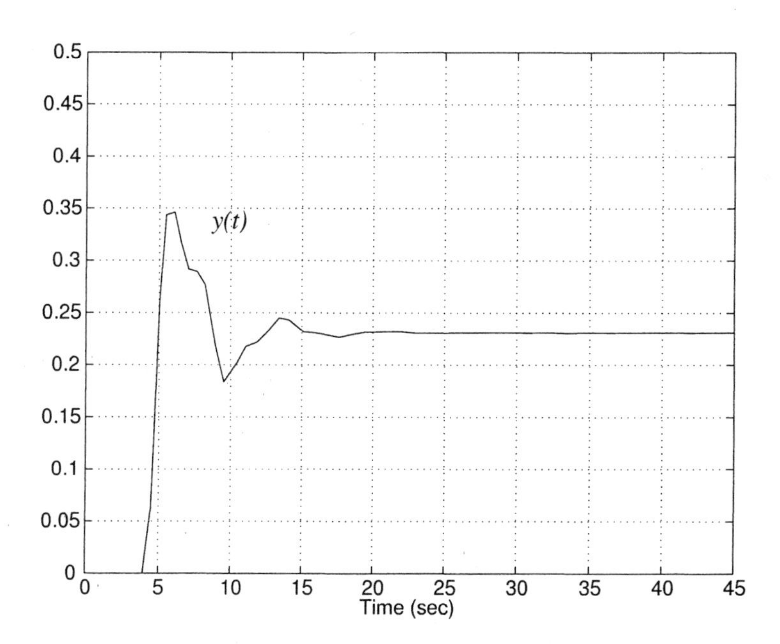

FIGURE 6.13. Time response of the closed-loop system for Example 6.3.

shows the time response of the closed-loop system to a unit step input r *applied at* $t = 1$ *sec.* △

6.4 Notes and References

The characterization of all stabilizing constant gain controllers for a given first-order plant with time delay was developed by Silva, Datta, and Bhattacharyya [41]. An alternative solution to the constant gain stabilization problem of a system with time delay, based on the Nyquist criterion, can be found in [11]. The solution to the constant gain stabilization problem for a second-order system with time delay is due to Silva, Datta, and Bhattacharyya [42].

7

PI Stabilization of First-Order Systems with Time Delay

In this chapter, we continue with the line of work presented in the last chapter and solve the problem of stabilizing a first-order plant with time delay using a PI controller. As before, the results in Chapter 5 will play a crucial role. Examples are included to clarify the detailed steps associated with the solution.

7.1 Introduction

In industrial control applications the performance requirements on a control system design include many factors. Some of these are response to command signals, insensitivity to measurement noise and process variations, and rejection of load disturbances. The design of a control system also involves the aspects of process dynamics and actuator saturation. It may seem surprising that controllers as simple in structure as the PI and PID controllers can perform so well in practice. It is also interesting to note that many industrial controllers only have the PI action. In some other cases the derivative action of the PID controller is switched off. In fact, it has been reported that 98% of the control loops in the pulp and paper industries are controlled by single-input single-output PI controllers. This is indicative of the popularity of PI controllers among industrial practitioners.

It has been pointed out that PI controllers are adequate for all processes where the dynamics are essentially of first order. Examples of these processes include level controls in single tanks and stirred tank reactors with perfect mixing. Even if the process has higher-order dynamics, if its step response looks like that of a first-order system, then PI control is usually sufficient. In such a case, the integral control will provide zero steady-state offset and the proportional action will provide an adequate transient response. Thus it is important to carefully select the parameters of the PI controller to achieve the desired performance specifications while maintaining closed-loop stability.

In this chapter we present a solution to the problem of stabilizing a first-order plant with time delay using a PI controller. Using the results developed in Chapter 5, a complete analytical characterization of all stabilizing PI gain values is provided. The chapter is organized as follows. In Section 7.2 we present the formal statement of the problem to be solved. In Section 7.3 we present the solution to the problem when the system is open-loop stable. In Section 7.4 we present a similar result for open-loop unstable plants. We also provide here a necessary condition on the time delay for the existence of stabilizing PI controllers. Throughout these sections, simulations and design examples are included to illustrate the applicability of the results.

7.2 The PI Stabilization Problem

As in Chapter 6, we consider the feedback control system shown in Fig. 6.2. The plant $G(s)$ is given by the following transfer function:

$$G(s) = \frac{k}{1+Ts}e^{-Ls} \tag{7.1}$$

where k represents the steady-state gain of the plant, L represents the time delay, and T represents the time constant of the plant. The controller $C(s)$ is of the PI type, i.e., it has a proportional term and an integral term:

$$C(s) = k_p + \frac{k_i}{s} .$$

Our objective is to analytically determine the region in the k_i—k_p parameter space for which the closed-loop system is stable.

When the time delay L of the plant model is zero, the characteristic equation of the closed-loop system is given by

$$\delta(s) = Ts^2 + (kk_p + 1)s + kk_i .$$

From the above equation, we conclude that for the closed-loop stability of the delay-free system, we must have either

$$kk_i > 0 \, , \; kk_p + 1 > 0 \, , \; T > 0 \tag{7.2}$$

or

$$kk_i < 0 \ , \ kk_p + 1 < 0 \ , \ T < 0 \ . \tag{7.3}$$

Clearly (7.2) must be satisfied for an open-loop stable plant while (7.3) must be satisfied for an open-loop unstable plant. Assuming that the steady-state gain k of the plant is positive we obtain the following conditions for closed-loop stability of the delay-free system:

$$k_p > -\frac{1}{k} \ , \ k_i > 0 \qquad \text{(open-loop stable plant, i.e., } T > 0\text{)} \tag{7.4}$$

$$k_p < -\frac{1}{k} \ , \ k_i < 0 \qquad \text{(open-loop unstable plant, i.e., } T < 0\text{)} \ . \tag{7.5}$$

We now bring in the time delay of the model. In this case the closed-loop characteristic equation of the system is given by

$$\delta(s) = (kk_i + kk_p s)e^{-Ls} + (1 + Ts)s \ .$$

As in Chapter 6, we will use Theorem 5.5 to find the set of stabilizing PI controllers. First we construct the quasi-polynomial $\delta^*(s)$, i.e.,

$$\delta^*(s) = e^{Ls}\delta(s) = kk_i + kk_p s + (1 + Ts)se^{Ls} \ .$$

Substituting $s = j\omega$, we have

$$\delta^*(j\omega) = \delta_r(\omega) + j\delta_i(\omega)$$

where

$$\begin{aligned} \delta_r(\omega) &= kk_i - \omega \sin(L\omega) - T\omega^2 \cos(L\omega) \\ \delta_i(\omega) &= \omega[kk_p + \cos(L\omega) - T\omega \sin(L\omega)] \ . \end{aligned}$$

In the following sections we present the analysis of the two different cases: open-loop stable plant and open-loop unstable plant.

7.3 Open-Loop Stable Plant

When the system described by (7.1) is open-loop stable, then we have $T > 0$. Furthermore, let us assume that $k > 0$ and $L > 0$. Clearly, the controller parameter k_i only affects the real part of $\delta^*(j\omega)$ whereas the controller parameter k_p affects the imaginary part of $\delta^*(j\omega)$. Moreover, we note that k_i, k_p appear affinely in $\delta_r(\omega)$, $\delta_i(\omega)$, respectively. Thus by sweeping over all real k_p and solving a constant gain stabilization problem at each stage (as in Section 6.2), we can determine the set of all stabilizing (k_p, k_i) values for the given plant.

The range of k_p values over which the sweeping needs to be carried out can be narrowed down by using the following result.

Theorem 7.1 *Under the above assumptions on k and L, the range of k_p values for which a solution to the PI stabilization problem of a given open-loop stable plant with transfer function $G(s)$ as in (7.1) exists is given by*

$$-\frac{1}{k} < k_p < \frac{T}{kL}\sqrt{\alpha_1^2 + \frac{L^2}{T^2}} \tag{7.6}$$

where α_1 is the solution of the equation

$$\tan(\alpha) = -\frac{T}{L}\alpha$$

in the interval $(\frac{\pi}{2}, \pi)$.

Proof. With the change of variables $z = L\omega$ the real and imaginary parts of $\delta^*(j\omega)$ can be rewritten as

$$\delta_r(z) = k[k_i - a(z)] \tag{7.7}$$

$$\delta_i(z) = \frac{z}{L}\left[kk_p + \cos(z) - \frac{T}{L}z\sin(z)\right] \tag{7.8}$$

where

$$a(z) \triangleq \frac{z}{kL}\left[\sin(z) + \frac{T}{L}z\cos(z)\right] . \tag{7.9}$$

According to Theorem 5.5, we need to check two conditions to ensure the stability of the quasi-polynomial $\delta^*(s)$.
Step 1. We start by checking condition 2 of Theorem 5.5:

$$E(\omega_o) = \delta_i'(\omega_o)\delta_r(\omega_o) - \delta_i(\omega_o)\delta_r'(\omega_o) > 0$$

for some ω_o in $(-\infty, \infty)$. Let us take $\omega_o = 0$, so $z_o = 0$. Thus $\delta_i(z_o) = 0$ and $\delta_r(z_o) = kk_i$. We also have

$$\begin{aligned} \delta_i'(z) &= \frac{kk_p}{L} + \left(\frac{1}{L} - \frac{T}{L^2}z^2\right)\cos(z) - \left(\frac{1}{L}z + \frac{2T}{L^2}z\right)\sin(z) \\ \Rightarrow E(z_o) &= \left(\frac{kk_p + 1}{L}\right)(kk_i) . \end{aligned}$$

By our initial assumption $k > 0$ and $L > 0$. Thus, if we pick $k_i > 0$ and $k_p > -\frac{1}{k}$, we have $E(z_o) > 0$. Notice that the case where $k_i < 0$ and $k_p < -\frac{1}{k}$ is ruled out since from (7.4) it is clear that this is not a stabilizing set for the delay-free case.
Step 2. We now check condition 1 of Theorem 5.5: the interlacing of the roots of $\delta_r(z)$ and $\delta_i(z)$. From (7.8) we can compute the roots of the imaginary part, i.e., $\delta_i(z) = 0$. This gives us the following equation

$$\frac{z}{L}\left[kk_p + \cos(z) - \frac{T}{L}z\sin(z)\right] = 0 .$$

Then,

$$z = 0 \ , \text{ or}$$
$$kk_p + \cos(z) - \frac{T}{L} z \sin(z) = 0 \ . \tag{7.10}$$

From this we see that one root of the imaginary part is $z_o = 0$. The other roots are difficult to find since we need to solve (7.10) analytically. However, we can plot the terms involved in (7.10) and graphically examine the nature of the solution. There are three different cases to consider. In each case, the positive real roots of (7.10) will be denoted by z_j, $j = 1, 2, ...$, arranged in increasing order of magnitude.

Case 1: $-\frac{1}{k} < k_p < \frac{1}{k}$. In this case, we graph $\frac{kk_p+\cos(z)}{\sin(z)}$ and $\frac{T}{L}z$ to obtain the plots shown in Fig. 7.1.

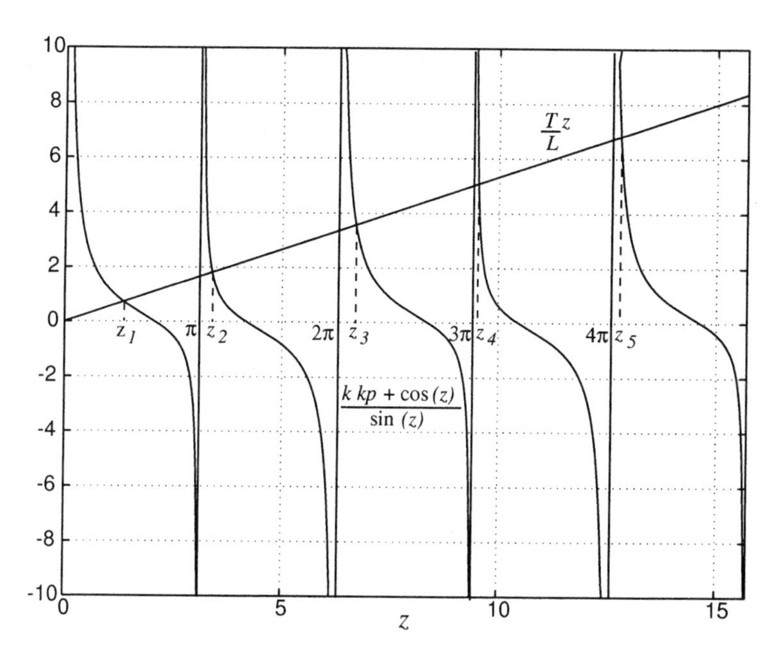

FIGURE 7.1. Plot of the terms involved in (7.10) for $-\frac{1}{k} < k_p < \frac{1}{k}$.

Case 2: $k_p = \frac{1}{k}$. In this case, we sketch $kk_p + \cos(z)$ and $\frac{T}{L}z\sin(z)$ to obtain the plots shown in Fig. 7.2.

Case 3: $\frac{1}{k} < k_p$. In this case, we sketch $\frac{kk_p+\cos(z)}{\sin(z)}$ and $\frac{T}{L}z$ to obtain the plots shown in Figs. 7.3(a) and 7.3(b). The plot in Fig. 7.3(a) corresponds to the case where $\frac{1}{k} < k_p < k_u$, and k_u is the largest number so that the plot of $\frac{kk_p+\cos(z)}{\sin(z)}$ intersects the line $\frac{T}{L}z$ twice in the interval $(0, \pi)$. The

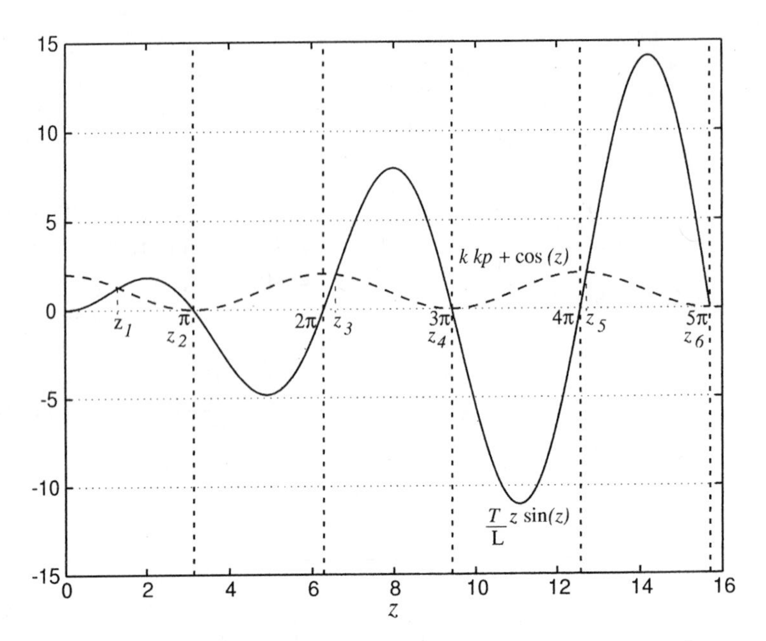

FIGURE 7.2. Plot of the terms involved in (7.10) for $k_p = \frac{1}{k}$.

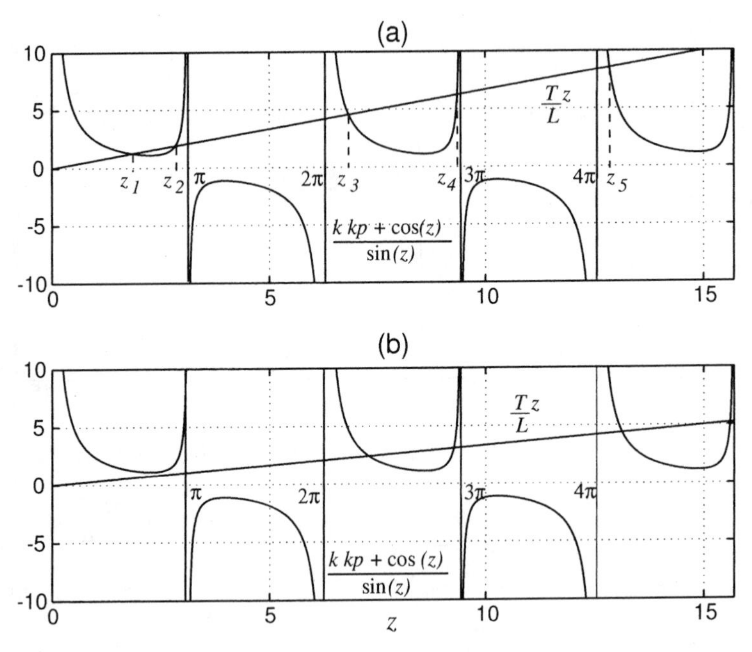

FIGURE 7.3. Plot of the terms involved in (7.10) for $\frac{1}{k} < k_p$.

plot in Fig. 7.3(b) corresponds to the case where $k_p \geq k_u$ and the plot of $\frac{kk_p+\cos(z)}{\sin(z)}$ does not intersect the line $\frac{T}{L}z$ twice in the interval $(0, \pi)$.

Let us now use the results presented in Section 5.5 to check if $\delta_i(z)$ has only real roots. Substituting $s_1 = Ls$ in the expression for $\delta^*(s)$, we see that for the new quasi-polynomial in s_1, $M = 2$ and $N = 1$. Next we choose $\eta = \frac{\pi}{4}$ to satisfy the requirement that $\sin(\eta) \neq 0$. Now from Figs. 7.1, 7.2, and 7.3(a), we see that in each of these cases, i.e., for $-\frac{1}{k} < k_p < k_u$, $\delta_i(z)$ has three real roots in the interval $[0, 2\pi - \frac{\pi}{4}] = [0, \frac{7\pi}{4}]$, including a root at the origin. Since $\delta_i(z)$ is an odd function of z, it follows that in the interval $[-\frac{7\pi}{4}, \frac{7\pi}{4}]$, $\delta_i(z)$ will have five real roots. Also observe from Figs. 7.1, 7.2, and 7.3(a) that $\delta_i(z)$ has a real root in the interval $(\frac{7\pi}{4}, \frac{9\pi}{4}]$. Thus $\delta_i(z)$ has $4N + M = 6$ real roots in the interval $[-2\pi + \frac{\pi}{4}, 2\pi + \frac{\pi}{4}]$. Moreover, it can be shown using Figs. 7.1, 7.2, and 7.3(a) that $\delta_i(z)$ has two real roots in each of the intervals $[2l\pi + \frac{\pi}{4}, 2(l+1)\pi + \frac{\pi}{4}]$ and $[-2(l+1)\pi + \frac{\pi}{4}, -2l\pi + \frac{\pi}{4}]$ for $l = 1, 2, \ldots$. Hence, it follows that $\delta_i(z)$ has exactly $4lN + M$ real roots in $[-2l\pi + \frac{\pi}{4}, 2l\pi + \frac{\pi}{4}]$ for $-\frac{1}{k} < k_p < k_u$. Hence from Theorem 5.6, we conclude that for $-\frac{1}{k} < k_p < k_u$, $\delta_i(z)$ has only real roots. Also note that the case $k_p \geq k_u$ corresponding to Fig. 7.3(b) does not merit any further consideration since using Theorem 5.6, we can easily argue that in this case, all the roots of $\delta_i(z)$ will not be real, thereby ruling out closed-loop stability.

We now evaluate $\delta_r(z)$ at the roots of the imaginary part $\delta_i(z)$. For $z_o = 0$, using (7.7) we obtain

$$\begin{aligned} \delta_r(z_o) &= k[k_i - a(0)] \\ &= kk_i \,. \end{aligned} \tag{7.11}$$

For z_j, where $j = 1, 2, 3, \ldots$, using (7.7) we obtain

$$\delta_r(z_j) = k[k_i - a(z_j)] \,. \tag{7.12}$$

Interlacing the roots of $\delta_r(z)$ and $\delta_i(z)$ is equivalent to $\delta_r(z_o) > 0$ (since $k_i > 0$ as derived in Step 1), $\delta_r(z_1) < 0$, $\delta_r(z_2) > 0$, $\delta_r(z_3) < 0$, and so on. Using this fact and (7.11)–(7.12) we obtain

$$\begin{aligned} \delta_r(z_o) > 0 &\Rightarrow k_i > 0 \\ \delta_r(z_1) < 0 &\Rightarrow k_i < a_1 \\ \delta_r(z_2) > 0 &\Rightarrow k_i > a_2 \\ \delta_r(z_3) < 0 &\Rightarrow k_i < a_3 \\ \delta_r(z_4) < 0 &\Rightarrow k_i > a_4 \\ &\vdots \end{aligned} \tag{7.13}$$

where the bounds a_j for $j = 1, 2, 3, \ldots$ are given by

$$a_j \stackrel{\Delta}{=} a(z_j) \,. \tag{7.14}$$

From this set of inequalities it is clear that we need the odd bounds (i.e., $a_1, a_3, \ldots$) to be strictly positive in order to obtain a feasible range for the controller parameter k_i. As we will see in the next lemma, for $k_p > -\frac{1}{k}$, this occurs if and only if

$$k_p < \frac{T}{kL}\sqrt{\alpha_1^2 + \frac{L^2}{T^2}} < k_u$$

where α_1 is the solution of the equation

$$\tan(\alpha) = -\frac{T}{L}\alpha$$

in the interval $(\frac{\pi}{2}, \pi)$. Moreover, from the same lemma we will see that the bounds a_j corresponding to even values of j are all negative for

$$k_p \in \left(-\frac{1}{k}, \frac{T}{kL}\sqrt{\alpha_1^2 + \frac{L^2}{T^2}}\right).$$

Thus, the conditions (7.13) reduce to

$$0 < k_i < \min_{j=1,3,5,\ldots} \{a_j\}. \tag{7.15}$$

One can make use of the interlacing property and the fact that $\delta_i(z)$ has only real roots to establish that for $-\frac{1}{k} < k_p < \frac{T}{kL}\sqrt{\alpha_1^2 + \frac{L^2}{T^2}}$, $\delta_r(z)$ also has only real roots. This completes the proof of the theorem. ■

Lemma 7.1 *For $k_p > -\frac{1}{k}$, a necessary and sufficient condition for a_j defined in (7.14) to be positive for odd values of j is that*

$$k_p < \frac{T}{kL}\sqrt{\alpha_1^2 + \frac{L^2}{T^2}}$$

where α_1 is the solution of the equation

$$\tan(\alpha) = -\frac{T}{L}\alpha \tag{7.16}$$

in the interval $(\frac{\pi}{2}, \pi)$.

Furthermore, for all $k_p \in \left(-\frac{1}{k}, \frac{T}{kL}\sqrt{\alpha_1^2 + \frac{L^2}{T^2}}\right)$, $a_j < 0$ for even values of j.

Proof. From Figs. 7.1, 7.2, and 7.3(a), we see that for $k_p \in (-\frac{1}{k}, k_u)$ the roots of (7.10) corresponding to odd values of j satisfy the following properties:

$$z_1 \in (0, \pi),\ z_3 \in (2\pi, 3\pi),\ z_5 \in (4\pi, 5\pi),$$

and so on, i.e., $z_j \epsilon ((j-1)\pi, j\pi)$. Thus in these three cases the roots of (7.10) corresponding to odd values of j are either in the first quadrant or in the second quadrant. Then,

$$\sin(z_j) > 0 \quad \text{for odd values of } j.$$

Recall from (7.14) that the parameter a_j was defined as

$$a_j = \frac{z_j}{kL}\left[\sin(z_j) + \frac{T}{L} z_j \cos(z_j)\right] .$$

Thus for $z_j \neq l\pi$, $l = 0, 1, 2, ...,$ we can write

$$\begin{aligned} a_j &= \frac{z_j}{kL}\left[\sin(z_j) + \frac{kk_p + \cos(z_j)}{\sin(z_j)} \cdot \cos(z_j)\right] \quad \text{[using (7.10)]} \\ &= \frac{z_j}{kL}\left[\frac{1 + kk_p \cos(z_j)}{\sin(z_j)}\right] . \end{aligned} \tag{7.17}$$

From this expression it is clear that if $z_j \neq l\pi$, then the parameter a_j is positive if and only if

$$\begin{aligned} \sin(z_j) > 0 \quad &\text{and} \quad 1 + kk_p \cos(z_j) > 0 \ \text{ or} \\ \sin(z_j) < 0 \quad &\text{and} \quad 1 + kk_p \cos(z_j) < 0 \end{aligned}$$

and it is negative otherwise. Figure 7.4 shows the k_p-z plane split into different regions according to the value of the parameter a_j. In those regions where $a_j > 0$ a plus sign (+) has been placed and in those regions where $a_j < 0$ a minus sign (-) has been placed. In this figure the dashed line corresponds to the function

$$1 + kk_p \cos(z) = 0$$

or equivalently

$$k_p = -\frac{1}{k \cos(z)} .$$

Although the plot here corresponds to the interval $z \epsilon [0, 2\pi]$, since the function is periodic, the plot repeats itself.

We will graph the solutions of (7.10) in the same k_p-z plane. Recall that the solutions of this equation represent the nonzero roots of the imaginary part $\delta_i(z)$. Now (7.10) can be rewritten as

$$k_p = \frac{1}{k}\left[\frac{T}{L} z \sin(z) - \cos(z)\right] . \tag{7.18}$$

Figure 7.5 shows the graph of this function along with the regions presented in Fig. 7.4. The intersection of (7.10) with the curve $1 + kk_p \cos(z) = 0$

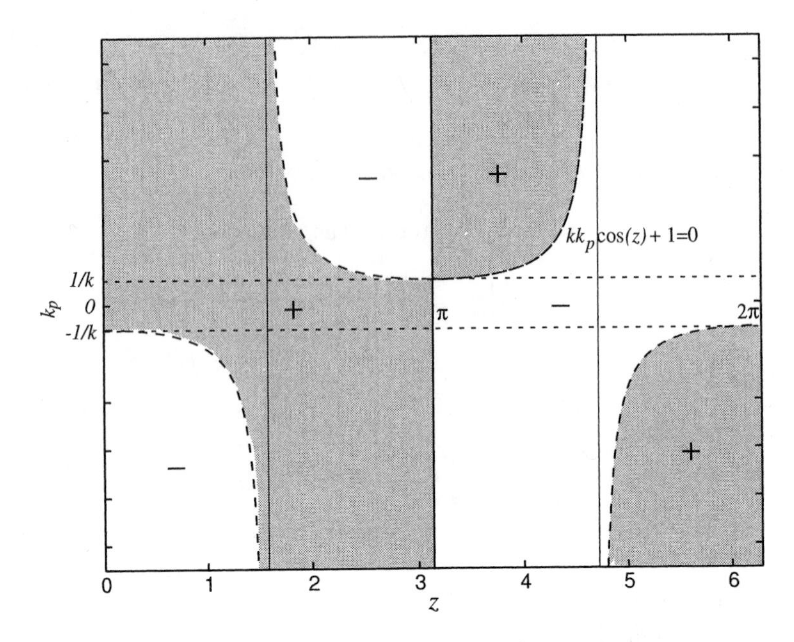

FIGURE 7.4. Regions associated with parameter a_j.

occurs at five values of the parameter z: 0, α_1, π, α_2, and 2π. Thus each of these values will satisfy the relationship

$$-\frac{1}{\cos(z)} = \frac{T}{L} z \sin(z) - \cos(z) \, .$$

Furthermore, if $z \neq l\pi$ then simplifying this equation, we obtain

$$\tan(z) = -\frac{T}{L} z \, .$$

Thus α_1 and α_2 will be solutions of the above equation.

For a given value of k_p, let $z_1(k_p)$ and $z_2(k_p)$ be the positive real roots of (7.18) arranged in ascending order of magnitude. From Fig. 7.5, it is clear that for $k_p \in (-\frac{1}{k}, -\frac{1}{k\cos(\alpha_1)}) - \{\frac{1}{k}\}$, i.e., excluding $\frac{1}{k}$ from this interval,

$$\begin{aligned} a_1 = a(z_1(k_p)) &> 0 \\ \text{and } a_2 = a(z_2(k_p)) &< 0 \, . \end{aligned}$$

For $k_p = \frac{1}{k}$, from Fig. 7.5, we once again conclude that $a_1 = a(z_1(k_p)) > 0$. Since $z_2 = \pi$, we cannot use Fig. 7.5 or (7.17) to determine the sign of $a(z_2(k_p))$. However from the original definition of $a(z)$ in (7.9), it follows that $a_2 = a(z_2(k_p)) < 0$. From Fig. 7.5, we also see that for $k_p > -\frac{1}{k\cos(\alpha_1)}$,

$$\begin{aligned} a_1 = a(z_1(k_p)) &< 0 \\ \text{and } a_2 = a(z_2(k_p)) &< 0 \, . \end{aligned}$$

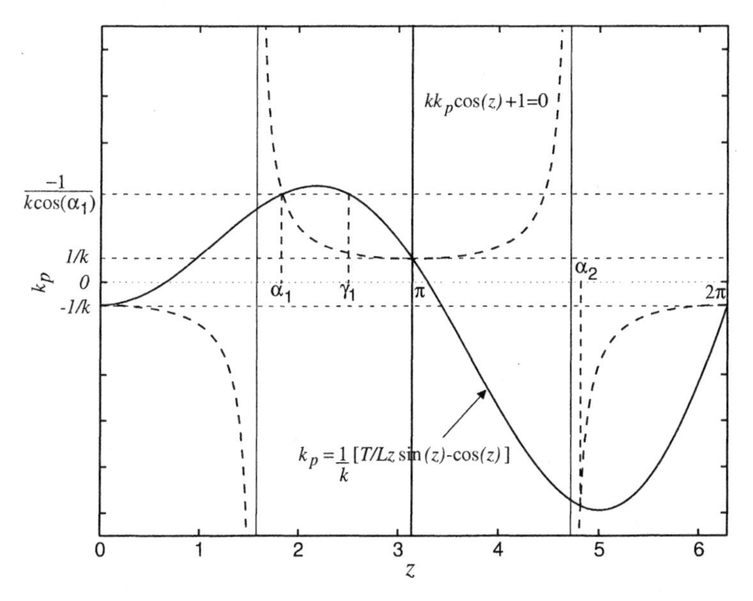

FIGURE 7.5. Study of (7.10) in the k_p-z plane when $T > 0$.

Thus we conclude that $a_1 > 0$ if and only if $k_p < -\frac{1}{k\cos(\alpha_1)}$. Since α_1 satisfies the relationship $\tan(\alpha_1) = -\frac{T}{L}\alpha_1$ we have that

$$\cos(\alpha_1) = -\frac{1}{\sqrt{1+\frac{T^2}{L^2}\alpha_1^2}}$$

so that

$$-\frac{1}{k\cos(\alpha_1)} = \frac{T}{kL}\sqrt{\alpha_1^2 + \frac{L^2}{T^2}}\,.$$

In view of the above discussion, we conclude that

$$\text{if } -\frac{1}{k} < k_p < \frac{T}{kL}\sqrt{\alpha_1^2 + \frac{L^2}{T^2}} \text{ then } a_1 > 0\,,\ a_2 < 0\,.$$

Notice from Fig. 7.5 that this upper bound on k_p is less than k_u, which in this figure corresponds to the maximum of the function

$$k_p = \frac{1}{k}\left[\frac{T}{L}z\sin(z) - \cos(z)\right]\,,$$

in the interval $z \in [0, \pi]$. As explained in the proof of Theorem 7.1, this is an important requirement for interlacing.

Using a similar approach we can show that

$$\text{if } -\frac{1}{k} < k_p < \frac{T}{kL}\sqrt{\alpha_3^2 + \frac{L^2}{T^2}} \quad \text{then} \quad a_3 > 0\,,\ a_4 < 0$$

$$\text{if } -\frac{1}{k} < k_p < \frac{T}{kL}\sqrt{\alpha_5^2 + \frac{L^2}{T^2}} \quad \text{then} \quad a_5 > 0\ ,\ a_6 < 0$$

$$\vdots$$

where α_j is the solution to $\tan(\alpha) = -\frac{T}{L}\alpha$ in the interval $((j-\frac{1}{2})\pi, j\pi)$. It is clear that all these upper bounds on k_p increase monotonically with α_j. Thus it suffices to take the upper bound corresponding to α_1 to guarantee that a_j for odd values of j are strictly positive and a_j for even values of j are strictly negative. This completes the proof. ■

Remark 7.1 *In the proof of the above lemma, we have not considered the case where the value of k_p is such that $k_p = \frac{1}{k}[\frac{T}{L}z\sin(z) - \cos(z)]$ does not have two zeros in the interval $z\epsilon[0, 2\pi]$. This is because we know from the proof of Theorem 7.1 that for such a value of k_p, closed-loop stability is ruled out.*

Remark 7.2 *As we can see from Figs. 7.1, 7.2, and 7.3(a), the odd roots of (7.10), i.e., z_j where $j = 1, 3, 5, \ldots$, are getting closer to $(j-1)\pi$ as j increases. So in the limit for odd values of j we have*

$$\lim_{j\to\infty} \cos(z_j) = 1\ .$$

Moreover, since the cosine function is monotonically decreasing between $(j-1)\pi$ and $j\pi$ for odd values of j, and because of the previous observation we have

$$\cos(z_1) < \cos(z_3) < \cos(z_5) < \cdots\ .$$

We now present a lemma that will be useful in the development of an algorithm for solving the PI stabilization problem.

Lemma 7.2 *If $\cos(z_j) > 0$ then $a_j < a_{j+2}$ for odd values of j.*

Proof. From (7.9), (7.14) we have

$$\begin{aligned} kLa_j &= z_j\sin(z_j) + \frac{T}{L}z_j^2\cos(z_j) \\ \Rightarrow Tka_j &= kk_p + \cos(z_j) + \frac{T^2}{L^2}z_j^2\cos(z_j) \;\; [\text{using } (7.10)] \\ \Rightarrow Tka_j - kk_p &= \cos(z_j)\left(1 + \frac{T^2}{L^2}z_j^2\right)\ . \end{aligned} \tag{7.19}$$

We know that the z_j, $j = 1, 3, 5...$ are arranged in increasing order of magnitude, i.e., $z_j < z_{j+2}$, so we have for odd values of j

$$1 + \frac{T^2}{L^2}z_j^2 < 1 + \frac{T^2}{L^2}z_{j+2}^2\ . \tag{7.20}$$

Because of Remark 7.2 we have

$$\cos(z_j) < \cos(z_{j+2}) \text{ for odd values of } j. \tag{7.21}$$

Since $\cos(z_j) > 0$, from (7.20) we have

$$\left(1 + \frac{T^2}{L^2} z_j^2\right) \cos(z_{j+2}) < \left(1 + \frac{T^2}{L^2} z_{j+2}^2\right) \cos(z_{j+2}) \,.$$

Since $1 + \frac{T^2}{L^2} z_j^2 > 0$, from (7.21) we have

$$\left(1 + \frac{T^2}{L^2} z_j^2\right) \cos(z_j) < \left(1 + \frac{T^2}{L^2} z_j^2\right) \cos(z_{j+2}) \,.$$

Combining these two latter inequalities:

$$\begin{aligned} \left(1 + \frac{T^2}{L^2} z_j^2\right) \cos(z_j) &< \left(1 + \frac{T^2}{L^2} z_{j+2}^2\right) \cos(z_{j+2}) \\ \Rightarrow Tka_j - kk_p &< Tka_{j+2} - kk_p \text{ [using (7.19)]} \\ \Rightarrow a_j &< a_{j+2} \end{aligned}$$

for odd values of j and this completes the proof. ∎

Remark 7.3 *Notice that for a fixed value of k_p inside the range proposed by Theorem 7.1, we can find the range of k_i such that the closed-loop system is stable. This range is given by (7.15) and depends on the bounds a_j corresponding to odd values of j. However, by Lemma 7.2, if $\cos(z_1) > 0$, then the bound a_1 is the minimum of all the odd bounds, and the range of stabilizing k_i is given by $0 < k_i < a_1$. If this is not the case, but we have that $\cos(z_3) > 0$, then the bound a_3 is less than all the other bounds a_j for $j = 5, 7, 9, \ldots$. Then, in this case the range of stabilizing k_i is given by $0 < k_i < \min\{a_1, a_3\}$. Note that since for odd values of j, $lim_{j \to \infty} \cos(z_j) = 1$, we are guaranteed that $\cos(z_j) > 0$, $\forall j \geq N$, where N is some* finite *integer.*

Theorem 7.1 and Lemma 7.2 together suggest a procedure for determining the set of all stabilizing (k_p, k_i) values for a given plant. This procedure is summarized in the following algorithm.

Algorithm for Determining Stabilizing PI Parameters.

- **Step 1:** Initialize $k_p = -\frac{1}{k}$, $step = \frac{1}{N+1}\left(\frac{T}{kL}\sqrt{\alpha_1^2 + \frac{L^2}{T^2}} + \frac{1}{k}\right)$ and $j = 1$, where N is the desired number of points;
- **Step 2:** Increase k_p as follows: $k_p = k_p + step$;

- **Step 3:** If $k_p < \frac{T}{kL}\sqrt{\alpha_1^2 + \frac{L^2}{T^2}}$ then go to Step 4. Else, terminate the algorithm;
- **Step 4:** Find the root z_j of (7.10);
- **Step 5:** Compute the parameter a_j associated with the z_j previously found by using (7.14);
- **Step 6:** If $\cos(z_j) > 0$ then go to Step 7. Else, increase $j = j + 2$ and go to Step 4;
- **Step 7:** Determine the lower and upper bounds for k_i as follows:

$$0 < k_i < \min_{l=1,3,5,\ldots,j} \{a_l\} ;$$

- **Step 8:** Go to Step 2;

We now present an example that illustrates the application of the results presented in this section.

Example 7.1 *Consider the problem of choosing stabilizing PI gains for the plant given in (7.1), where the plant parameters are* $k = 1$, $L = 1$ *sec, and* $T = 4$ *sec. Since the plant is open-loop stable, we first use Theorem 7.1 to obtain the range of* k_p *values over which the sweeping needs to be carried out. We compute* $\alpha_1 \epsilon (\frac{\pi}{2}, \pi)$ *satisfying (7.16),* i.e.,

$$\tan(\alpha) = -4\alpha .$$

Solving this equation we obtain $\alpha_1 = 1.7155$. *Thus, from (7.6) the range of* k_p *gains is given by*

$$-1 < k_p < 6.9345 .$$

We now sweep over this range of k_p *gains and use the previous algorithm to determine the range of* k_i *gains at each stage. Figure 7.6 shows the stabilizing region obtained in the* k_p*–*k_i *plane.*

We now set the controller parameters k_p *and* k_i *at 3 and 1, respectively. Clearly this point is inside the region sketched in Fig. 7.6. The step response of the closed-loop system with this PI controller is shown in Fig. 7.7. From this figure, we see that the closed-loop system is stable and the output* $y(t)$ *tracks the step input signal.* △

As we mentioned before, the purpose of the integral term in a PI controller is to achieve zero steady-state offset when tracking step inputs. Thus, we can employ several time domain performance specifications such as settling time, maximum overshoot, and minimum undershoot to quantify the performance of the closed-loop system. The characterization of all stabilizing (k_p, k_i) values provided in Fig. 7.6 enables us to graphically display

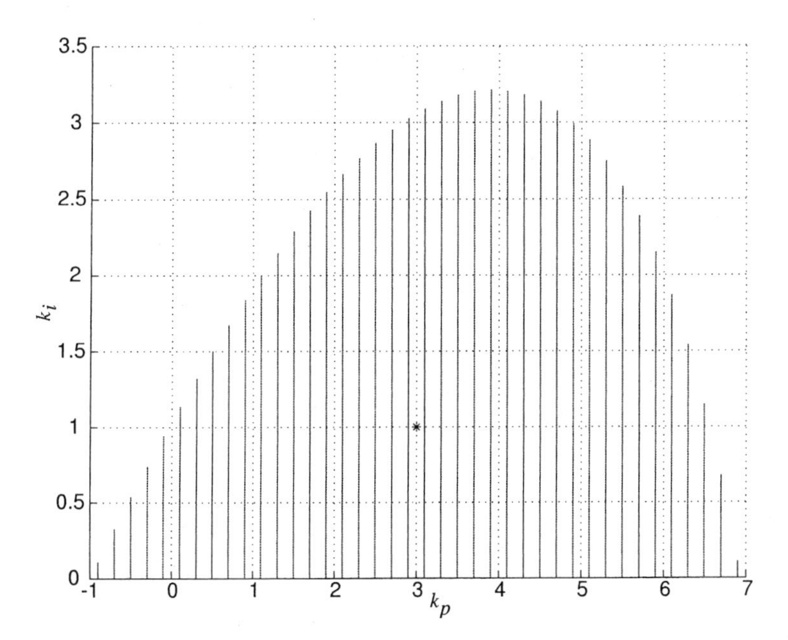

FIGURE 7.6. The stabilizing set of (k_p, k_i) values for Example 7.1.

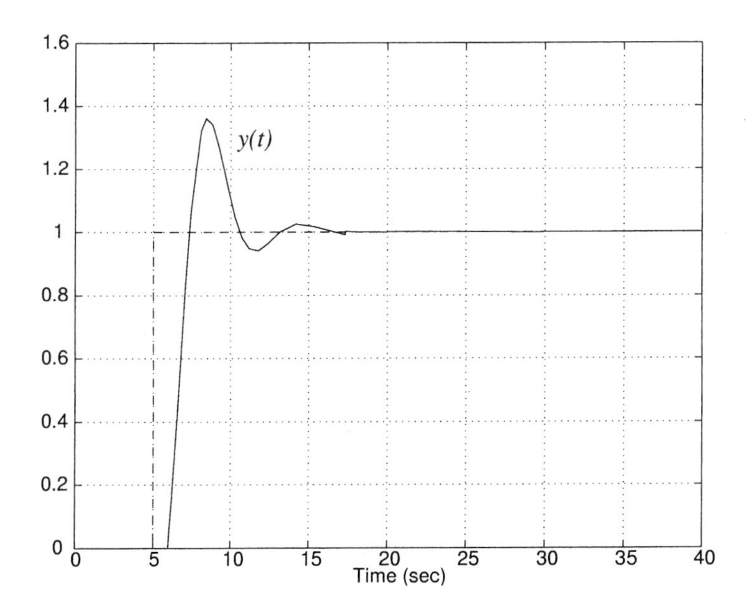

FIGURE 7.7. Time response of the closed-loop system for Example 7.1.

the variation of these performance indices over the entire stabilizing region in the parameter space. In this way, we can select the (k_p, k_i) values that satisfy the performance specifications. These ideas will be discussed further in Chapter 9.

Example 7.2 *Consider the following time-delay system*

$$G(s) = \frac{1}{1+4s} e^{-6s} .$$

We now approximate this transfer function by replacing the time-delay term e^{-6s} *with the first- and second-order Padé approximations. This results in the following approximated transfer functions:*

$$\begin{aligned} G_m^1(s) &= \frac{-s+0.3333}{4s^2+2.3333s+0.3333} \quad \textit{(first-order approx.)} \\ G_m^2(s) &= \frac{s^2-s+0.3333}{4s^3+5s^2+2.3333s+0.3333} \quad \textit{(second-order approx.)} . \end{aligned}$$

Using the results in Section 3.3 we can generate the set of stabilizing (k_p,k_i) *values for these transfer functions. Figure 7.8 shows the boundaries of the generated controller sets* C_m^1, C_m^2 *corresponding to* $G_m^1(s)$ *and* $G_m^2(s)$, *respectively, with dashed lines. Also illustrated in this figure as a shaded area is the true stabilizing set of* (k_p, k_i) *values. This set was obtained following the procedure presented in this section.*

As we can see from these plots, the second-order Padé approximation leads to an approximate set that matches the true set closely. On the other hand, the first-order Padé approximation provides not only an inaccurate set, but also contains controller gain values that lead to an unstable behavior of the closed-loop system. △

7.4 Open-Loop Unstable Plant

When the plant is open-loop unstable we have $T < 0$. Furthermore, as before, let us assume that $k > 0$ and $L > 0$. Recall from (7.5) that for the closed-loop stability of the delay-free system, we now require

$$k_p < -\frac{1}{k} , \; k_i < 0 .$$

The solution to the PI stabilization problem in this case also involves sweeping over all real k_p and solving a constant gain stabilization problem at each stage. The range of k_p values over which the sweeping needs to be carried out can be narrowed down by using the following theorem.

Theorem 7.2 *Under the above assumptions on* k *and* L, *a necessary condition for a PI controller to simultaneously stabilize the delay-free plant and*

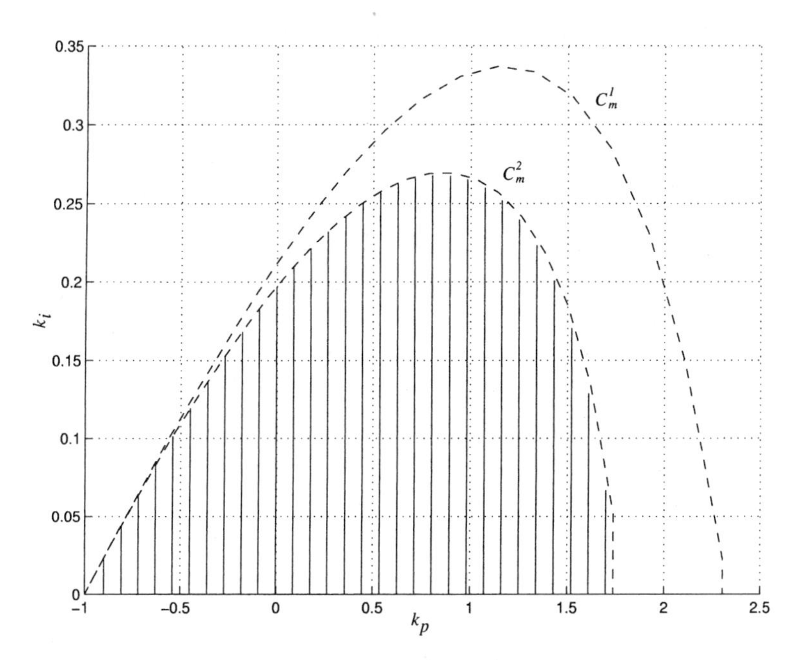

FIGURE 7.8. The set of stabilizing (k_p, k_i) values for Example 7.2.

the plant with delay is $|\frac{T}{L}| > 1$. *If this necessary condition is satisfied, then the range of* k_p *values for which a solution exists to the PI stabilization problem of a given open-loop unstable plant with transfer function* $G(s)$ *as in (7.1) is given by*

$$\frac{T}{kL}\sqrt{\alpha_1^2 + \frac{L^2}{T^2}} < k_p < -\frac{1}{k} \tag{7.22}$$

where α_1 *is the solution of the equation*

$$\tan(\alpha) = -\frac{T}{L}\alpha$$

in the interval $(0, \frac{\pi}{2})$.

Proof. The proof follows along the same lines as that of Theorem 7.1. We need to check the two conditions stated in Theorem 5.5 to ensure the stability of the quasi-polynomial $\delta^*(s)$.
Step 1. First we check condition 2 of Theorem 5.5:

$$E(\omega_o) = \delta_i^{'}(\omega_o)\delta_r(\omega_o) - \delta_i(\omega_o)\delta_r^{'}(\omega_o) > 0$$

for some ω_o in $(-\infty, \infty)$. Let us take $\omega_o = 0$, so $z_o = 0$. As in the proof of Theorem 7.1 we obtain

$$E(z_o) = \left(\frac{kk_p + 1}{L}\right)(kk_i) .$$

By our initial assumption $k > 0$ and $L > 0$. If we pick $k_i < 0$ and $k_p < -\frac{1}{k}$, we have $E(z_o) > 0$. Notice that the case where $k_i > 0$ and $k_p > -\frac{1}{k}$ is ruled out since from (7.5) it is clear that this is not a stabilizing set for the delay-free case.

Step 2. We now check the interlacing of the roots of $\delta_r(z)$ and $\delta_i(z)$ (condition 1 of Theorem 5.5). As in the proof of Theorem 7.1, we can compute the roots of the imaginary part, i.e., $\delta_i(z) = 0$ using (7.8). One root of the imaginary part is $z_o = 0$. The other positive real roots will be denoted by z_j, $j = 1, 2, 3...$, arranged in increasing order of magnitude. These roots are the solutions of (7.10). Since these roots are difficult to find analytically, we plot the terms involved in (7.10) and graphically examine the nature of the solution. Now, from Step 1 we only need to analyze the case $k_p < -\frac{1}{k}$. Thus, we sketch the terms $\frac{kk_p+\cos(z)}{\sin(z)}$ and $\frac{T}{L}z$ to obtain the plots shown in Figs. 7.9(a) and 7.9(b). The plot in Fig. 7.9(a) corresponds to the case where $k_l < k_p < -\frac{1}{k}$, and k_l is the smallest number so that the plot of $\frac{kk_p+\cos(z)}{\sin(z)}$ intersects the line $\frac{T}{L}z$ twice in the interval $(0, \pi)$. The plot in Fig. 7.9(b) corresponds to the case where $k_p \leq k_l$ and the plot of $\frac{kk_p+\cos(z)}{\sin(z)}$ does not intersect the line $\frac{T}{L}z$ twice in the interval $(0, \pi)$.

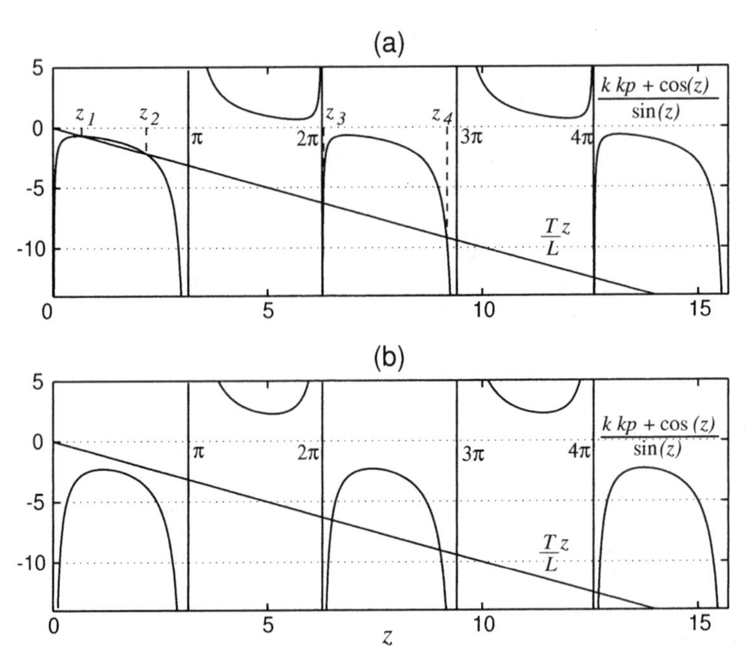

FIGURE 7.9. Plot of the terms involved in (7.10) for $k_p < -\frac{1}{k}$.

Let us now use Theorem 5.6 to check if $\delta_i(z)$ has only real roots. Substituting $s_1 = Ls$ in the expression for $\delta^*(s)$, we see that for the new quasi-polynomial in s_1, $M = 2$ and $N = 1$. Next we choose $\eta = \frac{\pi}{4}$ to satisfy the requirement that $\sin(\eta) \neq 0$. From Fig. 7.9(a) it can be shown that in this case, i.e., for $k_l < k_p < -\frac{1}{k}$, $\delta_i(z)$ has three real roots in the interval $[0, 2\pi - \frac{\pi}{4}] = [0, \frac{7\pi}{4}]$, including a root at the origin. Since $\delta_i(z)$ is an odd function of z, it follows that in the interval $[-\frac{7\pi}{4}, \frac{7\pi}{4}]$, $\delta_i(z)$ will have five real roots. Also observe from Fig. 7.9(a) that $\delta_i(z)$ has a real root in the interval $(\frac{7\pi}{4}, \frac{9\pi}{4}]$. Thus $\delta_i(z)$ has $4N + M = 6$ real roots in the interval $[-2\pi + \frac{\pi}{4}, 2\pi + \frac{\pi}{4}]$. Moreover, it can be shown using Fig. 7.9(a) that $\delta_i(z)$ has two real roots in each of the intervals $[2l\pi + \frac{\pi}{4}, 2(l+1)\pi + \frac{\pi}{4}]$ and $[-2(l+1)\pi + \frac{\pi}{4}, -2l\pi + \frac{\pi}{4}]$ for $l = 1, 2, \ldots$. Hence, it follows that $\delta_i(z)$ has exactly $4lN + M$ real roots in $[-2l\pi + \frac{\pi}{4}, 2l\pi + \frac{\pi}{4}]$ for $k_l < k_p < -\frac{1}{k}$. Thus from Theorem 5.6, we conclude that for $k_l < k_p < -\frac{1}{k}$, $\delta_i(z)$ has only real roots. Also note that the case $k_p \leq k_l$ corresponding to Fig. 7.9(b) does not merit any further consideration since, using Theorem 5.6, we can easily argue that in this case, all the roots of $\delta_i(z)$ will not be real, thereby ruling out closed-loop stability.

We now evaluate $\delta_r(z)$ at the roots of the imaginary part $\delta_i(z)$. For $z_o = 0$, using (7.7) we obtain

$$\delta_r(z_o) = kk_i \ . \tag{7.23}$$

For z_j, where $j = 1, 2, 3, \ldots$, using (7.7) we obtain

$$\delta_r(z_j) = k[k_i - a(z_j)] \ . \tag{7.24}$$

Interlacing the roots of $\delta_r(z)$ and $\delta_i(z)$ is equivalent to $\delta_r(z_o) < 0$ (since $k_i < 0$ as derived in Step 1), $\delta_r(z_1) > 0$, $\delta_r(z_2) < 0$, $\delta_r(z_3) > 0$, and so on. Using this fact and (7.23)–(7.24) we obtain

$$\begin{array}{lcl} \delta_r(z_o) < 0 & \Rightarrow & k_i < 0 \\ \delta_r(z_1) > 0 & \Rightarrow & k_i > a_1 \\ \delta_r(z_2) < 0 & \Rightarrow & k_i < a_2 \\ \delta_r(z_3) > 0 & \Rightarrow & k_i > a_3 \\ & \vdots & \end{array} \tag{7.25}$$

where the bounds a_j for $j = 1, 2, 3, \ldots$ are given by

$$a_j \stackrel{\Delta}{=} a(z_j) \ . \tag{7.26}$$

From this set of inequalities it is clear that we need the odd bounds (i.e., $a_1, a_3, \ldots$) to be strictly negative in order to obtain a feasible range for the controller parameter k_i. As we will see in the next lemma, to have $a_j < 0$ for odd values of j, we must have $1 + \frac{T}{L} < 0$ or

$$\left|\frac{T}{L}\right| > 1 \ .$$

If this necessary condition is satisfied, then for $k_p < -\frac{1}{k}$, the odd bounds are all negative if and only if

$$k_p > \frac{T}{kL}\sqrt{\alpha_1^2 + \frac{L^2}{T^2}} > k_l$$

where α_1 is the solution of the equation

$$\tan(\alpha) = -\frac{T}{L}\alpha$$

in the interval $(0, \frac{\pi}{2})$. Moreover, from the same lemma we will see that the bounds a_j corresponding to even values of j are all positive for $k_p \epsilon$ $(\frac{T}{kL}\sqrt{\alpha_1^2 + \frac{L^2}{T^2}}, -\frac{1}{k})$. Thus, the conditions (7.25) reduce to

$$\max_{j=1,3,5,\dots} \{a_j\} < k_i < 0\,. \tag{7.27}$$

As in the proof of Theorem 7.1, we can make use of the interlacing property and the fact that $\delta_i(z)$ has only real roots to establish that for $\frac{T}{kL}\sqrt{\alpha_1^2 + \frac{L^2}{T^2}} < k_p < -\frac{1}{k}$, $\delta_r(z)$ also has only real roots. This completes the proof of the theorem. ∎

Lemma 7.3 *A necessary condition for the bounds a_j defined in (7.26) corresponding to odd values of j to be negative is $|\frac{T}{L}| > 1$. If this necessary condition is satisfied, then for $k_p < -\frac{1}{k}$, the bounds a_j for odd values of j, are negative if and only if*

$$k_p > \frac{T}{kL}\sqrt{\alpha_1^2 + \frac{L^2}{T^2}}$$

where α_1 is the solution of the equation

$$\tan(\alpha) = -\frac{T}{L}\alpha$$

in the interval $(0, \frac{\pi}{2})$.

Furthermore, for all $k_p \epsilon \left(\frac{T}{kL}\sqrt{\alpha_1^2 + \frac{L^2}{T^2}}, -\frac{1}{k}\right)$, $a_j > 0$ for even values of j.

Proof. Recall from the proof of Lemma 7.1 that the parameter a_j can be rewritten as follows

$$a_j = \frac{z_j}{kL}\left[\frac{1 + kk_p\cos(z_j)}{\sin(z_j)}\right]$$

for $z_j \neq l\pi$, $l = 0, 1, 2, \dots$. From this expression it is clear that if $z_j \neq l\pi$, then the parameter a_j is negative if and only if

$$\begin{array}{lll} \sin(z_j) > 0 & \text{and} & 1 + kk_p\cos(z_j) < 0 \ \text{ or} \\ \sin(z_j) < 0 & \text{and} & 1 + kk_p\cos(z_j) > 0 \end{array}$$

and it is positive otherwise. As in the proof of Lemma 7.1 we will make use of Fig. 7.4. Recall that this figure shows the k_p-z plane split into different regions according to the value of parameter a_j. In those regions where $a_j > 0$ a plus sign (+) has been placed and in those regions where $a_j < 0$ a minus sign (-) has been placed. In this figure the dashed line corresponds to the function

$$k_p = -\frac{1}{k\cos(z)} .$$

We will now graph the solutions of (7.10) in the same k_p-z plane. Recall that the solutions of this equation represent the nonzero roots of the imaginary part $\delta_i(z)$. Now, (7.10) can be rewritten as

$$k_p = \frac{1}{k}\left[\frac{T}{L}z\sin(z) - \cos(z)\right] . \tag{7.28}$$

Figure 7.10 shows the graph of this function along with the regions presented in Fig. 7.4.

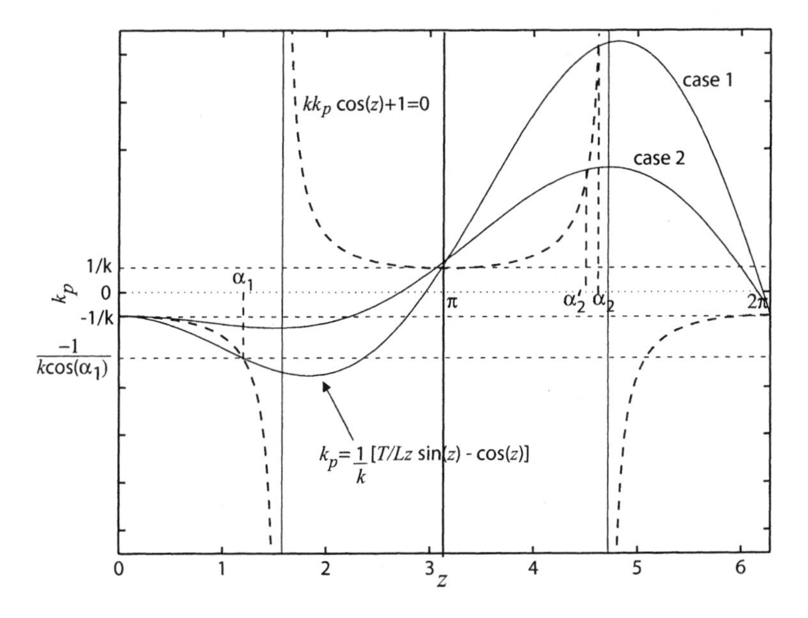

FIGURE 7.10. Study of (7.10) in the k_p-z plane when $T < 0$.

The behavior of the curve represented by (7.28) depends on the value of the parameter $\frac{T}{L}$. There are two different cases to analyze.

Case 1: The intersection of (7.28) with the curve $1 + kk_p\cos(z) = 0$ occurs at five values of the parameter z: 0, α_1, π, α_2, and 2π. In this case, each of these values will satisfy the relationship

$$-\frac{1}{\cos(z)} = \frac{T}{L}z\sin(z) - \cos(z) .$$

Furthermore, if $z \neq l\pi$ then simplifying this equation, we obtain

$$\tan(z) = -\frac{T}{L} z \,.$$

Thus α_1 and α_2 will be solutions of the above equation. Notice that α_1 is in the interval $(0, \frac{\pi}{2})$. We require $1 + \frac{T}{L} < 0$ in order to guarantee the existence of α_1.

Case 2: The intersection of (7.28) with the curve $1 + kk_p \cos(z) = 0$ occurs at four values of the parameter z: 0, π, α_2', and 2π. In this case, there is no intersection in the interval $(0, \frac{\pi}{2})$, which means that the equation $\tan(z) = -\frac{T}{L} z$ does not have a solution in this interval. This occurs when $1 + \frac{T}{L} \geq 0$. Moreover, notice from (7.9) that the bound a_1 corresponding to $z_1 = \pi$ is always positive for this case:

$$a_1 = -\frac{\pi^2 T}{kL^2} \quad \text{and} \quad T < 0 \,.$$

Hence, when $1 + \frac{T}{L} \geq 0$ the bound a_1 will not be negative.

From the previous discussion, we see that a necessary condition for the bound a_1 to be negative is $1 + \frac{T}{L} < 0$ or equivalently

$$\left| \frac{T}{L} \right| > 1 \,.$$

Thus, Case 2 will not be considered in the following analysis. For a given value of k_p, let $z_1(k_p)$ and $z_2(k_p)$ be the positive real roots of (7.28) arranged in ascending order of magnitude. Now from Fig. 7.10, it is clear that for $k_p \epsilon (-\frac{1}{k \cos(\alpha_1)}, -\frac{1}{k})$,

$$\begin{aligned} a_1 = a(z_1(k_p)) &< 0 \\ \text{and } a_2 = a(z_2(k_p)) &> 0 \,. \end{aligned}$$

Since α_1 satisfies the relationship $\tan(\alpha_1) = -\frac{T}{L}\alpha_1$ we have that

$$\cos(\alpha_1) = \frac{1}{\sqrt{1 + \frac{T^2}{L^2}\alpha_1^2}}$$

so that

$$-\frac{1}{k \cos(\alpha_1)} = \frac{T}{kL}\sqrt{\alpha_1^2 + \frac{L^2}{T^2}} \,.$$

In view of the above discussion, we conclude that

$$\text{if } \frac{T}{kL}\sqrt{\alpha_1^2 + \frac{L^2}{T^2}} < k_p < -\frac{1}{k} \text{ then } a_1 < 0 \,,\ a_2 > 0 \,.$$

Notice from Fig. 7.10 that this lower bound on k_p is bigger than k_l, which in this figure corresponds to the minimum of the function

$$k_p = \frac{1}{k}\left[\frac{T}{L} z \sin(z) - \cos(z)\right],$$

in the interval $z \in [0, \pi]$. As explained in the proof of Theorem 7.2, $k_p > k_l$ is an important requirement for interlacing.

Using a similar approach we can show that

$$\begin{array}{lll} \text{if } \frac{T}{kL}\sqrt{\alpha_3^2 + \frac{L^2}{T^2}} < k_p < -\frac{1}{k} & \text{then} & a_3 < 0\,, a_4 > 0 \\ \text{if } \frac{T}{kL}\sqrt{\alpha_5^2 + \frac{L^2}{T^2}} < k_p < -\frac{1}{k} & \text{then} & a_5 < 0\,, a_6 > 0 \\ & \vdots & \end{array}$$

where α_j is the solution to $\tan(\alpha) = -\frac{T}{L}\alpha$ in the interval $((j-1)\pi, (j-\frac{1}{2})\pi)$. It is clear that all these lower bounds on k_p decrease monotonically with α_j. Thus it suffices to take the lower bound corresponding to α_1 to guarantee that a_j for odd values of j are strictly negative and a_j for even values of j are strictly positive. This completes the proof. ■

Remark 7.4 *From Fig. 7.9(a) we see that the odd roots of (7.10), i.e., z_j where $j = 1, 3, 5, \ldots$, are getting closer to $(j-1)\pi$ as j increases. Also notice from Fig. 7.10 that z_1 is in the interval $(0, \alpha_1)$. Since α_1 is in the first quadrant we conclude that z_1 is also in the first quadrant. Moreover, since the cosine function is monotonically decreasing between $(j-1)\pi$ and $j\pi$ for odd values of j, and because of the previous observations we have*

$$0 < \cos(z_1) < \cos(z_3) < \cos(z_5) < \cdots .$$

The following lemma is useful for determining the range of stabilizing k_i gains for a fixed value of k_p inside the range proposed by Theorem 7.2.

Lemma 7.4 *For a_j defined in (7.26), the following holds:*

$$a_1 = \max_{j=1,3,5,\ldots} \{a_j\}.$$

Proof. Since the parameters z_j, $j = 1, 3, 5, \ldots$, are arranged in increasing order of magnitude we have

$$\begin{aligned} 0 < z_1 &< z_j \quad \text{for } j = 3, 5, 7, \ldots \\ \Rightarrow 1 + \frac{T^2}{L^2} z_1^2 &< 1 + \frac{T^2}{L^2} z_j^2 . \end{aligned} \tag{7.29}$$

Because of Remark 7.4 we have

$$0 < \cos(z_1) < \cos(z_j) \quad \text{for } j = 3, 5, 7, \ldots . \tag{7.30}$$

Using the fact that $\cos(z_1) > 0$ in (7.29) we have

$$\left(1+\frac{T^2}{L^2}z_1^2\right)\cos(z_1) < \left(1+\frac{T^2}{L^2}z_j^2\right)\cos(z_1)\,.$$

Since $1+\frac{T^2}{L^2}z_j^2 > 0$, from (7.30) we have

$$\left(1+\frac{T^2}{L^2}z_j^2\right)\cos(z_1) < \left(1+\frac{T^2}{L^2}z_j^2\right)\cos(z_j)\,.$$

Combining these two latter inequalities:

$$\begin{aligned}\left(1+\frac{T^2}{L^2}z_1^2\right)\cos(z_1) &< \left(1+\frac{T^2}{L^2}z_j^2\right)\cos(z_j)\\ \Rightarrow Tka_1 - kk_p &< Tka_j - kk_p \text{ [using (7.19)]}\\ \Rightarrow a_1 &> a_j \text{ for } j=3,5,7,\ldots \text{ [since } T<0].\end{aligned}$$

Thus we conclude that

$$a_1 = \max_{j=1,3,5,\ldots}\{a_j\}$$

and this completes the proof. ■

Notice that Theorem 7.2 and Lemma 7.4 can be used together for determining the set of all stabilizing (k_p, k_i) values for a given open-loop unstable plant. First, we fix the parameter k_p inside the range given by Theorem 7.2. For this value of the controller parameter k_p, we know that the range of stabilizing k_i is given by (7.27). However, from Lemma 7.4 this set is reduced to

$$a_1 < k_i < 0\,.$$

Thus we only need to find the root z_1 of (7.10) to obtain the range of stabilizing k_i.

Example 7.3 *Consider the problem of finding the set of stabilizing PI controllers for the plant given in (7.1), where the plant parameters are* $k=1$, $L=0.8$ *seconds, and* $T=-6$. *Since the plant is open-loop unstable we will use Theorem 7.2 to find the range of* k_p *values over which the sweeping needs to be carried out. Since* $|\frac{T}{L}| = 7.5 > 1$ *we can proceed to compute* $\alpha_1 \epsilon (0, \frac{\pi}{2})$ *satisfying the following equation*

$$\tan(\alpha) = 7.5\alpha\,.$$

Solving this equation we obtain $\alpha_1 = 1.4810$. *Thus, from (7.22) the range of* k_p *values is given by*

$$-11.1525 < k_p < -1\,.$$

By sweeping over the above range of k_p *values we can determine the range of* k_i *values at each stage. Figure 7.11 shows the stabilizing region obtained in the* k_p*–*k_i *plane.* △

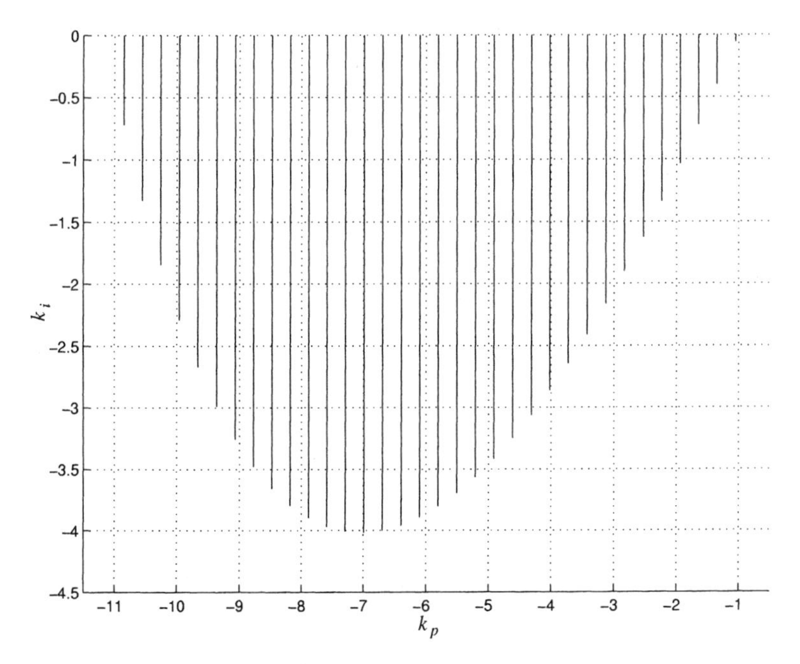

FIGURE 7.11. The stabilizing set of $(k_p,\ k_i)$ values for Example 7.3.

7.5 Notes and References

The results presented in Sections 7.3 and 7.4 are due to Silva, Datta, and Bhattacharyya [38], where the computation of the entire stabilizing set was first developed. A comprehensive study of the PI controller, its working principles, and tuning methods can be found in [2]. The importance of PI controllers in the pulp and paper industries has been stressed by Bialkowski in [6].

8
PID Stabilization of First-Order Systems with Time Delay

In this chapter we present a complete solution to the problem of characterizing all PID controllers that stabilize a given first-order system with time delay. As will be seen shortly, the PID stabilization problem is considerably more complicated than the P and PI cases considered in previous chapters. The solution presented here makes use of the results introduced in Chapter 5. The range of admissible proportional gains is first determined in closed form. Then for each proportional gain in this range the stabilizing set in the space of the integral and derivative gains is shown to be either a trapezoid, a triangle, or a quadrilateral.

8.1 Introduction

The PID controller is by far the most common control algorithm used in process control applications. The Japan Electric Measuring Instrument Manufacturers' Association conducted a survey of the state of process control systems in 1989. According to the survey more than 90% of the control loops were of the PID type. The popularity of the PID controller can be attributed to its different characteristic features: it provides feedback; it has the ability to eliminate steady-state offsets through integral action; and it can anticipate the future through derivative action. PID controllers come in many different forms. In some instances, the controller can be found as a stand-alone system in boxes for one or a few control loops. In other instances, PID control is combined with logic, sequential machines, transmitters, and simple function blocks to build complicated automation systems. These kinds of systems are often used for energy production, transportation, and manufacturing. Indeed, the PID controller can be considered to be the bread and butter of control engineering.

The general empirical observation is that most industrial processes can be controlled reasonably well with PID control provided that the demands on the performance specifications are not too high. A PID controller is sufficient when the process has dominant dynamics of first or second order. Most of the time, there are no significant benefits gained by using a more complex controller for such processes. With the derivative action, improved damping is provided. Hence, a higher proportional gain can be used to speed up the transient response. An example of this is temperature control inside a chamber. However, tuning of the derivative action should be carefully done because it can amplify high-frequency noise. Because of this, most of the commercially available PID controllers have a limitation on the gain of the derivative term.

Over the last four decades, several methods have been developed for setting the parameters of the PID controller. Some of these methods are based on characterizing the dynamic response of the plant to be controlled with a first-order model with time delay. It is interesting to note that even though most of these tuning techniques provide satisfactory results, the set of all stabilizing PID controllers for these first-order models with time delay has remained unknown until recently. Since this is the basic set in which every design must reside, it is important to determine. This fact constitutes the motivation for this chapter, which is to provide a complete solution to the problem of characterizing the set of all PID controllers that stabilize a given first-order plant with time delay.

The chapter is organized as follows. In Section 8.2 we present the formal statement of the PID stabilization problem. In Section 8.3 we present the solution to the problem when the system to be controlled is open-loop stable. Section 8.4 contains a similar result for the case of an open-loop unstable plant. We also provide here a necessary and sufficient condition on the time delay for the existence of stabilizing PID controllers. Simulations and examples are provided to illustrate the applicability of the results.

8.2 The PID Stabilization Problem

In this chapter we again study the problem of stabilizing a first-order system with time delay using a PID controller. As in Chapters 6 and 7, our feedback control system is as shown in Fig. 6.2, where $G(s)$ given by

$$G(s) = \frac{k}{1+Ts}e^{-Ls} \tag{8.1}$$

is the plant to be controlled, and $C(s)$ is the PID controller. The PID controller has a proportional term, an integral term, and a derivative term. There are different ways of representing the PID control algorithm. In our case, we will use the following representation:

$$C(s) = k_p + \frac{k_i}{s} + k_d s$$

where k_p is the proportional gain, k_i is the integral gain, and k_d is the derivative gain. Our objective is to analytically determine the set of controller parameters (k_p,k_i,k_d) for which the closed-loop system is stable.

We first analyze the system without the time delay, i.e., $L = 0$. In this case the closed-loop characteristic equation of the system is given by

$$\delta(s) = (T + kk_d)s^2 + (1 + kk_p)s + kk_i \ .$$

Since this is a second-order polynomial, closed-loop stability is equivalent to all the coefficients having the same sign. Assuming that the steady-state gain k of the plant is positive these conditions are

$$k_p > -\frac{1}{k} \ , k_i > 0 \quad \text{and} \quad k_d > -\frac{T}{k} \tag{8.2}$$

or

$$k_p < -\frac{1}{k} \ , k_i < 0 \quad \text{and} \quad k_d < -\frac{T}{k} \ . \tag{8.3}$$

A minimal requirement for any control design is that the delay-free closed-loop system be stable. Consequently, it will be henceforth assumed in this section that the PID gains used to stabilize the plant with delay always satisfy one of the conditions (8.2) or (8.3).

Next consider the case where the time delay of the plant model is different from zero. The closed-loop characteristic equation of the system is then

$$\delta(s) = (kk_i + kk_p s + kk_d s^2)e^{-Ls} + (1 + Ts)s \ .$$

As before, we can make use of Theorems 5.5 and 5.6 to solve the stability problem and find the set of stabilizing PID controllers.

We start by rewriting the quasi-polynomial $\delta(s)$ as

$$\delta^*(s) = e^{Ls}\delta(s) = kk_i + kk_p s + kk_d s^2 + (1 + Ts)se^{Ls} \ .$$

Substituting $s = j\omega$, we have

$$\delta^*(j\omega) = \delta_r(\omega) + j\delta_i(\omega)$$

where

$$\begin{aligned}
\delta_r(\omega) &= kk_i - kk_d\omega^2 - \omega\sin(L\omega) - T\omega^2\cos(L\omega) \\
\delta_i(\omega) &= \omega[kk_p + \cos(L\omega) - T\omega\sin(L\omega)] \ .
\end{aligned}$$

The following sections separately treat the two cases of an open-loop stable plant and an open-loop unstable plant.

8.3 Open-Loop Stable Plant

If the system is open-loop stable, then $T > 0$ in (8.1). Furthermore, we make the standing assumption that $k > 0$ and $L > 0$. From the expressions for $\delta_r(\omega)$ and $\delta_i(\omega)$, it is clear that the controller parameter k_p only affects the imaginary part of $\delta^*(j\omega)$ whereas the parameters k_i and k_d affect the real part of $\delta^*(j\omega)$. Moreover, these three controller parameters appear affinely in $\delta_r(\omega)$ and $\delta_i(\omega)$. These facts are exploited in applying Theorems 5.5 and 5.6 to determine the range of stabilizing PID gains.

Before stating the main result of this section, we will present a few preliminary results that will be useful in solving the PID stabilization problem.

Lemma 8.1 *The imaginary part of $\delta^*(j\omega)$ has only simple real roots if and only if*

$$-\frac{1}{k} < k_p < \frac{1}{k}\left[\frac{T}{L}\alpha_1 \sin(\alpha_1) - \cos(\alpha_1)\right] \tag{8.4}$$

where α_1 is the solution of the equation

$$\tan(\alpha) = -\frac{T}{T+L}\alpha$$

in the interval $(0, \pi)$.

Proof. With the change of variables $z = L\omega$ the real and imaginary parts of $\delta^*(j\omega)$ can be expressed as

$$\delta_r(z) = kk_i - \frac{kk_d}{L^2}z^2 - \frac{1}{L}z\sin(z) - \frac{T}{L^2}z^2\cos(z) \tag{8.5}$$

$$\delta_i(z) = \frac{z}{L}\left[kk_p + \cos(z) - \frac{T}{L}z\sin(z)\right]. \tag{8.6}$$

From (8.6) we can compute the roots of the imaginary part, i.e., $\delta_i(z) = 0$. This gives us the following equation:

$$\frac{z}{L}\left[kk_p + \cos(z) - \frac{T}{L}z\sin(z)\right] = 0\,.$$

Then either

$$z = 0 \text{ or}$$

$$kk_p + \cos(z) - \frac{T}{L}z\sin(z) = 0\,. \tag{8.7}$$

From this it is clear that one root of the imaginary part is $z_o = 0$. The other roots are difficult to find since we need to solve (8.7) analytically. However, we can plot the terms involved in (8.7) and graphically examine the nature of the solution. Let us denote the positive real roots of (8.7) by

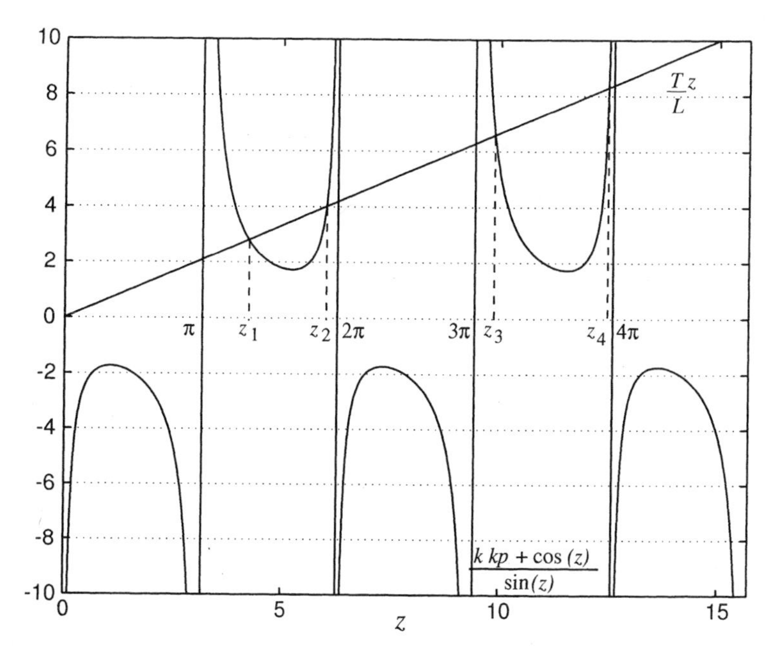

FIGURE 8.1. Plot of the terms involved in (8.7) for $k_p < -\frac{1}{k}$.

z_j, $j = 1, 2, ...$, arranged in increasing order of magnitude. There are now four different cases to consider.

Case 1: $k_p < -\frac{1}{k}$. In this case, we sketch $\frac{kk_p+\cos(z)}{\sin(z)}$ and $\frac{T}{L}z$ to obtain the plots shown in Fig. 8.1.

Case 2: $-\frac{1}{k} < k_p < \frac{1}{k}$. In this case, we graph $\frac{kk_p+\cos(z)}{\sin(z)}$ and $\frac{T}{L}z$ to obtain the plots shown in Fig. 8.2.

Case 3: $k_p = \frac{1}{k}$. In this case, we sketch $kk_p + \cos(z)$ and $\frac{T}{L}z\sin(z)$ to obtain the plots shown in Fig. 8.3.

Case 4: $\frac{1}{k} < k_p$. In this case, we sketch $\frac{kk_p+\cos(z)}{\sin(z)}$ and $\frac{T}{L}z$ to obtain the plots shown in Figs. 8.4(a) and 8.4(b). The plot in Fig. 8.4(a) corresponds to the case where $\frac{1}{k} < k_p < k_u$, and k_u is the largest number so that the plot of $\frac{kk_p+\cos(z)}{\sin(z)}$ intersects the line $\frac{T}{L}z$ twice in the interval $(0, \pi)$. The plot in Fig. 8.4(b) corresponds to the case where $k_p \geq k_u$ and the plot of $\frac{kk_p+\cos(z)}{\sin(z)}$ does not intersect the line $\frac{T}{L}z$ twice in the interval $(0, \pi)$.

Let us now use the results from Section 5.5 to check if $\delta_i(z)$ has only real roots. Substituting $s_1 = Ls$ in the expression for $\delta^*(s)$, we see that for the new quasi-polynomial in s_1, $M = 2$ and $N = 1$. Next we choose $\eta = \frac{\pi}{4}$ to satisfy the requirement that $\sin(\eta) \neq 0$. From Figs. 8.2 through 8.4(a) we see that in each of these cases, i.e., for $-\frac{1}{k} < k_p < k_u$, $\delta_i(z)$ has three real roots in the interval $[0, 2\pi - \frac{\pi}{4}] = [0, \frac{7\pi}{4}]$, including a root at the origin. Since $\delta_i(z)$ is an odd function of z, it follows that in the interval $[-\frac{7\pi}{4}, \frac{7\pi}{4}]$, $\delta_i(z)$ will have five real roots. Also observe from Figs. 8.2

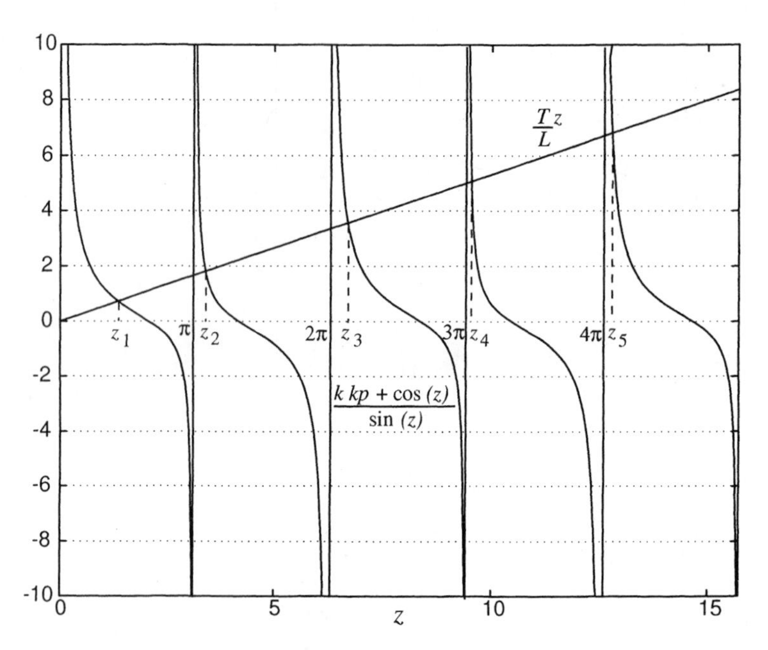

FIGURE 8.2. Plot of the terms involved in (8.7) for $-\frac{1}{k} < k_p < \frac{1}{k}$.

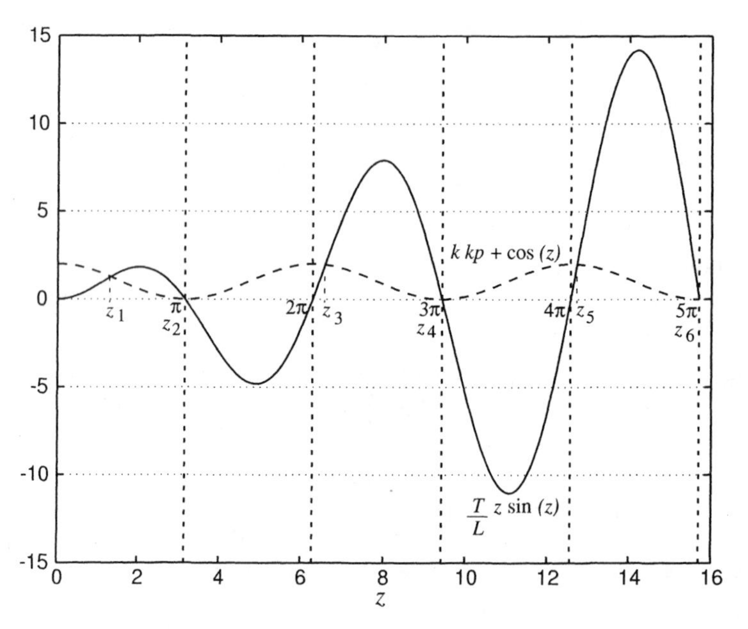

FIGURE 8.3. Plot of the terms involved in (8.7) for $k_p = \frac{1}{k}$.

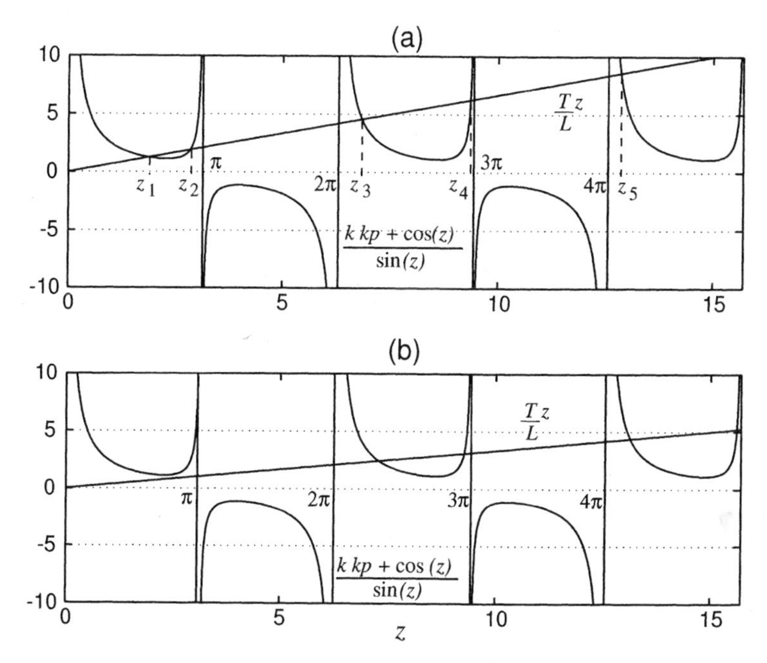

FIGURE 8.4. Plot of the terms involved in (8.7) for $\frac{1}{k} < k_p$.

through 8.4(a) that $\delta_i(z)$ has a real root in the interval $(\frac{7\pi}{4}, \frac{9\pi}{4}]$. Thus $\delta_i(z)$ has $4N + M = 6$ real roots in the interval $[-2\pi + \frac{\pi}{4}, 2\pi + \frac{\pi}{4}]$. Moreover, it can be shown using Figs. 8.2 through 8.4(a) that $\delta_i(z)$ has two real roots in each of the intervals $[2l\pi + \frac{\pi}{4}, 2(l+1)\pi + \frac{\pi}{4}]$ and $[-2(l+1)\pi + \frac{\pi}{4}, -2l\pi + \frac{\pi}{4}]$ for $l = 1, 2, \ldots$. It follows that $\delta_i(z)$ has exactly $4lN + M$ real roots in $[-2l\pi + \frac{\pi}{4}, 2l\pi + \frac{\pi}{4}]$ for $-\frac{1}{k} < k_p < k_u$. Hence from Theorem 5.6, we conclude that for $-\frac{1}{k} < k_p < k_u$, $\delta_i(z)$ has only real roots. Also note that the cases $k_p < -\frac{1}{k}$ and $k_p \geq k_u$ corresponding to Figs. 8.1 and 8.4(b), respectively, do not merit any further consideration since using Theorem 5.6, we can easily argue that in these cases, all the roots of $\delta_i(z)$ will not be real, thereby ruling out closed-loop stability.

It only remains to determine the upper bound k_u on the allowable value of k_p. From the definition of k_u, it follows that if $k_p = k_u$ the plot of $\frac{kk_p + \cos(z)}{\sin(z)}$ intersects the line $\frac{T}{L}z$ only once in the interval $(0, \pi)$. Let us denote by α_1 the value of z for which this intersection occurs. Then we know that for $z = \alpha_1 \in (0, \pi)$ we have

$$\frac{kk_u + \cos(\alpha_1)}{\sin(\alpha_1)} = \frac{T}{L}\alpha_1 \,. \tag{8.8}$$

Moreover, at $z = \alpha_1$, the line $\frac{T}{L}z$ is tangent to the plot of $\frac{kk_u + \cos(z)}{\sin(z)}$. Thus

$$
\begin{aligned}
\frac{d}{dz}\left[\frac{kk_u + \cos(z)}{\sin(z)}\right]_{z=\alpha_1} &= \frac{T}{L} \\
\Rightarrow 1 + kk_u \cos(\alpha_1) &= -\frac{T}{L}\sin^2(\alpha_1)\,. \qquad (8.9)
\end{aligned}
$$

Eliminating kk_u between (8.8) and (8.9) we conclude that $\alpha_1 \in (0, \pi)$ can be obtained as a solution of the following equation:

$$
\tan(\alpha_1) = -\frac{T}{T+L}\alpha_1\,.
$$

Once α_1 is determined the parameter k_u can be obtained using (8.8):

$$
k_u = \frac{1}{k}\left[\frac{T}{L}\alpha_1 \sin(\alpha_1) - \cos(\alpha_1)\right].
$$

This completes the proof of the lemma. ■

From (8.5), for $z \neq 0$, the real part $\delta_r(z)$ can be rewritten as

$$
\delta_r(z) = \frac{k}{L^2} z^2 [-k_d + m(z)k_i + b(z)] \qquad (8.10)
$$

where

$$
\begin{aligned}
m(z) &\triangleq \frac{L^2}{z^2} \qquad (8.11) \\
b(z) &\triangleq -\frac{L}{kz}\left[\sin(z) + \frac{T}{L}z\cos(z)\right]. \qquad (8.12)
\end{aligned}
$$

Lemma 8.2 *For each value of k_p in the range given by (8.4), the necessary and sufficient conditions on k_i and k_d for the roots of $\delta_r(z)$ and $\delta_i(z)$ to interlace are the following infinite set of inequalities:*

$$
\begin{aligned}
k_i &> 0 \\
k_d &> m_1 k_i + b_1 \\
k_d &< m_2 k_i + b_2 \\
k_d &> m_3 k_i + b_3 \\
k_d &< m_4 k_i + b_4 \qquad (8.13) \\
&\vdots
\end{aligned}
$$

where the parameters m_j and b_j for $j = 1, 2, 3, \ldots$ are given by

$$
\begin{aligned}
m_j &\triangleq m(z_j) \qquad (8.14) \\
b_j &\triangleq b(z_j)\,. \qquad (8.15)
\end{aligned}
$$

Proof. From condition 1 of Theorem 5.5, the roots of $\delta_r(z)$ and $\delta_i(z)$ have to interlace for the quasi-polynomial $\delta^*(s)$ to be stable. Thus we evaluate $\delta_r(z)$ at the roots of the imaginary part $\delta_i(z)$. For $z_o = 0$, using (8.5) we obtain

$$\delta_r(z_o) = kk_i \ . \tag{8.16}$$

For z_j, where $j = 1, 2, 3, \ldots$, using (8.10) we obtain

$$\delta_r(z_j) = \frac{k}{L^2} z_j^2 [-k_d + m(z_j)k_i + b(z_j)] \ . \tag{8.17}$$

Interlacing the roots of $\delta_r(z)$ and $\delta_i(z)$ is equivalent to $\delta_r(z_o) > 0$ (since Lemma 8.1 implies that k_p is necessarily greater than $-\frac{1}{k}$, which in view of the stability requirements (8.2) for the delay-free case implies that $k_i > 0$), $\delta_r(z_1) < 0$, $\delta_r(z_2) > 0$, $\delta_r(z_3) < 0$, and so on. Using this fact and (8.16)–(8.17) we obtain

$$\begin{aligned}
\delta_r(z_o) > 0 \quad &\Rightarrow \quad k_i > 0 \\
\delta_r(z_1) < 0 \quad &\Rightarrow \quad k_d > m_1 k_i + b_1 \\
\delta_r(z_2) > 0 \quad &\Rightarrow \quad k_d < m_2 k_i + b_2 \\
\delta_r(z_3) < 0 \quad &\Rightarrow \quad k_d > m_3 k_i + b_3 \\
\delta_r(z_4) > 0 \quad &\Rightarrow \quad k_d < m_4 k_i + b_4 \\
&\vdots
\end{aligned}$$

Thus intersecting all these regions in the k_i–k_d space, we obtain the set of (k_i,k_d) values for which the roots of $\delta_r(z)$ and $\delta_i(z)$ interlace for a given fixed value of k_p. Notice that all these regions are half planes with their boundaries being lines with positive slopes m_j. This completes the proof of the lemma. ■

Example 8.1 *Consider the transfer function (8.1) with the following plant parameters: $k = 1$, $T = 2$ seconds, and $L = 4$ seconds. Then, the quasi-polynomial $\delta^*(s)$ is given by*

$$\delta^*(s) = k_i + k_p s + k_d s^2 + (2s^2 + s)e^{4s} \ .$$

From Lemma 8.1 we need to find the solution of the following equation:

$$\tan(\alpha) = -\frac{1}{3}\alpha \ .$$

Solving this equation in the interval $(0, \pi)$ we get $\alpha_1 = 2.4556$. Then, from (8.4), the imaginary part of $\delta^*(s)$ has only simple real roots if and only if

$$-1 < k_p < 1.5515 \ .$$

We now set the controller parameter k_p to 0.8, which is inside the previous range. For this k_p value, (8.7) takes the form

$$0.8 + \cos(z) - 0.5z\sin(z) = 0 \, .$$

We next compute some of the positive real roots of this equation and arrange them in increasing order of magnitude:

$$z_1 = 1.5806 \, , \; z_2 = 3.2602 \, , \; z_3 = 6.7971 \, , \; z_4 = 9.4669 \, .$$

Using (8.14) and (8.15) we now calculate the parameters m_j and b_j for $j = 1, \cdots, 4$:

$$\begin{aligned} m_1 &= 6.4044 & b_1 &= -2.5110 \\ m_2 &= 1.5053 & b_2 &= 2.1311 \\ m_3 &= 0.3463 & b_3 &= -2.0309 \\ m_4 &= 0.1785 & b_4 &= 2.0160 \, . \end{aligned}$$

From Lemma 8.2, interlacing the roots of the real and imaginary parts of $\delta^*(j\omega)$ occurs for $k_p = 0.8$, if and only if the following set of inequalities are satisfied:

$$\begin{aligned} k_i &> 0 \\ k_d &> 6.4044k_i - 2.5110 \\ k_d &< 1.5053k_i + 2.1311 \\ k_d &> 0.3463k_i - 2.0309 \\ k_d &< 0.1785k_i + 2.0160 \\ &\vdots \end{aligned}$$

The boundaries of these regions are illustrated in Fig. 8.5. Notice that the boundaries corresponding to $z_2, z_4, \ldots$, converge to the line $k_d = \frac{T}{k} = 2$, whereas the boundaries corresponding to $z_1, z_3, \ldots$, converge to the line $k_d = -\frac{T}{k} = -2$. △

As pointed out in the proof of Lemma 8.2 and in Example 8.1, the inequalities given by (8.13) represent half planes in the space of k_i and k_d. Their boundaries are given by lines with the following equations:

$$k_d = m_j k_i + b_j \quad \text{for } j = 1, 2, 3, \ldots \, .$$

The focus of the remainder of this section will be to show that this intersection is nonempty. We will also determine the intersection of this *countably infinite number* of half planes in a computationally tractable way. To this end, let us denote by v_j the k_i-coordinate of the intersection of the line

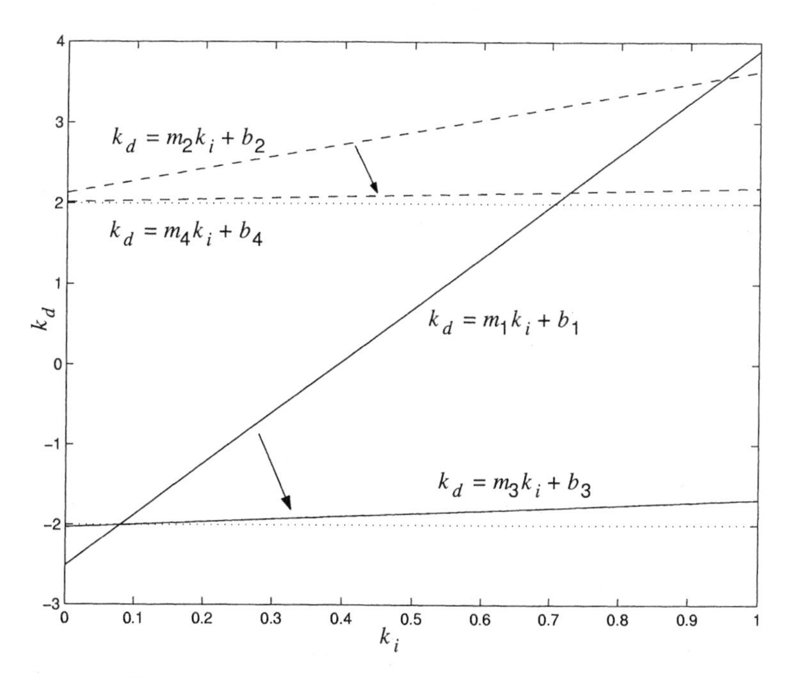

FIGURE 8.5. Boundaries of the regions of Example 8.1.

$k_d = m_j k_i + b_j$, $j = 1, 2, 3, \ldots$, with the line $k_d = -\frac{T}{k}$. From (8.14) and (8.15) it is not difficult to show that

$$v_j = \frac{z_j}{kL}\left[\sin(z_j) + \frac{T}{L} z_j (\cos(z_j) - 1)\right] . \tag{8.18}$$

In a similar fashion, let us now denote by w_j the k_i-coordinate of the intersection of the line $k_d = m_j k_i + b_j$, $j = 1, 2, 3, \ldots$, with the line $k_d = \frac{T}{k}$. Using (8.14) and (8.15) it can be once again shown that

$$w_j = \frac{z_j}{kL}\left[\sin(z_j) + \frac{T}{L} z_j (\cos(z_j) + 1)\right] . \tag{8.19}$$

We now state three important technical lemmas that will allow us to develop an algorithm for solving the PID stabilization problem of an open-loop stable plant ($T > 0$). These lemmas show the behavior of the parameters b_j, v_j, and w_j, $j = 1, 2, 3, \ldots$, for different values of the parameter k_p inside the range proposed by Lemma 8.1. The proofs of these lemmas are long and will be omitted here. They are given in Appendix A.

Lemma 8.3 *If* $-\frac{1}{k} < k_p < \frac{1}{k}$ *then*

(*i*) $\quad b_j < b_{j+2} < -\frac{T}{k}$ *for odd values of* j

$$(ii) \quad b_j > \frac{T}{k} \text{ and } b_j \to \frac{T}{k} \text{ as } j \to \infty \text{ for even values of } j$$

$$(iii) \quad 0 < v_j < v_{j+2} \text{ for odd values of } j.$$

Lemma 8.4 *If $k_p = \frac{1}{k}$ then*

$$(i) \quad b_j = -\frac{T}{k} \text{ for odd values of } j$$

$$(ii) \quad b_j = \frac{T}{k} \text{ for even values of } j.$$

Lemma 8.5 *If $\frac{1}{k} < k_p < \frac{1}{k}\left[\frac{T}{L}\alpha_1 \sin(\alpha_1) - \cos(\alpha_1)\right]$ where α_1 is the solution of the equation*

$$\tan(\alpha) = -\frac{T}{T+L}\alpha$$

in the interval $(0, \pi)$, then

$$(i) \quad b_j > b_{j+2} > -\frac{T}{k} \text{ for odd values of } j$$

$$(ii) \quad b_j < b_{j+2} < \frac{T}{k} \text{ for even values of } j$$

$$(iii) \quad w_j > w_{j+2} > 0 \text{ for even values of } j$$

$$(iv) \quad b_1 < b_2 \,, \; w_1 > w_2 \,.$$

We are now ready to state the main result of this section.

Theorem 8.1 (Main Result) *The range of k_p values for which a given open-loop stable plant, with transfer function $G(s)$ as in (8.1), can be stabilized using a PID controller is given by*

$$-\frac{1}{k} < k_p < \frac{1}{k}\left[\frac{T}{L}\alpha_1 \sin(\alpha_1) - \cos(\alpha_1)\right] \tag{8.20}$$

where α_1 is the solution of the equation

$$\tan(\alpha) = -\frac{T}{T+L}\alpha \tag{8.21}$$

in the interval $(0, \pi)$. For k_p values outside this range, there are no stabilizing PID controllers. The complete stabilizing region is given by (see Fig. 8.6):

1. *For each $k_p \in (-\frac{1}{k}, \frac{1}{k})$, the cross-section of the stabilizing region in the (k_i, k_d) space is the trapezoid T;*

2. *For $k_p = \frac{1}{k}$, the cross-section of the stabilizing region in the (k_i, k_d) space is the triangle Δ;*

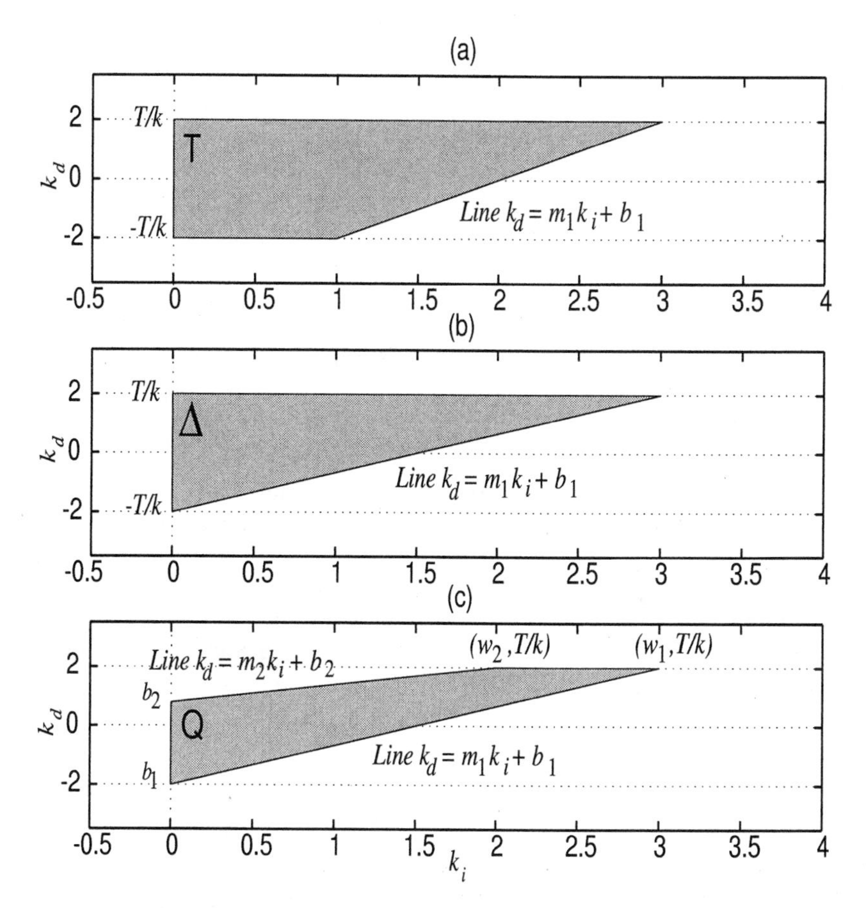

FIGURE 8.6. The stabilizing region of (k_i, k_d) for (a) $-\frac{1}{k} < k_p < \frac{1}{k}$, (b) $k_p = \frac{1}{k}$, (c) $\frac{1}{k} < k_p < k_u$.

3. For each $k_p \in \left(\frac{1}{k}, k_u := \frac{1}{k}\left[\frac{T}{L}\alpha_1 \sin(\alpha_1) - \cos(\alpha_1)\right]\right)$*, the cross-section of the stabilizing region in the* (k_i, k_d) *space is the quadrilateral* Q*.*

Proof. To ensure the stability of the quasi-polynomial $\delta^*(s)$ we need to check the two conditions given in Theorem 5.5.

Step 1. We first check condition 2 of Theorem 5.5:

$$E(\omega_o) = \delta_i^{'}(\omega_o)\delta_r(\omega_o) - \delta_i(\omega_o)\delta_r^{'}(\omega_o) > 0$$

for some ω_o in $(-\infty, \infty)$. Let us take $\omega_o = 0$, so $z_o = 0$. Thus $\delta_i(z_o) = 0$ and $\delta_r(z_o) = kk_i$. We also have

$$\begin{aligned} \delta_i^{'}(z) &= \frac{kk_p}{L} + \left(\frac{1}{L} - \frac{T}{L^2}z^2\right)\cos(z) - \left(\frac{1}{L}z + \frac{2T}{L^2}z\right)\sin(z) \\ \Rightarrow E(z_o) &= \left(\frac{kk_p + 1}{L}\right)(kk_i)\,. \end{aligned}$$

Recall that $k > 0$ and $L > 0$. Thus for

$$k_i > 0 \text{ and } k_p > -\frac{1}{k} \tag{8.22}$$

or

$$k_i < 0 \text{ and } k_p < -\frac{1}{k}$$

we have $E(z_o) > 0$. Notice that from these conditions we can safely discard $k_p = -\frac{1}{k}$ from the set of k_p values for which a stabilizing PID controller can be found.

Step 2. Next we check condition 1 of Theorem 5.5, i.e., $\delta_r(z)$ and $\delta_i(z)$ have only simple real roots and these interlace. From Lemma 8.1 we know that the roots of $\delta_i(z)$ are all real if and only if the parameter k_p lies inside the range

$$\left(-\frac{1}{k}, \frac{1}{k}\left[\frac{T}{L}\alpha_1 \sin(\alpha_1) - \cos(\alpha_1)\right]\right)$$

where α_1 is the solution of the equation

$$\tan(\alpha) = -\frac{T}{T+L}\alpha$$

in the interval $(0, \pi)$. From the proof of Lemma 8.2, we see that interlacing of the roots of $\delta_r(z)$ and $\delta_i(z)$ leads to the following set of inequalities

$$\begin{aligned} k_i &> 0 \\ k_d &> m_1 k_i + b_1 \\ k_d &< m_2 k_i + b_2 \\ k_d &> m_3 k_i + b_3 \\ k_d &< m_4 k_i + b_4 \\ &\vdots \end{aligned}$$

We now show that for $-\frac{1}{k} < k_p < k_u$, where $k_u = \frac{1}{k}[\frac{T}{L}\alpha_1 \sin(\alpha_1) - \cos(\alpha_1)]$, all these regions have a nonempty intersection. Notice first that the slopes m_j of the boundary lines of these regions decrease with z_j. Moreover, in the limit we have

$$\lim_{j\to\infty} m_j = 0 \,.$$

Using this fact we have the following observations:

1. When $-\frac{1}{k} < k_p < \frac{1}{k}$, the intersection is given by the trapezoid T sketched in Fig. 8.6(a). This region can be found using the properties stated in Lemma 8.3.

2. When $k_p = \frac{1}{k}$, the intersection is given by the triangle Δ sketched in Fig. 8.6(b). This region can be found using the properties stated in Lemma 8.4.

3. When $\frac{1}{k} < k_p < k_u$, the intersection is given by the quadrilateral Q sketched in Fig. 8.6(c). This region can be found using the properties stated in Lemma 8.5.

For values of k_p in $(-\frac{1}{k}, k_u)$, the interlacing property and the fact that the roots of $\delta_i(z)$ are all real can be used in Theorem 5.6 to guarantee that $\delta_r(z)$ also has only real roots. Thus for values of k_p inside this range there is a solution to the PID stabilization problem for a first-order open-loop stable plant with time delay. For values of k_p outside this range the aforementioned problem does not have a solution. This completes the proof of the theorem. ■

Remark 8.1 *For $L = 0$, i.e., no time delay, we can solve (8.21) analytically to obtain $\alpha_1 = 2.0288$. Using this value, the upper bound in (8.20) evaluates out to ∞, which is consistent with the condition imposed on k_p by (8.2), for one of the scenarios arising in the delay-free case. Also, by plotting the graphs of $\tan(\alpha)$ and $-\frac{T}{T+L}\alpha$ versus α, it is easy to see that as $\frac{T}{L}$ decreases, the intersection α_1 approaches π. By substituting for $\frac{T}{L}$ from (8.21) into the upper bound in (8.20) and differentiating with respect to α_1, it can be shown that as α_1 increases from 2.0288 and approaches π, the upper bound in (8.20) monotonically decreases to $\frac{1}{k}$ (see also Fig. 8.7). This shows that as $\frac{L}{T}$ increases, the range of k_p values shrinks, which is consistent with the empirical observation of Astrom and Hagglund [1].*

In view of Theorem 8.1, we now propose an algorithm to determine the set of stabilizing parameters for the plant (8.1) with $T > 0$.

Algorithm for Determining Stabilizing PID Parameters.

- **Step 1:** Initialize $k_p = -\frac{1}{k}$ and $step = \frac{1}{N+1}\left(k_u + \frac{1}{k}\right)$, where N is the desired number of points;

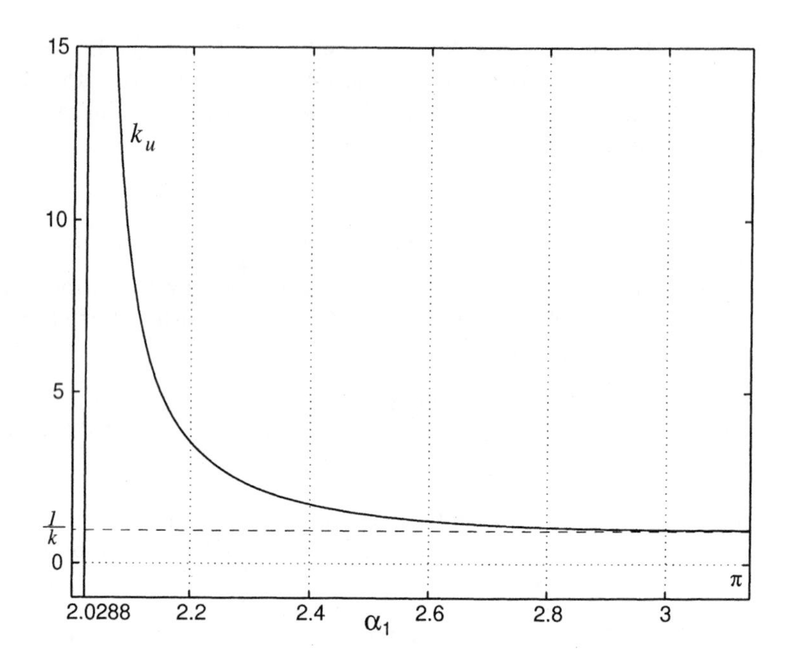

FIGURE 8.7. Plot of the upper bound k_u in (8.20) as a function of α_1.

- **Step 2:** Increase k_p as follows: $k_p = k_p + step$;
- **Step 3:** If $k_p < k_u$ then go to Step 4. Else, terminate the algorithm;
- **Step 4:** Find the roots z_1 and z_2 of (8.7);
- **Step 5:** Compute the parameters m_j and b_j, $j = 1, 2$ associated with the previously found z_j by using (8.14) and (8.15);
- **Step 6:** Determine the stabilizing region in the k_i–k_d space using Fig. 8.6;
- **Step 7:** Go to Step 2.

We now present two examples that illustrate the procedure involved in solving the PID stabilization problem using the results of this section.

Example 8.2 *Consider the PID controller design for the first-order process with deadtime using the Ziegler-Nichols step response method. The process model is given as*

$$G(s) = \frac{k}{Ts+1}e^{-Ls} \tag{8.23}$$

where $k = 0.1$, $T = 0.01$ *seconds, and* $L = 0.1$ *seconds. Using the Ziegler-Nichols step response method, we obtain the controller parameter values* $k_p = 1.2$, $k_i = 6.0$, *and* $k_d = 0.06$.

We now use the results of this section to determine the set of all stabilizing (k_i, k_d) *values when* k_p *is kept fixed at 1.2 (the Ziegler-Nichols value). First we compute the roots* z_1 *and* z_2 *of the imaginary part* $\delta_i(z)$ *of the characteristic equation of the closed-loop system, i.e., we solve the following equation*

$$0.12 + \cos(z) - 0.1z\sin(z) = 0 \, .$$

The roots obtained are $z_1 = 1.537$ *and* $z_2 = 4.204$*. Then from Fig. 8.6 we only need to compute the boundary line corresponding to* z_1*. This line is given by* $k_d = 0.00423k_i - 0.6535$*. Thus the set of stabilizing* (k_i, k_d) *values when* $k_p = 1.2$ *is the one sketched in Fig. 8.8.*

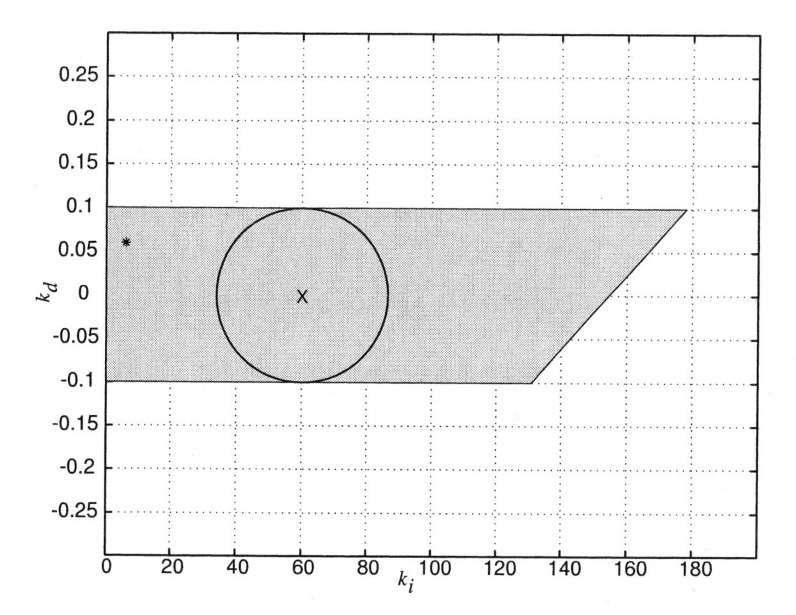

FIGURE 8.8. The stabilizing region of (k_i, k_d) when $k_p = 1.2$ for Example 8.2.

*From Fig. 8.8, it is clear that the PID controller obtained by the Ziegler-Nichols step response method (denoted by *) is very close to the stability boundary. So this example shows that a PID controller design obtained using the Ziegler-Nichols step response method may suffer from "fragility." Also shown in Fig. 8.8 is the circle of largest radius inscribed inside the stabilizing region. This circle has a radius of* $r = 0.1$ *and its center can be placed anywhere on the line* $k_d = 0$, $0 < k_i < 130.26$*. By choosing the* (k_i, k_d) *value at the center of this circle, we can obtain the largest* l_2 *parametric stability margin in the space of* k_i *and* k_d*, thereby alleviating the controller fragility problem.* △

Example 8.3 *Let us revisit the design problem presented in Example 8.2. The plant parameters are* $k = 0.1$, $T = 0.01$ *seconds, and* $L = 0.1$ *sec-*

onds. We will now use a different approach to solve this problem. We first approximate the deadtime of (8.23) by the first-order Padé approximation introduced in Section 5.3. The approximated process model is given by

$$\begin{aligned} G_m^1(s) &= \frac{k}{Ts+1} \cdot \frac{-\frac{L}{2}s+1}{\frac{L}{2}s+1} \\ &= \frac{0.1(-0.05s+1)}{(0.01s+1)(0.05s+1)} . \end{aligned}$$

Since $G_m^1(s)$ is a rational transfer function we use the PID controller design procedure presented in Section 4.2. Using this procedure we obtained the set of all stabilizing (k_p, k_i, k_d) values. The set of all stabilizing (k_i, k_d) values corresponding to $k_p = 1.2$ is sketched in Fig. 8.9 with a continuous line. We next compare this set with the one obtained in Example 8.2. This latter set is superimposed on Fig. 8.9 using a dashed line. As we can see from this figure, the set obtained by the Padé approximation includes settings of the PID controller that lead to an unstable closed-loop system. This shows that the Padé approximation may indeed be unsatisfactory when designing a PID controller for systems with deadtime.

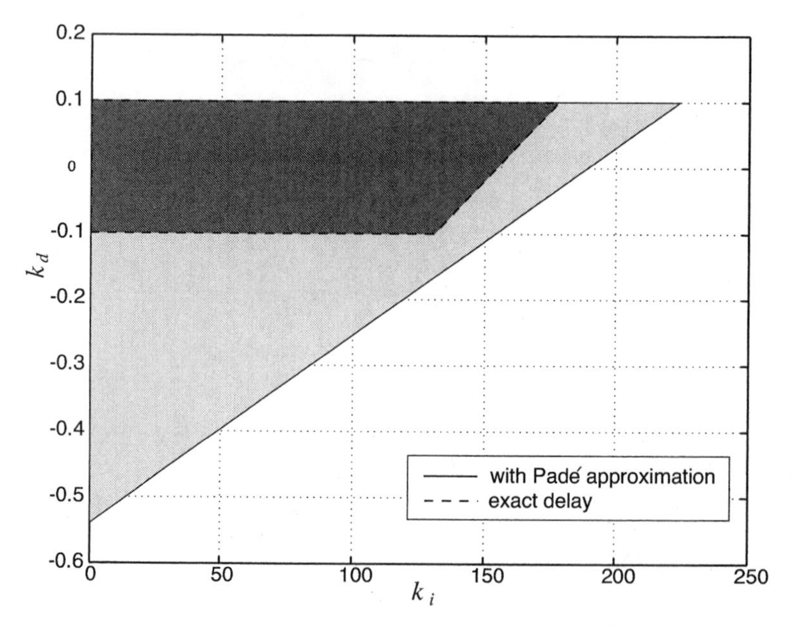

FIGURE 8.9. The stabilizing region of (k_i, k_d) when $k_p = 1.2$ for Example 8.3.

Finally let us use the results of this section to determine the entire set of stabilizing PID parameters. The range of k_p values specified by Theorem 8.1 is given by

$$-10 < k_p < 10.4048 .$$

By sweeping over this range and using the algorithm presented earlier, we obtain the stabilizing set of (k_p, k_i, k_d) values sketched in Fig. 8.10. △

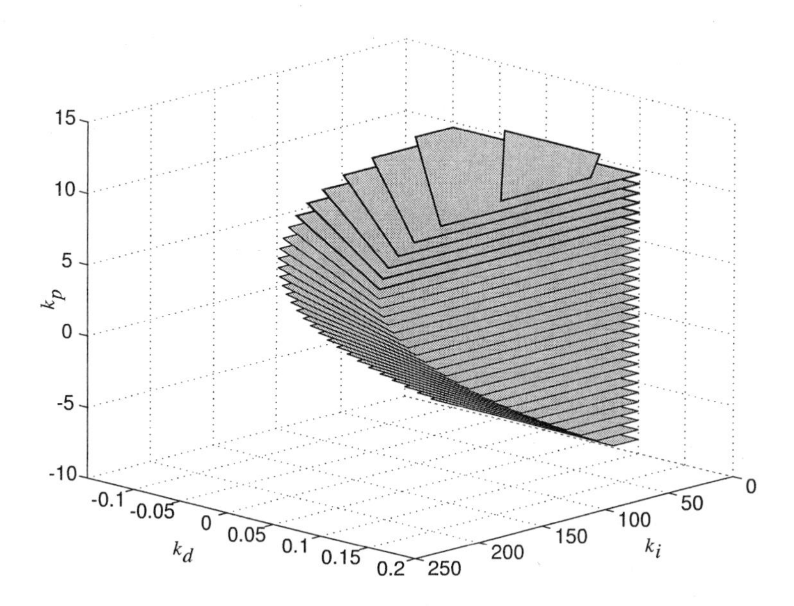

FIGURE 8.10. The stabilizing region of (k_p, k_i, k_d) values for the PID controller in Example 8.3.

8.4 Open-Loop Unstable Plant

In this case $T < 0$ in (8.1). Furthermore, let us assume that $k > 0$ and $L > 0$. The same procedure used in the last section will be used here to solve the problem of stabilizing an open-loop unstable plant using a PID controller. In other words, we will find the set of all stabilizing (k_p,k_i,k_d) values by repeatedly using Theorem 5.5.

Before stating the main result of this section, we will present a few preliminary results leading up to it. As the next lemma shows, the range of k_p values for which stabilization is possible can be determined exactly.

Lemma 8.6 *For $|\frac{T}{L}| > 0.5$, the imaginary part of $\delta^*(j\omega)$ has only simple real roots if and only if*

$$\frac{1}{k}\left[\frac{T}{L}\alpha_1 \sin(\alpha_1) - \cos(\alpha_1)\right] < k_p < -\frac{1}{k} \tag{8.24}$$

where α_1 is the solution of the equation

$$\tan(\alpha) = -\frac{T}{T+L}\alpha$$

in the interval $(0,\pi)$. *In the special case of* $|\frac{T}{L}|=1$, *we have* $\alpha_1=\frac{\pi}{2}$. *For* $|\frac{T}{L}|\leq 0.5$, *the roots of the imaginary part of* $\delta^*(j\omega)$ *are not all real.*

Proof. As in the proof of Lemma 8.1 we make use of the change of variables $z=L\omega$. With this change of variables, the real and imaginary parts of $\delta^*(j\omega)$ can be expressed as

$$\begin{aligned}\delta_r(z) &= kk_i-\frac{kk_d}{L^2}z^2-\frac{1}{L}z\sin(z)-\frac{T}{L^2}z^2\cos(z) && (8.25)\\ \delta_i(z) &= \frac{z}{L}\left[kk_p+\cos(z)-\frac{T}{L}z\sin(z)\right]. && (8.26)\end{aligned}$$

We can compute the roots of the imaginary part from (8.26), i.e., $\delta_i(z)=0$. This gives us the following equation:

$$\frac{z}{L}\left[kk_p+\cos(z)-\frac{T}{L}z\sin(z)\right]=0\,.$$

Then either

$$\begin{aligned}z &= 0 \text{ or}\\ kk_p+\cos(z)-\frac{T}{L}z\sin(z) &= 0\,. && (8.27)\end{aligned}$$

From this expression one root of the imaginary part is $z_o=0$. As in Section 8.3, we will plot the terms involved in (8.27) and graphically examine the nature of the solution. Let us denote the positive real roots of (8.27) by z_j, $j=1,2,...,$ arranged in increasing order of magnitude. There are now four different cases to consider.

Case 1: $k_p<-\frac{1}{k}$. In this case, we sketch $\frac{kk_p+\cos(z)}{\sin(z)}$ and $\frac{T}{L}z$. It can be shown (see Lemma 8.7) that if $\frac{T}{L}\geq -0.5$, then the curves $\frac{kk_p+\cos(z)}{\sin(z)}$ and $\frac{T}{L}z$ do not intersect at all in the interval $(0,\pi)$ *regardless of the value of* k_p *in* $(-\infty,-\frac{1}{k})$. As will be shortly shown, in such a case the roots of the imaginary part are not all real. Accordingly, let us focus on the case $\frac{T}{L}<-0.5$ in which case again there are two possibilities, depending on the value of k_p. These two possibilities are shown in Figs. 8.11(a) and 8.11(b). The plot in Fig. 8.11(a) corresponds to the case where $k_l<k_p<-\frac{1}{k}$, and k_l is the smallest number so that the plot of $\frac{kk_p+\cos(z)}{\sin(z)}$ intersects the line $\frac{T}{L}z$ twice in the interval $(0,\pi)$. The plot in Fig. 8.11(b) corresponds to the case where $k_p\leq k_l$ and the plot of $\frac{kk_p+\cos(z)}{\sin(z)}$ does not intersect the line $\frac{T}{L}z$ twice in the interval $(0,\pi)$.

Case 2: $-\frac{1}{k}<k_p<\frac{1}{k}$. As in the previous case, we graph $\frac{kk_p+\cos(z)}{\sin(z)}$ and $\frac{T}{L}z$ to obtain the plots shown in Fig. 8.12.

Case 3: $k_p=\frac{1}{k}$. In this case, we sketch $kk_p+\cos(z)$ and $\frac{T}{L}z\sin(z)$ to obtain the plots shown in Fig. 8.13.

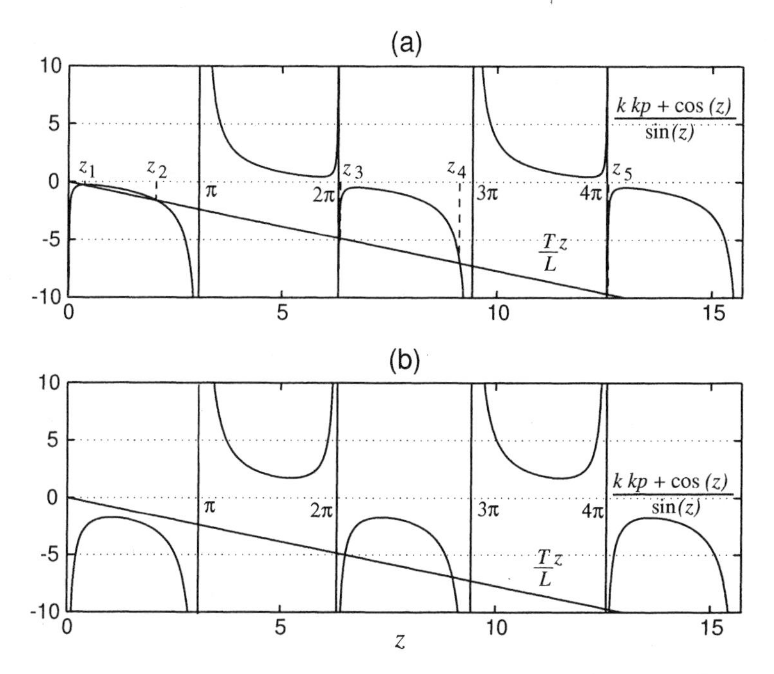

FIGURE 8.11. Plot of the terms involved in (8.27) for $k_p < -\frac{1}{k}$.

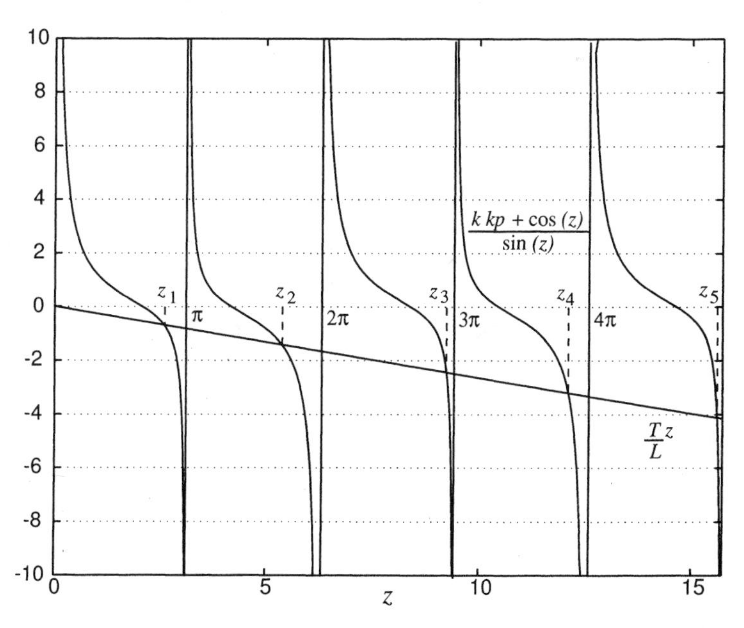

FIGURE 8.12. Plot of the terms involved in (8.27) for $-\frac{1}{k} < k_p < \frac{1}{k}$.

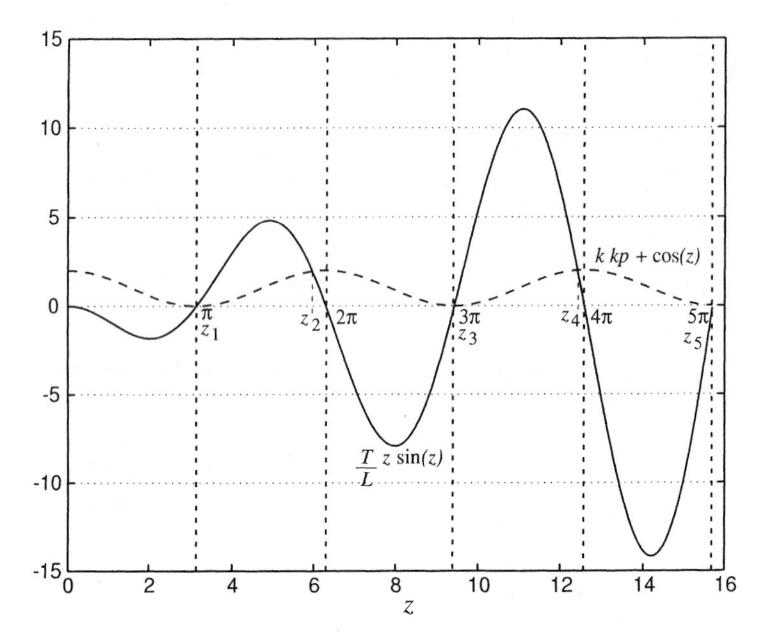

FIGURE 8.13. Plot of the terms involved in (8.27) for $k_p = \frac{1}{k}$.

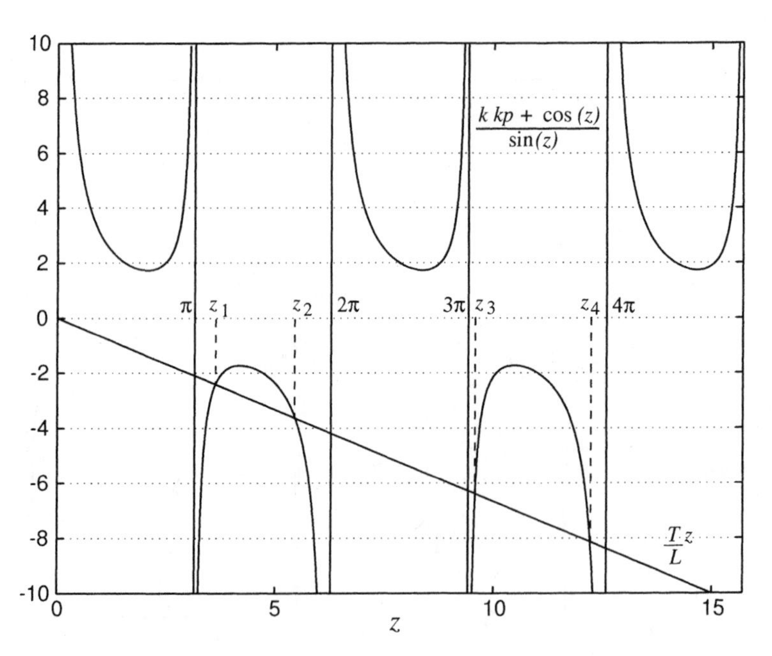

FIGURE 8.14. Plot of the terms involved in (8.27) for $\frac{1}{k} < k_p$.

Case 4: $\frac{1}{k} < k_p$. In this case, we sketch $\frac{kk_p+\cos(z)}{\sin(z)}$ and $\frac{T}{L}z$ to obtain the plots shown in Fig. 8.14.

We will now use Theorem 5.6 to check if $\delta_i(z)$ has only real roots. Substituting $s_1 = Ls$ in the expression for $\delta^*(s)$, we see that for the new quasi-polynomial in s_1, $M = 2$ and $N = 1$. Next we choose $\eta = \frac{\pi}{4}$ to satisfy the requirement that $\sin(\eta) \neq 0$. From Fig. 8.11(a), we see that in this case, i.e., for $k_l < k_p < -\frac{1}{k}$, $\delta_i(z)$ has three real roots in the interval $[0, 2\pi - \frac{\pi}{4}] = [0, \frac{7\pi}{4}]$, including a root at the origin. Since $\delta_i(z)$ is an odd function of z, it follows that in the interval $[-\frac{7\pi}{4}, \frac{7\pi}{4}]$, $\delta_i(z)$ will have five real roots. Also observe from Fig. 8.11(a) that $\delta_i(z)$ has a real root in the interval $(\frac{7\pi}{4}, \frac{9\pi}{4}]$. Thus $\delta_i(z)$ has $4N + M = 6$ real roots in the interval $[-2\pi + \frac{\pi}{4}, 2\pi + \frac{\pi}{4}]$. Moreover, it can be shown using Fig. 8.11(a) that $\delta_i(z)$ has two real roots in each of the intervals $[2l\pi + \frac{\pi}{4}, 2(l+1)\pi + \frac{\pi}{4}]$ and $[-2(l+1)\pi + \frac{\pi}{4}, -2l\pi + \frac{\pi}{4}]$ for $l = 1, 2, \ldots$. Hence, it follows that $\delta_i(z)$ has exactly $4lN + M$ real roots in $[-2l\pi + \frac{\pi}{4}, 2l\pi + \frac{\pi}{4}]$ for $k_l < k_p < -\frac{1}{k}$. Hence from Theorem 5.6, we conclude that for $k_l < k_p < -\frac{1}{k}$, $\delta_i(z)$ has only real roots. Also note that the cases $k_p \leq k_l$ and $k_p > -\frac{1}{k}$ corresponding to Figs. 8.11(b) through 8.14, respectively, do not merit any further consideration since using Theorem 5.6, we can easily argue that in these cases all the roots of $\delta_i(z)$ will not be real. The same argument can also be used in conjunction with Lemma 8.7 to conclude that for all the roots of $\delta_i(z)$ to be real, the condition $\frac{T}{L} < -0.5$ must necessarily be satisfied.

We now need to determine the lower bound k_l on the allowable value for k_p. From the definition of k_l, it follows that for $k_p = k_l$ the plot of $\frac{kk_p+\cos(z)}{\sin(z)}$ intersects the line $\frac{T}{L}z$ only once in the interval $(0, \pi)$. Let us denote by α_1 the value of z for which this intersection occurs. Then we know that for $z = \alpha_1 \in (0, \pi)$ we have

$$\frac{kk_l + \cos(\alpha_1)}{\sin(\alpha_1)} = \frac{T}{L}\alpha_1 . \tag{8.28}$$

Now, at $z = \alpha_1$, the line $\frac{T}{L}z$ is tangent to the plot of $\frac{kk_l+\cos(z)}{\sin(z)}$. Thus

$$\begin{aligned} \frac{d}{dz}\left[\frac{kk_l + \cos(z)}{\sin(z)}\right]_{z=\alpha_1} &= \frac{T}{L} \\ \Rightarrow 1 + kk_l\cos(\alpha_1) &= -\frac{T}{L}\sin^2(\alpha_1) . \end{aligned} \tag{8.29}$$

Eliminating kk_l between (8.28) and (8.29) we conclude that $\alpha_1 \in (0, \pi)$ can be obtained as a solution of the following equation:

$$\sin(\alpha_1)\left(\frac{T}{L} + 1\right) = -\frac{T}{L}\alpha_1\cos(\alpha_1) .$$

If $\frac{T}{L} \neq -1$, then this expression can be rewritten as follows:

$$\tan(\alpha_1) = -\frac{T}{T+L}\alpha_1 . \tag{8.30}$$

If $\frac{T}{L} = -1$, then $\alpha_1 = \frac{\pi}{2}$. In either case, the parameter k_l is given by

$$k_l = \frac{1}{k}\left[\frac{T}{L}\alpha_1 \sin(\alpha_1) - \cos(\alpha_1)\right] \quad \text{[from (8.28)]}$$

and this completes the proof. ■

In Lemma 8.6, it was stated that if $|\frac{T}{L}| \leq 0.5$, then the roots of the imaginary part $\delta_i(z)$ are not all real. The following lemma forms the basis of this claim. The proof is given in Appendix B.

Lemma 8.7 *If $-0.5 \leq \frac{T}{L} < 0$, then the curves $\frac{kk_p+\cos(z)}{\sin(z)}$ and $\frac{T}{L}z$ do not intersect in the interval $(0,\pi)$ regardless of the value of k_p in $(-\infty, -\frac{1}{k})$.*

For $z \neq 0$, the real part $\delta_r(z)$ can be rewritten as

$$\delta_r(z) = \frac{k}{L^2}z^2[-k_d + m(z)k_i + b(z)] \tag{8.31}$$

where

$$m(z) \triangleq \frac{L^2}{z^2} \tag{8.32}$$

$$b(z) \triangleq -\frac{L}{kz}\left[\sin(z) + \frac{T}{L}z\cos(z)\right]. \tag{8.33}$$

Lemma 8.8 *For each value of k_p in the range given by (8.24), the necessary and sufficient conditions on k_i and k_d for the roots of $\delta_r(z)$ and $\delta_i(z)$ to interlace are the following infinite set of inequalities:*

$$\begin{aligned}
k_i &< 0 \\
k_d &< m_1k_i + b_1 \\
k_d &> m_2k_i + b_2 \\
k_d &< m_3k_i + b_3 \\
k_d &> m_4k_i + b_4 \\
&\vdots
\end{aligned} \tag{8.34}$$

where the parameters m_j and b_j for $j = 1, 2, 3, \ldots$ are given by

$$m_j \triangleq m(z_j) \tag{8.35}$$

$$b_j \triangleq b(z_j). \tag{8.36}$$

Proof. From condition 1 of Theorem 5.5, the roots of $\delta_r(z)$ and $\delta_i(z)$ have to interlace in order for the quasi-polynomial $\delta^*(s)$ to stable. Thus we evaluate the real part $\delta_r(z)$ at the roots of the imaginary part $\delta_i(z)$. For $z_o = 0$, using (8.25) we obtain

$$\delta_r(z_o) = kk_i. \tag{8.37}$$

For z_j, where $j = 1, 2, 3, ...$, using (8.31) we obtain

$$\delta_r(z_j) = \frac{k}{L^2} z_j^2 [-k_d + m(z_j) k_i + b(z_j)] \,. \tag{8.38}$$

Interlacing the roots of $\delta_r(z)$ and $\delta_i(z)$ is equivalent to $\delta_r(z_o) < 0$ (since Lemma 8.6 implies that k_p is necessarily less than $-\frac{1}{k}$, which in view of the stability requirements (8.3) for the delay-free case, implies that $k_i < 0$), $\delta_r(z_1) > 0$, $\delta_r(z_2) < 0$, $\delta_r(z_3) > 0$, and so on. Using this fact and (8.37)–(8.38) we obtain

$$\begin{aligned}
\delta_r(z_o) < 0 \quad &\Rightarrow \quad k_i < 0 \\
\delta_r(z_1) > 0 \quad &\Rightarrow \quad k_d < m_1 k_i + b_1 \\
\delta_r(z_2) < 0 \quad &\Rightarrow \quad k_d > m_2 k_i + b_2 \\
\delta_r(z_3) > 0 \quad &\Rightarrow \quad k_d < m_3 k_i + b_3 \\
\delta_r(z_4) < 0 \quad &\Rightarrow \quad k_d > m_4 k_i + b_4 \\
&\vdots
\end{aligned}$$

Thus intersecting all these regions in the k_i–k_d space, we obtain the set of (k_i, k_d) values for which the roots of $\delta_r(z)$ and $\delta_i(z)$ interlace for a given fixed value of k_p. Notice that all these regions are half planes with their boundaries being lines with positive slopes m_j. This completes the proof of the lemma. ■

The inequalities given by (8.34) represent half planes in the space of k_i and k_d. Their boundaries are given by lines with the following equations:

$$k_d = m_j k_i + b_j \quad \text{for } j = 1, 2, 3, \ldots .$$

The focus of the remainder of this section will be to determine the intersection of this countably infinite number of half planes in a computationally tractable way. To this end, let us denote by w_j the k_i-coordinate of the intersection of the line $k_d = m_j k_i + b_j$, $j = 1, 2, 3, ...$, with the line $k_d = \frac{T}{k}$. Using (8.35) and (8.36), it can be shown that

$$w_j = \frac{z_j}{kL} \left[\sin(z_j) + \frac{T}{L} z_j (\cos(z_j) + 1) \right] \,. \tag{8.39}$$

We now state a lemma that describes the behavior of the parameters b_j defined in (8.36) and w_j defined in (8.39) for $k_p \in (k_l, -\frac{1}{k})$. The proof of this lemma is long and technical and is relegated to Appendix B.

Lemma 8.9 *If* $\frac{1}{k} \left[\frac{T}{L} \alpha_1 \sin(\alpha_1) - \cos(\alpha_1) \right] < k_p < -\frac{1}{k}$ *where* α_1 *is the solution of the equation*

$$\tan(\alpha) = -\frac{T}{T + L} \alpha$$

in the interval $(0,\pi)$ *or* $\alpha_1 = \frac{\pi}{2}$ *if* $|\frac{T}{L}| = 1$, *then*

$$\begin{array}{ll} (i) & b_j < b_{j+2} < -\frac{T}{k} \text{ for odd values of } j \\ (ii) & b_j > b_{j+2} > \frac{T}{k} \text{ for even values of } j \\ (iii) & w_j < w_{j+2} < 0 \text{ for even values of } j \\ (iv) & b_1 > b_2 \ , \ w_1 < w_2 \ . \end{array}$$

We are now ready to state the main result of this section.

Theorem 8.2 *A necessary and sufficient condition for the existence of a stabilizing PID controller for the open-loop unstable plant (8.1) is* $|\frac{T}{L}| > 0.5$. *If this condition is satisfied, then the range of* k_p *values for which a given open-loop unstable plant, with transfer function* $G(s)$ *as in (8.1), can be stabilized using a PID controller is given by*

$$\frac{1}{k}\left[\frac{T}{L}\alpha_1 \sin(\alpha_1) - \cos(\alpha_1)\right] < k_p < -\frac{1}{k} \tag{8.40}$$

where α_1 *is the solution of the equation*

$$\tan(\alpha) = -\frac{T}{T+L}\alpha \tag{8.41}$$

in the interval $(0,\pi)$. *In the special case of* $|\frac{T}{L}| = 1$, *we have* $\alpha_1 = \frac{\pi}{2}$. *For* k_p *values outside this range, there are no stabilizing PID controllers. Moreover, the complete stabilizing region is given by (see Fig. 8.15):*

> *For each* $k_p \in \left(k_l := \frac{1}{k}\left[\frac{T}{L}\alpha_1 \sin(\alpha_1) - \cos(\alpha_1)\right], -\frac{1}{k}\right)$, *the cross-section of the stabilizing region in the* (k_i, k_d) *space is the quadrilateral* Q.

Proof. To ensure the stability of the quasi-polynomial $\delta^*(s)$, we need to check the two conditions given in Theorem 5.5.

Step 1. We first check condition 2 of Theorem 5.5:

$$E(\omega_o) = \delta_i^{'}(\omega_o)\delta_r(\omega_o) - \delta_i(\omega_o)\delta_r^{'}(\omega_o) > 0$$

for some ω_o in $(-\infty,\infty)$. Again we take $\omega_o = 0$, so $z_o = 0$. Thus $\delta_i(z_o) = 0$ and $\delta_r(z_o) = kk_i$ and we have

$$E(z_o) = \left(\frac{kk_p + 1}{L}\right)(kk_i) \ .$$

Recall $k > 0$ and $L > 0$. Thus if we pick

$$k_i > 0 \text{ and } k_p > -\frac{1}{k}$$

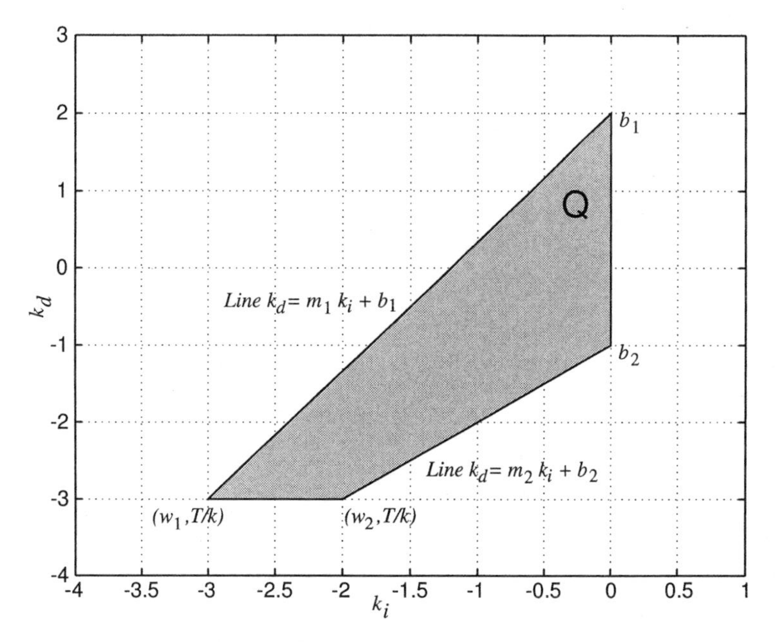

FIGURE 8.15. The stabilizing region of (k_i, k_d) for $k_l < k_p < -\frac{1}{k}$.

or

$$k_i < 0 \text{ and } k_p < -\frac{1}{k}$$

we have $E(z_o) > 0$. Notice that from these conditions we can safely discard $k_p = -\frac{1}{k}$ from the set of k_p values for which a stabilizing PID controller can be found.

Step 2. The second step is to check condition 1 of Theorem 5.5: $\delta_r(z)$ and $\delta_i(z)$ have only simple real roots and they interlace. From Lemma 8.6 we know that for $\frac{T}{L} < -0.5$, the roots of $\delta_i(z)$ are all real if and only if the parameter k_p is inside the range $\left(\frac{1}{k}\left[\frac{T}{L}\alpha_1 \sin(\alpha_1) - \cos(\alpha_1)\right], -\frac{1}{k}\right)$, where α_1 is the solution of the equation

$$\tan(\alpha) = -\frac{T}{T+L}\alpha$$

in the interval $(0, \pi)$. Since for $\frac{T}{L} \geq -0.5$ the roots of the imaginary part are not all real, we conclude that for the existence of a stabilizing PID controller, the condition $\frac{T}{L} < -0.5$ must necessarily be satisfied. From Lemma 8.8, interlacing the roots of $\delta_r(z)$ and $\delta_i(z)$ leads to the following set of inequalities:

$$\begin{aligned} k_i &< 0 \\ k_d &< m_1 k_i + b_1 \\ k_d &> m_2 k_i + b_2 \end{aligned}$$

$$
\begin{aligned}
k_d &< m_3 k_i + b_3 \\
k_d &> m_4 k_i + b_4 \\
&\vdots
\end{aligned}
$$

We now show that for $k_l < k_p < -\frac{1}{k}$ all these regions have a nonempty intersection. Notice first that the slopes m_j of the boundary lines of these regions decrease with z_j. Moreover, in the limit, we have

$$\lim_{j \to \infty} m_j = 0 \, .$$

Using this fact and Lemma 8.9 we get the intersection shown in Fig. 8.15. Finally we note that for values of k_p in the range $(k_l, -\frac{1}{k})$, the interlacing property and the fact that all the roots of $\delta_i(z)$ are real can be used in Theorem 5.6 to guarantee that $\delta_r(z)$ also has only real roots. Thus for values of k_p inside this range there is a solution to the PID stabilization problem for a first-order open-loop unstable plant with time delay. ■

Remark 8.2 *It is not difficult to see that when $\frac{T}{L} \leq -1$ there is always a solution to (8.41) in the interval $(0, \pi)$. However, when $\frac{T}{L} > -1$ we have two situations to consider. These situations are illustrated in Fig. 8.16 where the terms $\tan(\alpha)$ and $-\frac{T}{T+L}\alpha$ involved in (8.41) are plotted. Figure 8.16(a) corresponds to the case where $-1 < \frac{T}{L} < -0.5$. Figure 8.16(b) corresponds to the case where $-0.5 \leq \frac{T}{L} < 0$. As can be seen from Fig. 8.16(a), there is always a solution to (8.41) in the interval $(0, \pi)$. However, in the case of Fig. 8.16(b), we see that there is no solution in this open interval. Thus for this situation, the parameter α_1 does not exist and neither does k_l. However, as pointed out earlier, this corresponds to the case where no stabilizing PID controller exists.*

A similar algorithm to the one presented in the previous section can now be developed to solve the PID stabilization problem of an open-loop unstable plant. We only need to sweep the parameter k_p over the interval proposed by Theorem 8.2 and use Fig. 8.15 to find the stabilizing region of (k_i, k_d) values at each admissible value of k_p.

Example 8.4 *Consider a process described by the differential equation*

$$\frac{dy(t)}{dt} = 0.25y(t) - 0.25u(t - 0.8) \, .$$

This process can be described by the transfer function $G(s)$ in (8.23) with the following parameters: $k = 1$, $T = -4$, and $L = 0.8$ seconds. Since the system is open-loop unstable we use Theorem 8.2 to find the range of k_p values for which a solution to the PID stabilization problem exists. Since $|\frac{T}{L}| = 5 > 0.5$, we can proceed to compute $\alpha_1 \in (0, \pi)$ satisfying the following equation:

$$\tan(\alpha) = -1.25\alpha \, .$$

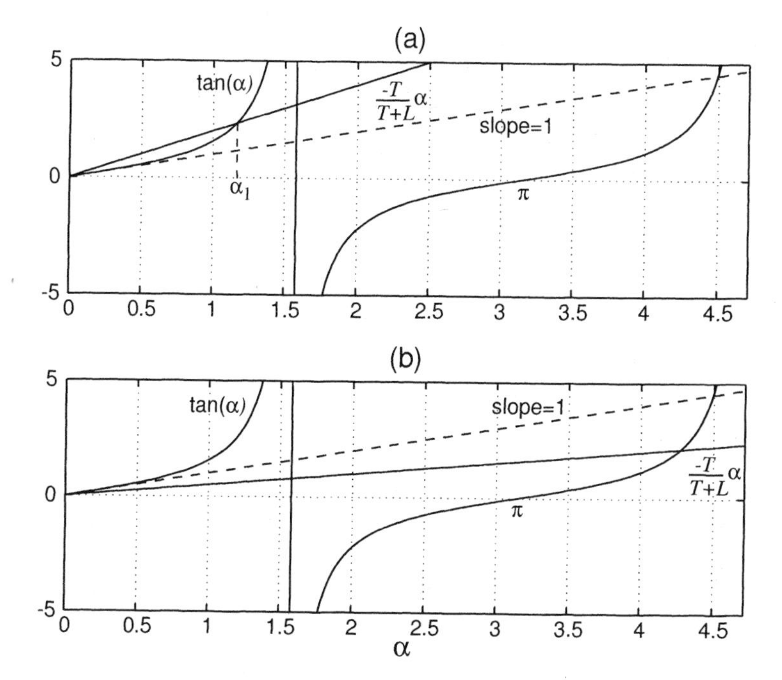

FIGURE 8.16. Cases involved in determining the parameter α_1 when $\frac{T}{L} > -1$.

Solving this equation we obtain $\alpha_1 = 1.9586$. *Thus from (8.40) the range of* k_p *values is given by*

$$-8.6876 < k_p < -1 \,.$$

We now sweep over the above range of k_p *values and determine the stabilizing set of* (k_i, k_d) *values at each stage. These regions are sketched in Fig. 8.17.* △

8.5 Notes and References

The characterization of all stabilizing PID controllers for a given first-order plant with time delay was developed by Silva, Datta, and Bhattacharyya [39]. To the best of the authors' knowledge, this is the first time that such a characterization was provided in the literature. An excellent account of PID theory and design for first-order plants with time delay can be found in [2]. The importance of PID controllers in modern industry is documented in the survey conducted by the Japan Electric Measuring Instrument Manufacturers' Association [52].

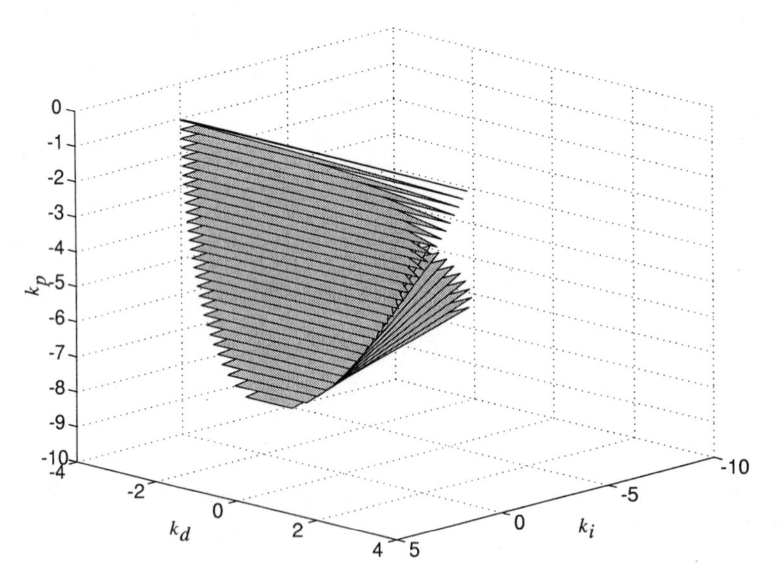

FIGURE 8.17. The stabilizing region of (k_p, k_i, k_d) values for the PID controller in Example 8.4.

9

Control System Design Using the PID Controller

In this chapter we present some tools that are useful when designing a PI or a PID controller for a first-order system with time delay. These tools show the importance of knowing the set of controller parameter values that stabilize the closed-loop system derived in the previous chapters. The chapter also provides a solution to the problem of robustly stabilizing a given delay-free interval plant family using the PID controller.

9.1 Introduction

In general, the task of a controller is to maintain a desired system performance while coping with possible system disturbances. In the case of the PID controller, to achieve a desired performance, the user needs to select carefully the amount of each control action: proportional, integral, and derivative. Oscillation and instability can occur if the parameters are chosen incorrectly. In an industrial situation this can lead to material loss and destruction of equipment. Setting the parameters of a PID controller is an important step in the design of PID control systems.

In this chapter we discuss different approaches for designing PI and PID controllers for first-order plants with time delay. As will be seen later, it is important for the designer to know *a priori* the set of stabilizing controller parameter values. We also discuss a technique for synthesizing PID controllers that simultaneously stabilize a given delay-free interval plant family. In Section 9.2, we recall some relevant results from the area of parametric robust control and use them to provide a constructive procedure for obtaining all P, PI, and PID controllers that stabilize a given delay-free interval plant family. Section 9.3 considers the case of a system with unknown delay that requires the design of a robust PID controller. In Section 9.4 we study the problem of designing a resilient or nonfragile controller. Section 9.5 introduces time domain performance specifications into the PID design.

9.2 Robust Controller Design: Delay-Free Case

We start this section by briefly reviewing some results from the area of parametric robust control. We will only highlight those results that are most relevant to the subsequent development. For an exhaustive treatment, the reader is referred to [5].

Definition 9.1 *Consider the set $\mathcal{F}$ of all real polynomials of degree n of the form*

$$p(s) = p_0 + p_1 s + p_2 s^2 + \cdots + p_n s^n$$

where the coefficients vary in independent intervals

$$p_0 \in [x_0,\ y_0],\ p_1 \in [x_1,\ y_1],\ \ldots,\ p_n \in [x_n,\ y_n],\ 0 \notin [x_n,\ y_n].$$

Such a set of polynomials is called an interval polynomial.

We next state Kharitonov's celebrated theorem, which provides a necessary and sufficient condition for the Hurwitz stability of such an interval polynomial.

Theorem 9.1 (Kharitonov's Theorem) *Every polynomial in the interval family $\mathcal{F}$ is Hurwitz if and only if the following four polynomials called Kharitonov polynomials are Hurwitz:*

$$\begin{aligned}
K^1(s) &= x_0 + x_1 s + y_2 s^2 + y_3 s^3 + x_4 s^4 + x_5 s^5 + y_6 s^6 + \cdots \\
K^2(s) &= x_0 + y_1 s + y_2 s^2 + x_3 s^3 + x_4 s^4 + y_5 s^5 + y_6 s^6 + \cdots \\
K^3(s) &= y_0 + x_1 s + x_2 s^2 + y_3 s^3 + y_4 s^4 + x_5 s^5 + x_6 s^6 + \cdots \\
K^4(s) &= y_0 + y_1 s + x_2 s^2 + x_3 s^3 + y_4 s^4 + y_5 s^5 + x_6 s^6 + \cdots .
\end{aligned}$$

We now proceed to state a generalization of Kharitonov's Theorem that will play an important role in the sequel. However, before stating this theorem, we need to introduce some notation. Let m be an arbitrary integer and let $\bar{P}(s) = [P_1(s),\ P_2(s),\ \ldots,\ P_m(s)]$ be an m-tuple of real polynomials where each component $P_i(s)$ $i = 1, 2, \ldots, m$ is an interval polynomial. Consequently, with each $P_i(s)$, we can associate its four Kharitonov polynomials $K_i^1(s)$, $K_i^2(s)$, $K_i^3(s)$, and $K_i^4(s)$. We denote by $\mathcal{K}_m$ the set of m-tuples obtained as follows. For every fixed integer i between 1 and m set

$$P_i(s) = K_i^k(s),\ \text{for some } k = 1,\ 2,\ 3,\ 4.$$

Clearly there are at most 4^m distinct elements in $\mathcal{K}_m$. In addition, we define a family of m-tuples called *generalized Kharitonov segments* as follows. For any fixed integer l between 1 and m, set

$$P_i(s) = K_i^k(s), \text{ for } i \neq l \text{ and for some } k = 1, 2, 3, 4$$

and for $i = l$, suppose that $P_l(s)$ varies in one of the four segments

$$\begin{gathered}[K_l^1(s),\ K_l^2(s)]\\ [K_l^1(s),\ K_l^3(s)]\\ [K_l^2(s),\ K_l^4(s)]\\ [K_l^3(s),\ K_l^4(s)]\,.\end{gathered}$$

By the segment $[K_l^1(s),\ K_l^2(s)]$, we mean the set of all convex combinations of the form

$$(1-\lambda)K_l^1(s) + \lambda K_l^2(s),\ \lambda \in [0,\ 1].$$

There are at most $m4^m$ distinct generalized Kharitonov segments and we will denote by S_m the family of all these m-tuples. With these preliminaries, we now state the theorem of Chapellat and Bhattacharyya.

Theorem 9.2 (Generalized Kharitonov Theorem) [5]

(1) Given an m-tuple of fixed real or complex polynomials $[F_1(s),\ F_2(s),\ \ldots,\ F_m(s)]$, the polynomial family

$$P_1(s)F_1(s) + P_2(s)F_2(s) + \cdots + P_m(s)F_m(s) \qquad (9.1)$$

is Hurwitz stable if and only if all the one-parameter polynomial families that result from replacing $[P_1(s),\ P_2(s),\ \ldots, P_m(s)]$ in the above expression by the elements of S_m are Hurwitz stable.

(2) If the polynomials $F_i(s)$ are real and of the form $F_i(s) = s^{t_i}(a_i s + b_i)U_i(s)Q_i(s)$ where $t_i \geq 0$ is an arbitrary integer, a_i and b_i are arbitrary real numbers, $U_i(s)$ is an anti-Hurwitz polynomial, and $Q_i(s)$ is an even or odd polynomial, then it is sufficient that (9.1) be Hurwitz stable with the P_is replaced by the elements of $\mathcal{K}_m$.

(3) If the F_is are complex and

$$\frac{d}{d\omega}\arg[F_i(j\omega)]\ \leq\ 0\ \ \forall i = 1,\ 2,\ \ldots,\ m$$

then it is sufficient that (9.1) be Hurwitz stable with the P_is replaced by the elements of $\mathcal{K}_m$.

9.2.1 Robust Stabilization Using a Constant Gain

In this subsection, we consider the problem of characterizing all constant gain controllers that stabilize a given delay-free interval plant. The key idea is to use the results of Chapter 3 in conjunction with the Generalized Kharitonov Theorem.

Now let $\mathcal{G}(s)$ be an interval plant:

$$\mathcal{G}(s) = \frac{\mathcal{N}(s)}{\mathcal{D}(s)}$$

$$\begin{aligned}\mathcal{N}(s) &= a_0 + a_1 s + a_2 s^2 + \cdots + a_m s^m \\ \mathcal{D}(s) &= b_0 + b_1 s + b_2 s^2 + \cdots + b_n s^n\end{aligned}$$

where $n \geq m$, $a_m \neq 0$, $b_n \neq 0$, and the coefficients of $\mathcal{N}(s)$ and $\mathcal{D}(s)$ vary in independent intervals, i.e.,

$$\begin{aligned}&a_0 \in [\underline{a}_0,\, \overline{a}_0],\ a_1 \in [\underline{a}_1,\, \overline{a}_1], \ldots,\ a_m \in [\underline{a}_m,\, \overline{a}_m] \\ &b_0 \in [\underline{b}_0,\, \overline{b}_0],\ b_1 \in [\underline{b}_1,\, \overline{b}_1], \ldots,\ b_n \in [\underline{b}_n,\, \overline{b}_n].\end{aligned}$$

The controller $C(s)$ in question is a constant gain, i.e.,

$$C(s) = k_c.$$

The family of closed-loop characteristic polynomials $\Delta(s,\, k_c)$ is given by

$$\Delta(s,\, k_c) = \mathcal{D}(s) + k_c \mathcal{N}(s).$$

The problem of characterizing all constant gain stabilizers for an interval plant is to determine all the values of k_c for which the family of closed-loop characteristic polynomials $\Delta(s,\, k_c)$ is Hurwitz.

Let $\mathcal{N}^i(s)$, $\mathcal{D}^i(s)$, $i = 1, 2, 3, 4$, be the Kharitonov polynomials corresponding to $N(s)$ and $D(s)$, respectively, and let $\mathcal{G}_K(s)$ denote the family of 16 vertex plants defined as

$$\mathcal{G}_K(s) = \left\{ G_{ij}(s) \mid G_{ij}(s) = \frac{\mathcal{N}^i(s)}{\mathcal{D}^j(s)}, i = 1, 2, 3, 4,\ j = 1, 2, 3, 4 \right\}.$$

The closed-loop characteristic polynomial for each of these vertex plants $G_{ij}(s)$ is denoted by $\delta_{ij}(s, k_c)$ and is defined as

$$\delta_{ij}(s, k_c) = \mathcal{D}^j(s) + k_c \mathcal{N}^i(s).$$

We can now state the following theorem, which characterizes all constant gain stabilizers for a delay-free interval plant.

Theorem 9.3 *Let $\mathcal{G}(s)$ be a delay-free interval plant as defined above. Then the entire family $\mathcal{G}(s)$ is stabilizable by a constant gain k_c if and only if the following conditions hold:*

(i) Each $G_{ij}(s) \in \mathcal{G}_K(s)$ is stabilizable by a constant gain

(ii) $\mathcal{K} \neq \emptyset$, where $\mathcal{K} = \cap_{i=1,\dots,4,\ j=1,\dots,4} K_{ij}$, and K_{ij} is the set of all stabilizing gain values for $G_{ij}(s)$[1].

Furthermore the set of all stabilizing gain values for the entire family $\mathcal{G}(s)$ is precisely given by $\mathcal{K}$.

Proof. Now we have

$$\begin{aligned}\Delta(s,\, k_c) &= \mathcal{D}(s) + k_c \mathcal{N}(s) \\ &= F_1(s)\mathcal{D}(s) + F_2(s)\mathcal{N}(s)\end{aligned}$$

where $F_1(s) = 1$ and $F_2(s) = k_c$. Using Theorem 9.2 (2), it follows that the entire family $\Delta(s, k_c)$ is Hurwitz stable if and only if $\delta_{ij}(s, k_c)$, $i = 1, 2, 3, 4$, $j = 1, 2, 3, 4$ are all Hurwitz. Therefore, the entire family $\mathcal{G}(s)$ is stabilized by a constant gain k_c if and only if every element of $\mathcal{G}_K(s)$ is simultaneously stabilized by k_c. Thus we can use the results of Section 3.2 to find out all the constant gain stabilizers for each member of $\mathcal{G}_K(s)$ and then take their intersection to obtain all constant gain stabilizers for the given interval plant. This completes the proof. ■

We now present a simple example to illustrate the detailed calculations involved in coming up with all constant gain stabilizers for an interval plant.

Example 9.1 *Consider the interval plant*

$$\mathcal{G}(s) = \frac{\mathcal{N}(s)}{\mathcal{D}(s)},$$

$$\begin{aligned}\textit{where } \mathcal{N}(s) &= a_0 + a_1 s + a_2 s^2 \\ \mathcal{D}(s) &= b_0 + b_1 s + b_2 s^2 + b_3 s^3 + b_4 s^4\end{aligned}$$

with

$$\begin{aligned}& a_2 \in [1,\ 1],\ a_1 \in [1,\ 2],\ a_0 \in [1,\ 2] \\ \textit{and}\quad & b_4 \in [1,\ 1],\ b_3 \in [3,\ 4],\ b_2 \in [4,\ 4],\ b_1 \in [5,\ 8],\ b_0 \in [6,\ 7].\end{aligned}$$

Then the Kharitonov polynomials corresponding to $\mathcal{N}(s)$ and $\mathcal{D}(s)$ are

$$\begin{aligned}&\mathcal{N}^1(s) = s^2 + s + 1,\ \mathcal{N}^2(s) = s^2 + 2s + 1 \\ &\mathcal{N}^3(s) = s^2 + s + 2,\ \mathcal{N}^4(s) = s^2 + 2s + 2 \\ &\mathcal{D}^1(s) = s^4 + 4s^3 + 4s^2 + 5s + 6,\ \mathcal{D}^2(s) = s^4 + 3s^3 + 4s^2 + 8s + 6 \\ &\mathcal{D}^3(s) = s^4 + 4s^3 + 4s^2 + 5s + 7,\ \mathcal{D}^4(s) = s^4 + 3s^3 + 4s^2 + 8s + 7.\end{aligned}$$

[1]Note that for any particular $G_{ij}(s)$, the set of all stabilizing feedback gains K_{ij} can be determined using the results of Section 3.2, specifically Theorem 3.1.

Furthermore,

$$\mathcal{G}_K(s) = \left\{ G_{ij}(s) \mid G_{ij}(s) = \frac{\mathcal{N}^i(s)}{\mathcal{D}^j(s)}, i = 1,2,3,4,\ j = 1,2,3,4 \right\}.$$

Using the results of Section 3.2, specifically Theorem 3.1, the sets K_{ij} *corresponding to* $G_{ij}(s)$, $i = 1,2,3,4,\ j = 1,2,3,4$ *are*

$$\begin{aligned}
&K_{11} = (2.3885, \infty),\ K_{12} = (1.5584, \infty),\\
&K_{13} = (3.0, \infty),\ K_{14} = (2.0523, \infty),\\
&K_{21} = (1.7749, \infty),\ K_{22} = (2.0, \infty),\\
&K_{23} = (2.2720, \infty),\ K_{24} = (2.5584, \infty),\\
&K_{31} = (-3.0, -2.8297) \cup (4.8297, \infty),\ K_{32} = (2.8541, \infty),\\
&K_{33} = (-3.5, -3.4721) \cup (5.4721, \infty),\ K_{34} = (3.4686, \infty),\\
&K_{41} = (3.2016, \infty),\ K_{42} = (-3.0, -2.8541) \cup (3.8541, \infty),\\
&K_{43} = (3.7749, \infty),\ K_{44} = (-3.5, -3.4686) \cup (4.4686, \infty).
\end{aligned}$$

Therefore the set of all k_c *values that stabilize the entire family* $\mathcal{G}(s)$ *is given by*

$$\begin{aligned}
\mathcal{K} &= \cap_{i=1,\ldots,4,\ j=1,\ldots,4} K_{ij}\\
&= (5.4721, \infty).
\end{aligned}$$

△

9.2.2 Robust Stabilization Using a PI Controller

In this subsection, we consider the problem of characterizing all PI controllers that stabilize a given delay-free interval plant $\mathcal{G}(s) = \frac{\mathcal{N}(s)}{\mathcal{D}(s)}$ where $\mathcal{N}(s)$, $\mathcal{D}(s)$ are as defined in Section 9.2.1. Since the controller $C(s)$ is now given by

$$C(s) = k_p + \frac{k_i}{s},$$

the family of closed-loop characteristic polynomials $\Delta(s, k_p, k_i)$ becomes

$$\Delta(s, k_p, k_i) = s\mathcal{D}(s) + (k_i + k_p s)\mathcal{N}(s).$$

The problem of characterizing all stabilizing PI controllers is to determine all the values of k_p and k_i for which the entire family of closed-loop characteristic polynomials $\Delta(s, k_p, k_i)$ is Hurwitz.

Let $\mathcal{N}^i(s)$, $i = 1,2,3,4$ and $\mathcal{D}^j(s)$, $j = 1,2,3,4$ be the Kharitonov polynomials corresponding to $\mathcal{N}(s)$ and $\mathcal{D}(s)$, respectively, and let $\mathcal{G}_K(s)$ denote the family of 16 vertex plants:

$$\mathcal{G}_K(s) = \left\{ G_{ij}(s) \mid G_{ij}(s) = \frac{\mathcal{N}^i(s)}{\mathcal{D}^j(s)}, i = 1,2,3,4,\ j = 1,2,3,4 \right\}.$$

The closed-loop characteristic polynomial for each of these vertex plants $G_{ij}(s)$ is denoted by $\delta_{ij}(s, k_p, k_i)$ and is defined as

$$\delta_{ij}(s, k_p, k_i) = s\mathcal{D}^j(s) + (k_i + k_p s)\mathcal{N}^i(s).$$

We can now state the following theorem on stabilizing a delay-free interval plant using a PI controller. The proof is essentially the same as that of Theorem 9.3 and is therefore omitted.

Theorem 9.4 *Let $\mathcal{G}(s)$ be an interval plant. Then the entire family $\mathcal{G}(s)$ is stabilized by a particular PI controller if and only if each $G_{ij}(s) \in \mathcal{G}_K(s)$ is stabilized by that same PI controller.*

In view of the above theorem, we can now use the results of Section 3.3 to obtain a characterization of all PI controllers that stabilize the delay-free interval family $\mathcal{G}(s)$. As in Section 3.3, for any fixed k_p, we can solve the constant gain stabilization problem for each $G_{ij}(s)$ to determine the stabilizing set of k_i for that particular $G_{ij}(s)$. We denote $KI_{ij}(k_p)$ to be the set of stabilizing k_i corresponding to $G_{ij}(s)$ and a fixed k_p. With such a fixed k_p, the set of all stabilizing $(k_p,\ k_i)$ values for the entire $\mathcal{G}_K(s)$, denoted by $\mathcal{S}_{k_p}$ is given by

$$\mathcal{S}_{k_p} = \{(k_p,\ k_i) | k_i \in \cap_{i=1,\dots,4,\ j=1,\dots,4} KI_{ij}(k_p)\}.$$

The set of all stabilizing $(k_p,\ k_i)$ values for the entire family $\mathcal{G}(s)$ can now be found by simply sweeping over k_p. From the results of Section 3.3, we know that for a given fixed plant, the range of k_p values for which a stabilizing k_i may exist can usually be narrowed down using some necessary conditions derived from roots locus ideas. Let KP_{ij} be the set of k_p values satisfying such a necessary condition for $G_{ij}(s)$. Define

$$\mathcal{KP} = \cap_{i=1,\dots,4,\ j=1,\dots,4} KP_{ij}.$$

Then $k_p \in \mathcal{KP}$ is a necessary condition that must be satisfied for any $(k_p,\ k_i)$ that stabilizes the entire family $\mathcal{G}(s)$. Thus by sweeping over all $k_p \in \mathcal{KP}$ and solving constant gain stabilization problems for the 16 vertex plants at each stage, we can determine the set of all stabilizing (k_p, k_i) values for the entire family $\mathcal{G}(s)$.

We now present a simple example to illustrate the detailed calculations involved in determining all the stabilizing (k_p, k_i) values for a given delay-free interval plant $\mathcal{G}(s)$.

Example 9.2 *Consider the interval plant*

$$\mathcal{G}(s) = \frac{\mathcal{N}(s)}{\mathcal{D}(s)},$$

$$\begin{aligned} \textit{where } \mathcal{N}(s) &= a_0 + a_1 s + a_2 s^2 + a_3 s^3 + a_4 s^4 \\ \mathcal{D}(s) &= b_0 + b_1 s + b_2 s^2 + b_3 s^3 + b_4 s^4 + b_5 s^5 \end{aligned}$$

with

$$a_4 \in [1,\ 1],\ a_3 \in [2,\ 3],\ a_2 \in [39,\ 41],\ a_1 \in [48,\ 50],\ a_0 \in [-6,\ -3]$$

and $b_5 \in [1,\ 1],\ b_4 \in [2,\ 3],\ b_3 \in [31,\ 32],\ b_2 \in [35,\ 38],\ b_1 \in [49,\ 51],$ $b_0 \in [97,\ 101].$

Then the Kharitonov polynomials corresponding to $\mathcal{N}(s)$ and $\mathcal{D}(s)$ are

$$\begin{aligned}
\mathcal{N}^1(s) &= s^4 + 3s^3 + 41s^2 + 48s - 6 \\
\mathcal{N}^2(s) &= s^4 + 2s^3 + 41s^2 + 50s - 6 \\
\mathcal{N}^3(s) &= s^4 + 3s^3 + 39s^2 + 48s - 3 \\
\mathcal{N}^4(s) &= s^4 + 2s^3 + 39s^2 + 50s - 3 \\
\mathcal{D}^1(s) &= s^5 + 2s^4 + 32s^3 + 38s^2 + 49s + 97 \\
\mathcal{D}^2(s) &= s^5 + 2s^4 + 31s^3 + 38s^2 + 51s + 97 \\
\mathcal{D}^3(s) &= s^5 + 3s^4 + 32s^3 + 35s^2 + 49s + 101 \\
\mathcal{D}^4(s) &= s^5 + 3s^4 + 31s^3 + 35s^2 + 51s + 101.
\end{aligned}$$

For a fixed k_p, for instance $k_p = 2$, we can solve constant gain stabilization problems to determine the set of stabilizing $KI_{ij}(2)$ corresponding to each $G_{ij}(s)$.

$$\begin{aligned}
&KI_{11}(2) = (-1.4993, 0),\ KI_{12}(2) = (-1.5049, 0) \\
&KI_{13}(2) = (-1.5673, 0),\ KI_{14}(2) = (-1.5734, 0) \\
&KI_{21}(2) = (-1.4610, 0),\ KI_{22}(2) = (-1.4656, 0) \\
&KI_{23}(2) = (-1.5281, 0),\ KI_{24}(2) = (-1.5332, 0) \\
&KI_{31}(2) = (-1.7356, 0),\ KI_{32}(2) = (-1.7394, 0) \\
&KI_{33}(2) = (-1.8100, 0),\ KI_{34}(2) = (-1.8142, 0) \\
&KI_{41}(2) = (-1.6807, 0),\ KI_{42}(2) = (-1.6837, 0) \\
&KI_{43}(2) = (-1.7534, 0),\ KI_{44}(2) = (-1.7567, 0).
\end{aligned}$$

Since

$$\cap_{i=1,\dots,4,\ j=1,\dots,4} [KI_{ij}(2)] = (-1.4610,\ 0),$$

it follows that

$$\mathcal{S}_2 = \{(k_p,\ k_i) | kp = 2,\ k_i \in (-1.4610,\ 0)\}.$$

Furthermore, $KP_{ij}, i = 1, 2, 3, 4,\ j = 1, 2, 3, 4$ are given by

$$\begin{aligned}
&KP_{11} = (0.0010, 16.1667),\ KP_{12} = (-0.0055, 16.1667), \\
&KP_{13} = (0, 16.8333),\ KP_{14} = (0, 16.8333), \\
&KP_{21} = (0, 16.1667),\ KP_{22} = (0, 16.1667),
\end{aligned}$$

$$KP_{23} = (-1.7382, 16.8333),\ KP_{24} = (-1.6128, 17.9000),$$
$$KP_{31} = (0.0022, 32.3333),\ KP_{32} = (0.0011, 32.3333),$$
$$KP_{33} = (0, 33.6667),\ KP_{34} = (0, 33.6667),$$
$$KP_{41} = (0, 32.3333),\ KP_{42} = (0, 32.3333),$$
$$KP_{43} = (-2.0923, 33.6667),\ KP_{44} = (-1.9073, 33.6667).$$

Thus

$$\begin{aligned}\mathcal{KP} &= \cap_{i=1,\ldots,4,\ j=1,\ldots,4} KP_{ij} \\ &= (0.0022, 16.1667).\end{aligned}$$

Now by sweeping over all $k_p \in (0.0022, 16.1667)$ *and solving the constant gain stabilization problems for each of the* 16 *vertex plants at each stage, we obtained the stabilizing* (k_p, k_i) *values for the entire family* $\mathcal{G}(s)$*. This is sketched in Fig. 9.1.*

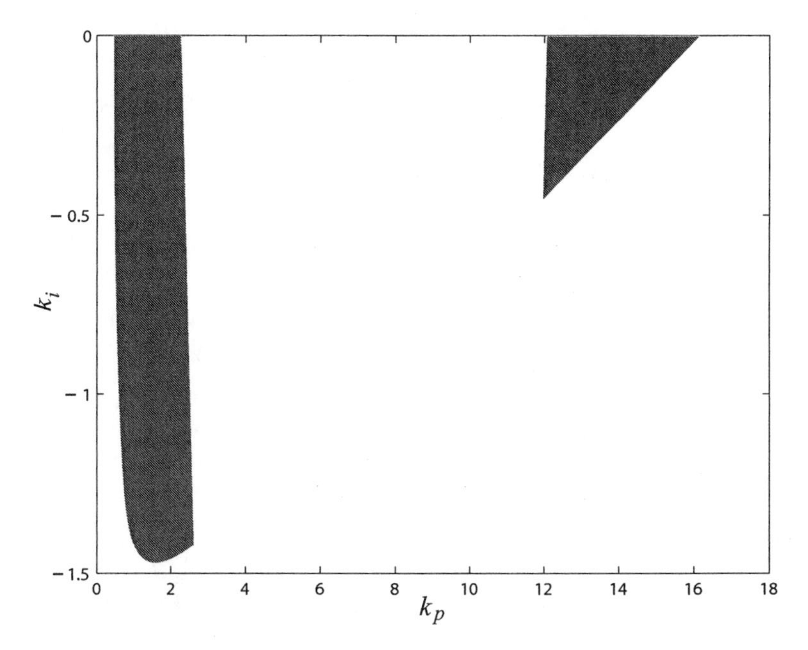

FIGURE 9.1. The stabilizing set of (k_p, k_i) values (Example 9.2).

△

9.2.3 *Robust Stabilization Using a PID Controller*

In this subsection, we consider the problem of characterizing all PID controllers that stabilize a given delay-free interval plant $\mathcal{G}(s) = \frac{\mathcal{N}(s)}{\mathcal{D}(s)}$ where

$\mathcal{N}(s)$, $\mathcal{D}(s)$ are as defined in Section 9.2.1. Since the controller $C(s)$ in question is now a PID controller, i.e., $C(s) = \frac{k_d s^2 + k_p s + k_i}{s}$, the family of closed-loop characteristic polynomials $\Delta(s, k_p, k_i, k_d)$ becomes

$$\Delta(s, k_p, k_i, k_d) = s\mathcal{D}(s) + (k_d s^2 + k_p s + k_i)\mathcal{N}(s).$$

The problem of characterizing all stabilizing PID controllers for the entire family $\mathcal{G}(s)$ is to determine all the values of k_p, k_i, and k_d for which the entire family of closed-loop characteristic polynomials $\Delta(s, k_p, k_i, k_d)$ is Hurwitz.

Let $\mathcal{N}^i(s)$, $i = 1, 2, 3, 4$ and $\mathcal{D}^j(s)$, $j = 1, 2, 3, 4$ be the Kharitonov polynomials corresponding to $\mathcal{N}(s)$ and $\mathcal{D}(s)$, respectively. Furthermore, let $\mathcal{NS}^i(s)$, $i = 1, 2, 3, 4$ be the four Kharitonov segments of $\mathcal{N}(s)$, where

$$\begin{aligned} \mathcal{NS}^1(s, \lambda) &= (1-\lambda)\mathcal{N}^1(s) + \lambda\mathcal{N}^2(s) \\ \mathcal{NS}^2(s, \lambda) &= (1-\lambda)\mathcal{N}^1(s) + \lambda\mathcal{N}^3(s) \\ \mathcal{NS}^3(s, \lambda) &= (1-\lambda)\mathcal{N}^2(s) + \lambda\mathcal{N}^4(s) \\ \mathcal{NS}^4(s, \lambda) &= (1-\lambda)\mathcal{N}^3(s) + \lambda\mathcal{N}^4(s) \end{aligned}$$

and $\lambda \in [0,\ 1]$. Let $\mathcal{G}_S(s)$ denote the family of 16 segment plants:

$$\begin{aligned} \mathcal{G}_S(s) &= \left\{ G_{ij}(s,\ \lambda) \mid G_{ij}(s,\ \lambda) = \frac{\mathcal{NS}^i(s,\ \lambda)}{\mathcal{D}^j(s)}, i = 1, 2, 3, 4,\ j = 1, 2, 3, 4, \right. \\ &\qquad \left. \lambda \in [0,\ 1] \right\}. \end{aligned}$$

The family of closed-loop characteristic polynomials for each segment plant $G_{ij}(s,\ \lambda)$ is denoted by $\delta_{ij}(s, k_p, k_i, k_d, \lambda)$ and is given by

$$\delta_{ij}(s, k_p, k_i, k_d, \lambda) = s\mathcal{D}^j(s) + (k_i + k_p s + k_d s^2)\mathcal{NS}^i(s, \lambda).$$

We can now state the following result on stabilizing a delay-free interval plant using a PID controller.

Theorem 9.5 *Let $\mathcal{G}(s)$ be a delay-free interval plant. Then the entire family $\mathcal{G}(s)$ is stabilized by a particular PID controller, if and only if each segment plant $G_{ij}(s, \lambda) \in \mathcal{G}_S(s)$ is stabilized by that same PID controller.*

Proof. We have

$$\begin{aligned} \Delta(s, k_p, k_i, k_d) &= s\mathcal{D}(s) + (k_i + k_p s + k_d s^2)\mathcal{N}(s) \\ &= F_1(s)\mathcal{D}(s) + F_2(s)\mathcal{N}(s) \end{aligned}$$

where

$$\begin{aligned} F_1(s) &= s \\ F_2(s) &= k_i + k_p s + k_d s^2. \end{aligned}$$

Using Theorem 9.2, it follows that the entire family $\Delta(s, k_p, k_i, k_d)$ is Hurwitz stable if and only if the entire family $\delta_{ij}(s, k_p, k_i, k_d, \lambda)$, $i = 1, 2, 3, 4$, $j = 1, 2, 3, 4$, $\lambda \in [0,\ 1]$ is Hurwitz. Hence we can conclude that the entire family $\mathcal{G}(s)$ is stabilized by a PID controller if and only if every element of $\mathcal{G}_S(s)$ is simultaneously stabilized by such a PID controller. ∎

In view of the above theorem, we can now use the results of Section 4.2 to obtain a computational characterization of all PID controllers that stabilize the interval family $\mathcal{G}(s)$. As in Section 4.2, for any fixed k_p and any fixed $\lambda^* \in [0, 1]$, we can solve a linear programming problem to determine the stabilizing set of (k_i, k_d) values for $G_{ij}(s,\ \lambda^*)$. Let this set be denoted by $\mathcal{S}_{(i,\ j,\ k_p,\ \lambda^*)}$. By keeping k_p fixed and letting λ^* vary in $[0, 1]$ we can determine the set of stabilizing (k_i, k_d) values for the entire segment plant $G_{ij}(s,\ \lambda)$. This set is denoted by $\mathcal{S}_{(i,\ j,\ k_p)}$ and is defined as

$$\mathcal{S}_{(i,\ j,\ k_p)} = \cap_{\lambda \in [0,\ 1]} \mathcal{S}_{(i,\ j,\ k_p,\ \lambda)}.$$

For a fixed k_p, the set of all stabilizing $(k_i,\ k_d)$ values for the entire set $\mathcal{G}_S(s)$, denoted by $\mathcal{S}_{k_p}$ is given by

$$\mathcal{S}_{k_p} = \cap_{i=1,2,3,4,\ j=1,2,3,4} \mathcal{S}_{(i,\ j,\ k_p)}\ .$$

The set of all stabilizing (k_p, k_i, k_d) values for the delay-free interval plant $\mathcal{G}(s)$ can now be found by simply sweeping over k_p and determining $\mathcal{S}_{k_p}$ at each stage. Once again the range of k_p values over which the sweeping has to be carried out for the entire $\mathcal{G}_S(s)$ can be *a priori* narrowed down by using root locus ideas. We now present a simple example to illustrate the detailed calculations involved in determining all the stabilizing (k_p, k_i, k_d) values for a given delay-free interval plant $\mathcal{G}(s)$.

Example 9.3 *Consider the interval plant*

$$\mathcal{G}(s) = \frac{\mathcal{N}(s)}{\mathcal{D}(s)},$$

$$\begin{aligned} \textit{where } \mathcal{N}(s) &= a_0 + a_1 s + a_2 s^2 \\ \mathcal{D}(s) &= b_0 + b_1 s + b_2 s^2 + b_3 s^3 + b_4 s^4 + b_5 s^5 \end{aligned}$$

with

$$\begin{aligned} & a_2 \in [1,\ 1],\ a_1 \in [-5,\ -4],\ a_0 \in [2,\ 4] \\ \textit{and}\quad & b_5 \in [1,\ 1],\ b_4 \in [3,\ 4],\ b_3 \in [5,\ 5],\ b_2 \in [7,\ 9],\ b_1 \in [8,\ 9] \\ & b_0 \in [-2,\ -1]. \end{aligned}$$

The Kharitonov polynomials corresponding to $\mathcal{N}(s)$ *and* $\mathcal{D}(s)$ *are*

$$\mathcal{N}^1(s) = s^2 - 5s + 2\mathcal{N}^2(s) = s^2 - 4s + 2$$

$$\begin{aligned}
\mathcal{N}^3(s) &= s^2 - 5s + 4\mathcal{N}^4(s) = s^2 - 4s + 4 \\
\mathcal{D}^1(s) &= s^5 + 3s^4 + 5s^3 + 9s^2 + 8s - 2 \\
\mathcal{D}^2(s) &= s^5 + 3s^4 + 5s^3 + 9s^2 + 9s - 2 \\
\mathcal{D}^3(s) &= s^5 + 4s^4 + 5s^3 + 7s^2 + 8s - 1 \\
\mathcal{D}^4(s) &= s^5 + 4s^4 + 5s^3 + 7s^2 + 9s - 1.
\end{aligned}$$

The Kharitonov segments and the family of 16 segment plants are defined as at the beginning of this subsection.

Now for a fixed k_p, for instance $k_p = 1.05$, sweeping over $\lambda \in [0,1]$ and using the results of Section 4.2, we obtained the stabilizing (k_i, k_d) values for $G_{11}(s,\ \lambda)$, i.e., $\mathcal{S}_{(1,\ 1,\ 1.05)}$ sketched in Fig. 9.2. In Fig. 9.2, for different values of $\lambda \in [0,1]$, the boundaries of the stabilizing regions $\mathcal{S}_{(1,\ 1,\ 1.05,\ \lambda)}$ are indicated using solid lines. The shaded portion is $\mathcal{S}_{(1,\ 1,\ 1.05)}$ which is the intersection of $\mathcal{S}_{(1,\ 1,\ 1.05,\ \lambda)}$, as λ varies over the interval $[0,\ 1]$. Repeating

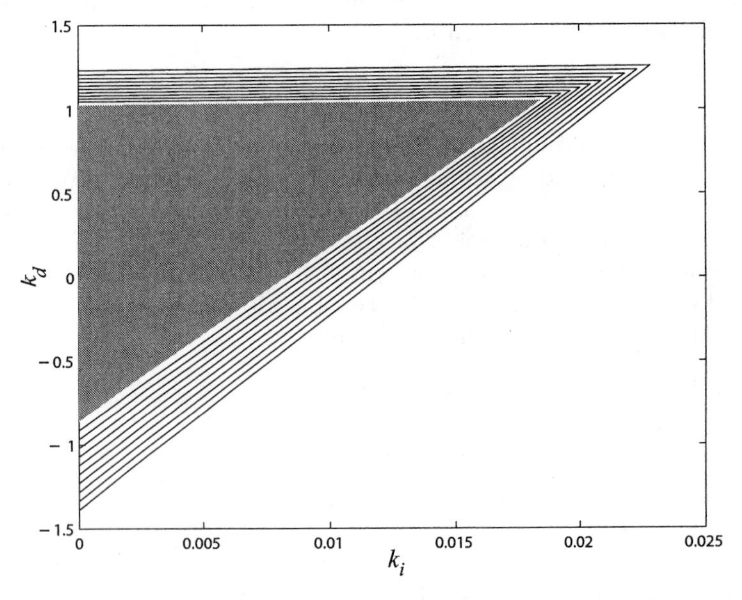

FIGURE 9.2. The stabilizing set of (k_i, k_d) values for $G_{11}(s,\ \lambda)$, $\lambda \in [0,\ 1]$ with $k_p = 1.05$ (Example 9.3).

the same procedure for the remaining 15 segment plants, we obtained the regions $\mathcal{S}_{(i,\ j,\ 1.05)}$, $i = 1,2,3,4$, $j = 1,2,3,4$ as indicated by the solid lines in Fig. 9.3. The shaded portion, which is their intersection, is the region $\mathcal{S}_{1.05}$.

Using root locus ideas, it was determined that a necessary condition for the existence of stabilizing (k_i, k_d) values for the entire family was that $k_p \in (1,\ 1.0869)$. Thus, by sweeping over $k_p \in (1,\ 1.0869)$ and repeating

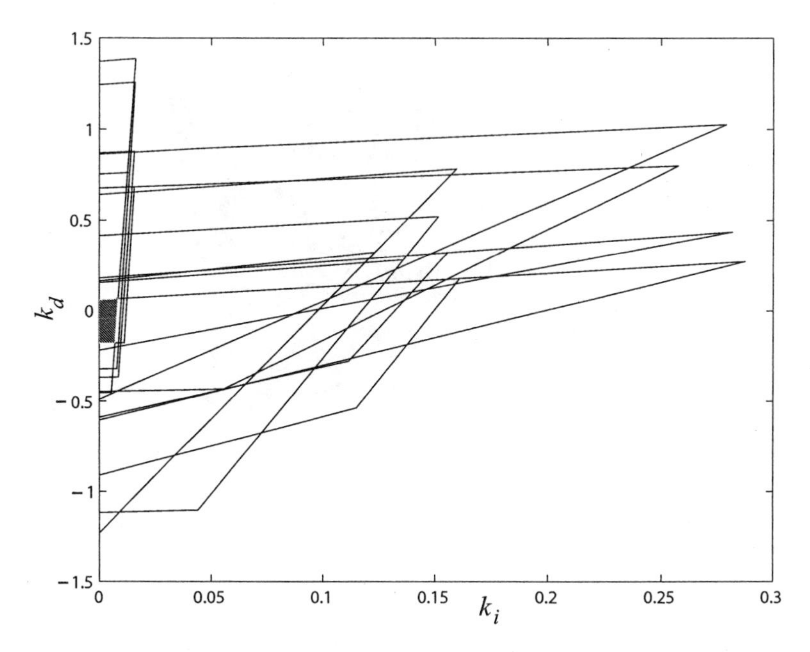

FIGURE 9.3. The stabilizing set of (k_i, k_d) values for $G_{ij}(s, \lambda)$, $\lambda \in [0, 1]$, $i = 1, 2, 3, 4$ and $j = 1, 2, 3, 4$ with $k_p = 1.05$ (Example 9.3).

the above procedure, we obtained the stabilizing set of (k_p, k_i, k_d) values sketched in Fig. 9.4.

$\triangle$

9.3 Robust Controller Design: Time-Delay Case

In this section we discuss the problem of stabilizing a first-order system with time delay, where the time delay of the system is unknown but lies inside a known interval. This generates a one-dimensional family of plants. As we will shortly see, it is sufficient to design a P, PI, or PID controller that stabilizes the system with the time delay equal to its upper bound. This guarantees that the entire plant family is stable.

We start by considering the plant family $\mathcal{G}(s)$:

$$\mathcal{G}(s) = \frac{k}{Ts+1} e^{-Ls} \tag{9.2}$$

where L is unknown and

$$L \in [L_1, L_2] \ .$$

For convenience we will focus on the case of an open-loop stable system, i.e., a system where $T > 0$.

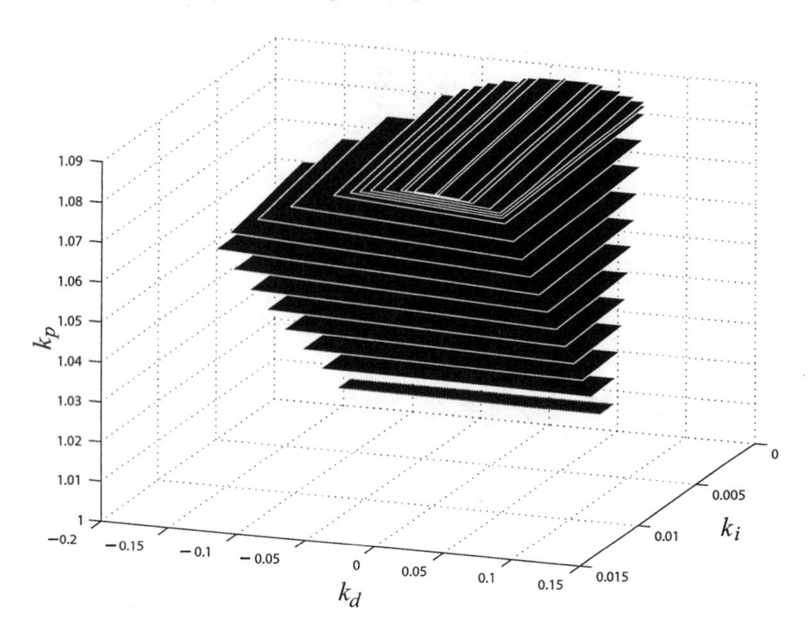

FIGURE 9.4. The stabilizing set of (k_p, k_i, k_d) values (Example 9.3).

9.3.1 Robust Stabilization Using a Constant Gain

In this case, the controller $C(s)$ is a constant gain, i.e.,

$$C(s) = k_c \ .$$

The problem of characterizing all constant gain controllers for the plant family $\mathcal{G}(s)$ is to determine all the values of the controller parameter k_c for which the family of first-order systems with time delay $L \in [L_1, L_2]$ is stable. From Theorem 6.1, for a fixed value $L^* \in [L_1, L_2]$, the range of stabilizing constant gain values is given by

$$-\frac{1}{k} < k_c < \frac{T}{kL^*}\sqrt{\alpha_1^{*2} + \frac{L^{*2}}{T^2}}$$

where α_1 is the solution of the equation

$$\tan(\alpha) = -\frac{T}{L^*}\alpha \tag{9.3}$$

in the interval $(\frac{\pi}{2}, \pi)$. Moreover, Lemma 6.1 asserts that the upper bound in the previous expression is a monotonically decreasing function of the time delay L of the system. Thus

$$L^* < L_2 \Rightarrow \frac{T}{kL^*}\sqrt{\alpha_1^{*2} + \frac{L^{*2}}{T^2}} > \frac{T}{kL_2}\sqrt{\alpha_1^2 + \frac{L_2^2}{T^2}}$$

for any $L^* \in [L_1, L_2]$. Notice that in the previous inequality, the two parameters α_1 appearing on the two sides are not the same as can be seen from (9.3) where L appears explicitly.
Thus it is not difficult to see that if we design for L_2, the set of resulting stabilizing constant gain controllers will stabilize any element of the plant family $\mathcal{G}(s)$.

9.3.2 Robust Stabilization Using a PI Controller

In this subsection, we consider the problem of determining the set of PI controllers that stabilize the plant family $\mathcal{G}(s)$ in (9.2). We consider here robustness with respect to the time-delay parameter, not with respect to the plant parameters. Since the controller $C(s)$ is now given by

$$C(s) = k_p + \frac{k_i}{s} \, ,$$

the family of closed-loop characteristic polynomials $\Delta(s, k_p, k_i)$ becomes

$$\Delta(s, k_p, k_i) = Ts^2 + s + (kk_p s + kk_i)e^{-Ls}$$

where $L \in [L_1, L_2]$. As in the previous subsection, our objective is to determine the values of k_p and k_i for which the entire family of closed-loop characteristic polynomials $\Delta(s, k_p, k_i)$ is stable.

We start by presenting the following result, which is based on the material introduced in Section 5.6.

Lemma 9.1 *Consider the system with transfer function (9.2). If a given PI controller stabilizes the delay-free system and the system with $L = L^* >$ 0, then the same PI controller stabilizes the system $\forall \ L \in [0, \ L^*]$.*

Proof. The idea of the proof is to follow the three-step procedure introduced in Section 5.6. This procedure allows us to analyze the behavior of the roots of $\Delta(s, k_p, k_i) = 0$ when the time delay L of the system increases from 0 to $+\infty$.

Let us denote by k_p^* and k_i^* some controller parameter values that stabilize the delay-free system and the system described by (9.2) with the time delay set to L^*. Also, we rewrite $\Delta(s, k_p, k_i)$ as follows:

$$\Delta(s, k_p, k_i) = d(s) + n(s)e^{-Ls}$$

where

$$\begin{aligned} d(s) &= Ts^2 + s \\ n(s) &= kk_p^* s + kk_i^* \, . \end{aligned}$$

Step 1. Stability at $L = 0$. This follows from our assumption concerning k_p^* and k_i^*. Note that in Theorem 7.1, we have imposed this (reasonable) requirement on any stabilizing PI controller.

Step 2. Increment L from 0 to an infinitesimally small and positive number. Since the degree of $d(s)$ is greater than the degree of $n(s)$ we conclude that all the new roots lie in the open LHP (see Section 5.6).
Step 3. Potential crossing points on the imaginary axis. First we determine $W(\omega^2)$ from (5.31):

$$W(\omega^2) = T^2\omega^4 + (1 - k^2k_p^{*2})\omega^2 - k^2k_i^{*2}\,. \tag{9.4}$$

Then

$$W'(\omega^2) = 2T^2\omega^2 + 1 - k^2k_p^{*2}\,.$$

The roots of $W(\omega^2)$ are given by the following expression:

$$\omega_1^2, \omega_2^2 = \frac{-(1 - k^2k_p^{*2}) \mp \sqrt{(1 - k^2k_p^{*2})^2 + 4T^2k^2k_i^{*2}}}{2T^2}\,,$$

which can be rewritten as

$$\omega_1^2, \omega_2^2 = -\frac{(1 - k^2k_p^{*2})}{2T^2}\left[1 \pm \sqrt{1 + \frac{4T^2k^2k_i^{*2}}{(1 - k^2k_p^{*2})^2}}\right]\,.$$

Clearly, the expression inside the square root is always greater than 1. Thus, we have

$$\begin{aligned} \omega_1^2 &= -\frac{(1 - k^2k_p^{*2})}{2T^2}\left[1 + \sqrt{1 + \frac{4T^2k^2k_i^{*2}}{(1 - k^2k_p^{*2})^2}}\right] \\ \Rightarrow \text{sgn}[\omega_1^2] &= -\text{sgn}\left[1 - k^2k_p^{*2}\right] \end{aligned} \tag{9.5}$$

and

$$\begin{aligned} \omega_2^2 &= -\frac{(1 - k^2k_p^{*2})}{2T^2}\left[1 - \sqrt{1 + \frac{4T^2k^2k_i^{*2}}{(1 - k^2k_p^{*2})^2}}\right] \\ \Rightarrow \text{sgn}[\omega_2^2] &= \text{sgn}\left[1 - k^2k_p^{*2}\right]\,. \end{aligned}$$

We are only interested in the positive roots of $W(\omega^2)$. From the previous expressions, it is clear that for a fixed value of the controller parameter $k_p \neq \frac{1}{k}$, there is only one positive root of $W(\omega^2)$ since ω_1^2 and ω_2^2 have opposite signs.

Suppose that k_p^* is such that ω_1^2 is the only positive root of $W(\omega^2)$. We now determine the value of S given by

$$\begin{aligned} S &= \text{sgn}[W'(\omega_1^2)] \\ &= \text{sgn}\left[-(1 - k^2k_p^{*2})\left(1 + \sqrt{1 + \frac{4T^2k^2k_i^{*2}}{(1 - k^2k_p^{*2})^2}}\right) + 1 - k^2k_p^{*2}\right] \\ &= \text{sgn}\left[-(1 - k^2k_p^{*2})\sqrt{1 + \frac{4T^2k^2k_i^{*2}}{(1 - k^2k_p^{*2})^2}}\right] \\ &= -\text{sgn}\left[1 - k^2k_p^{*2}\right] \end{aligned}$$

and from (9.5) we conclude that $S = 1$. This root is therefore destabilizing. For the case when k_p^* is such that ω_2^2 is the only positive root of $W(\omega^2)$, a similar analysis yields

$$S = \text{sgn}[W'(\omega_2^2)] = 1 .$$

Moreover, if $k_p^* = \frac{1}{k}$, then (9.4) reduces to the form

$$W(\omega^2) = T^2\omega^4 - k^2 k_i^{*2} .$$

In this case there is only one positive root at $\omega_1^2 = \frac{kk_i^*}{T}$ and it is not difficult to see that

$$S = \text{sgn}[W'(\omega_1^2)] = \text{sgn}[2Tkk_i^*] .$$

Since k_p^*, k_i^* stabilize the delay-free system, it follows that $k_i^* > 0$. Hence, we conclude that $S = 1$ for $k_p^* = \frac{1}{k}$.

Thus in any case, there is only one positive root of $W(\omega^2)$, and this root is always destabilizing. The corresponding values of L are given by (5.33):

$$\begin{aligned}\cos(L\omega) &= \frac{(Tkk_i^* - kk_p^*)\omega^2}{k^2k_i^{*2} + k^2k_p^{*2}\omega^2} \\ \sin(L\omega) &= \frac{(kk_i^* + Tkk_p^*\omega^2)\omega}{k^2k_i^{*2} + k^2k_p^{*2}\omega^2} .\end{aligned}$$

Solving for L we obtain $L = b_1, b_2, \ldots$, where $0 < b_1 < b_2 < \cdots$ are real numbers. This means that at $L = b_1$, two roots of $\Delta(s, k_p, k_i) = 0$ cross from the left to the right of the imaginary axis. Then two more cross at $L = b_2$ and so on. We conclude that the *only* region of stability is $0 \le L < L_{max}$, where $L_{max} = b_1$.

Since the closed-loop system is stable for L^*, it follows from the previous discussion that L^* is inside the interval $(0, L_{max})$. Hence, the closed-loop system is stable for $L \in [0, L^*]$ and this concludes the proof of the lemma. ■

In view of Lemma 9.1, any controller taken from the set of stabilizing PI controllers obtained for the given first-order system with the time delay set to L_2 (known upper bound of the time delay) will stabilize the plant family $\mathcal{G}(s)$. This set can be obtained following the algorithm presented in Section 7.3. The following example illustrates this property.

Example 9.4 *Consider the plant family*

$$\mathcal{G}(s) = \frac{1}{s+1}e^{-Ls}$$

where $L \in [1, 3]$ *seconds. By using the algorithm in Section 7.3 we can find the set of stabilizing PI controllers for different values of the time delay* $L \in [1, 3]$. *Figure 9.5 shows these sets for* $L = 1, 1.5, 2, 3$. *As can be seen*

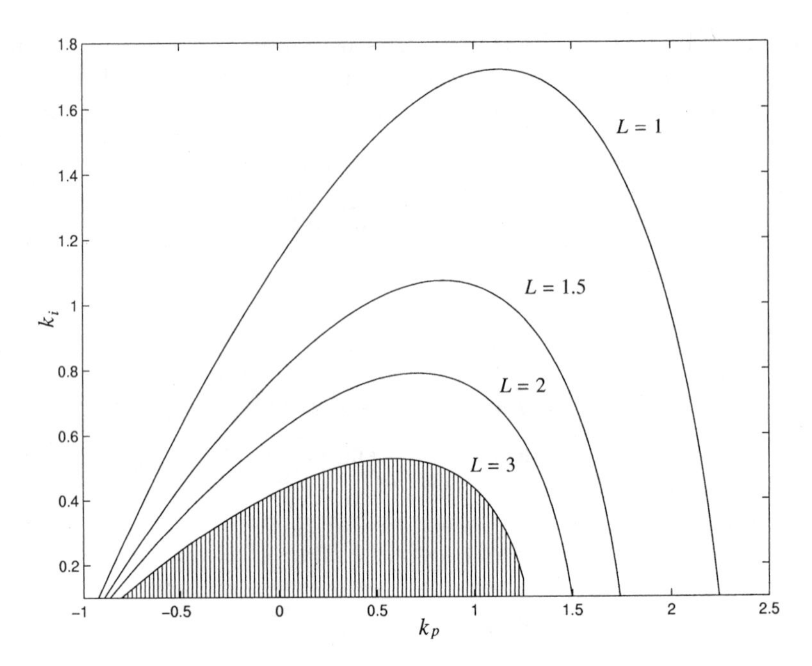

FIGURE 9.5. Sets of stabilizing PI controllers for Example 9.4

from this figure the intersection of all these sets is the set corresponding to $L = 3$ (dashed area). Thus, any PI controller from this set will stabilize the entire family of plants described by $\mathcal{G}(s)$. △

9.3.3 Robust Stabilization Using a PID Controller

In this subsection, we consider the problem of finding the set of PID controllers that stabilize the plant family $\mathcal{G}(s)$ described by (9.2). Since the controller $C(s)$ in question is now a PID controller, i.e.,

$$C(s) = k_p + \frac{k_i}{s} + k_d s \ ,$$

the family of closed-loop characteristic polynomials $\Delta(s, k_p, k_i, k_d)$ becomes

$$\Delta(s, k_p, k_i, k_d) = Ts^2 + s + (kk_d s^2 + kk_p s + kk_i)e^{-Ls}$$

where $L \in [L_1, L_2]$. We want to determine the values of the controller parameters k_p, k_i, k_d, for which the entire family of closed-loop characteristic polynomials $\Delta(s, k_p, k_i, k_d)$ is stable.

As in the last subsection, we start by presenting an interesting lemma based on the results introduced in Section 5.6. This lemma states that if a given PID controller stabilizes the delay-free system and the system in (9.2) with time delay L^*, then it also stabilizes the same system for all time delay $L \in [0,\ L^*]$.

Lemma 9.2 *Consider the system with transfer function (9.2). If a given PID controller stabilizes the delay-free system and the system with* $L = L^* > 0$*, then the same PID controller stabilizes the system* $\forall\ L \in [0,\ L^*]$.

Proof. The proof of this lemma is similar to that of Lemma 9.1 and follows the three-step procedure introduced in Section 5.6.

Let us denote by k_p^*, k_i^*, and k_d^* some controller parameter values that stabilize the delay-free system and the system described by (9.2) with $L = L^*$. Also, we rewrite $\Delta(s, k_p, k_i, k_d)$ as follows:

$$\Delta(s, k_p, k_i, k_d) = d(s) + n(s)e^{-Ls}$$

where

$$\begin{aligned} d(s) &= Ts^2 + s \\ n(s) &= kk_d^* s^2 + kk_p^* s + kk_i^* \,. \end{aligned}$$

Step 1. Stability at $L = 0$. This follows from our assumption concerning k_p^*, k_i^*, and k_d^*. Note that in Theorem 8.1 we have imposed this (reasonable) requirement on any stabilizing PID controller.

Step 2. Increment L from 0 to an infinitesimally small and positive number. Since the degree of $d(s)$ is equal to the degree of $n(s)$ we need to consider the behavior of $W(\omega^2)$ for large ω^2 (see Section 5.6). From (5.31) we have

$$W(\omega^2) = (T^2 - k^2 k_d^{*2})\omega^4 + (1 + 2k^2 k_d^* k_i^* - k^2 k_p^{*2})\omega^2 - k^2 k_i^{*2} \,.$$

For large ω^2 we have

$$\lim_{\omega^2 \to +\infty} W(\omega^2) = (T^2 - k^2 k_d^{*2}) \lim_{\omega^2 \to +\infty} \omega^4 \,.$$

It follows from Lemmas 8.3, 8.4, and 8.5 that $-\frac{T}{k} < k_d^* < \frac{T}{k}$ (see also Fig. 8.6). Thus $T^2 - k^2 k_d^{*2} > 0$ so $W(\omega^2) > 0$ for large ω^2. The infinite number of new roots therefore occurs in the LHP.

Step 3. Potential crossing points on the imaginary axis. From the expression for $W(\omega^2)$ derived in the previous step we have

$$W'(\omega^2) = 2(T^2 - k^2 k_d^{*2})\omega^2 + (1 + 2k^2 k_d^* k_i^* - k^2 k_p^{*2}) \,.$$

The roots of $W(\omega^2)$ are given by

$$\begin{aligned} \omega_1^2, \omega_2^2 &= \frac{-(1 + 2k^2 k_d^* k_i^* - k^2 k_p^{*2})}{2(T^2 - k^2 k_d^{*2})} \\ &\mp \frac{\sqrt{(1 + 2k^2 k_d^* k_i^* - k^2 k_p^{*2})^2 + 4(T^2 - k^2 k_d^{*2})k^2 k_i^{*2}}}{2(T^2 - k^2 k_d^{*2})} \end{aligned}$$

which can be rewritten as

$$\omega_1^2, \omega_2^2 = -\frac{(1 + 2k^2 k_d^* k_i^* - k^2 k_p^{*2})}{2(T^2 - k^2 k_d^{*2})} \left[1 \pm \sqrt{1 + \gamma}\right]$$

where

$$\gamma = \frac{4(T^2 - k^2 k_d^{*2}) k^2 k_i^{*2}}{(1 + 2k^2 k_d^* k_i^* - k^2 k_p^{*2})^2} \,.$$

Clearly the expression inside the square root is always greater than 1. Thus we have

$$\begin{aligned} \omega_1^2 &= -\frac{(1 + 2k^2 k_d^* k_i^* - k^2 k_p^{*2})}{2(T^2 - k^2 k_d^{*2})} \left[1 + \sqrt{1+\gamma}\right] \\ \Rightarrow \operatorname{sgn}[\omega_1^2] &= -\operatorname{sgn}\left[1 + 2k^2 k_d^* k_i^* - k^2 k_p^{*2}\right] \end{aligned} \tag{9.6}$$

since $T^2 - k^2 k_d^{*2} > 0$ as pointed out in Step 2. Moreover,

$$\begin{aligned} \omega_2^2 &= -\frac{(1 + 2k^2 k_d^* k_i^* - k^2 k_p^{*2})}{2(T^2 - k^2 k_d^{*2})} \left[1 - \sqrt{1+\gamma}\right] \\ \Rightarrow \operatorname{sgn}[\omega_2^2] &= \operatorname{sgn}\left[1 + 2k^2 k_d^* k_i^* - k^2 k_p^{*2}\right] . \end{aligned}$$

Since ω_1^2 and ω_2^2 have opposite signs, we conclude that for $k_p = k_p^*$, $k_i = k_i^*$, $k_d = k_d^*$, there is only one positive root of $W(\omega^2)$.

Consider now that the controller parameter values are such that ω_1^2 is the only positive root of $W(\omega^2)$. Then, the value of S is given by

$$\begin{aligned} S &= \operatorname{sgn}[W'(\omega_1^2)] \\ &= \operatorname{sgn}\Big[-(1 + 2k^2 k_d^* k_i^* - k^2 k_p^{*2})(1 + \sqrt{1+\gamma}\,) \\ &\qquad +1 + 2k^2 k_d^* k_i^* - k^2 k_p^{*2}\Big] \\ &= \operatorname{sgn}\left[-(1 + 2k^2 k_d^* k_i^* - k^2 k_p^{*2})\sqrt{1+\gamma}\,\right] \\ &= -\operatorname{sgn}\left[1 + 2k^2 k_d^* k_i^* - k^2 k_p^{*2}\,\right] \end{aligned}$$

and from (9.6) we conclude that $S = 1$. Thus this root is destabilizing. For the case when the controller parameters are such that ω_2^2 is the only positive root of $W(\omega^2)$, a similar analysis yields

$$S = \operatorname{sgn}[W'(\omega_2^2)] = 1 \,.$$

In any case, there is only one positive root of $W(\omega^2)$, and this root is always destabilizing. The corresponding values of L are given by (5.33):

$$\cos(L\omega) = Re\left[-\frac{d(j\omega)}{n(j\omega)}\right] \,, \ \sin(L\omega) = Im\left[\frac{d(j\omega)}{n(j\omega)}\right] .$$

Solving for L we obtain $L = b_1, b_2, \ldots$, where $0 < b_1 < b_2 < \cdots$ are real numbers. This means that at $L = b_1$, two roots of $\Delta(s, k_p, k_i, k_d) = 0$ cross from the left to the right of the imaginary axis. Then two more cross at $L = b_2$ and so on. We conclude that the *only* region of stability is $0 \leq L < L_{max}$, where $L_{max} = b_1$.

Since the closed-loop system is stable for L^*, it follows from the previous discussion that L^* is inside the interval $(0, L_{max})$. Hence the closed-loop system is stable for $L \in [0, L^*]$ and this concludes the proof of the lemma. ■

In view of Lemma 9.2, we conclude that when solving the PID stabilization problem for the plant family $\mathcal{G}(s)$ in (9.2), it is sufficient to solve the PID stabilization problem for the case of $L = L_2$ (known upper bound on the time delay). This will generate a set of PID controllers that stabilizes the family of plants with time delay in $[0, L_2]$, which is a superset of $[L_1, L_2]$. We now present a simple example to illustrate these observations.

Example 9.5 *Consider the same plant family as in Example 9.4:*

$$\mathcal{G}(s) = \frac{1}{s+1} e^{-Ls}$$

where $L \in [1, 3]$ seconds. By using Theorem 8.1 we can study the behavior of the upper bound of the controller parameter k_p as a function of the time delay L. This upper bound is given by the following expression:

$$k_{upp} = \frac{1}{k} \left[\frac{T}{L} \alpha_1 \sin(\alpha_1) - \cos(\alpha_1) \right]$$

where α_1 is the solution of the equation

$$\tan(\alpha) = -\frac{T}{T+L} \alpha$$

in the interval $(0, \pi)$. By sweeping over the time delay L, we obtained the plot shown in Fig. 9.6. This plot shows the behavior of the upper bound k_{upp} as a function of the time delay of the system.

As we can see from this plot, the upper bound k_{upp} is a monotonically decreasing function of the time delay of the system. This implies that if we design, for example, for $L = 1$, we will obtain PID controllers that produce an unstable behavior for some of the members of the plant family $\mathcal{G}(s)$. Thus we take $k_{upp} = 1.3045$ (value corresponding to $L = 3$) as a safe upper bound for the controller parameter k_p.

Next we fix the controller parameter k_p at 1.2. Using Fig. 8.6 we now determine the stabilizing region of (k_i, k_d) values for different values of $L \in [1, 3]$. Figure 9.7 shows these sets for $L = 1, 1.5, 2, 3$. As can be seen from this figure the intersection of all these sets is the set corresponding to $L = 3$ (shaded area). Thus when k_p is set equal to 1.2, any (k_i, k_d) from this set will stabilize the entire family of plants described by $\mathcal{G}(s)$. △

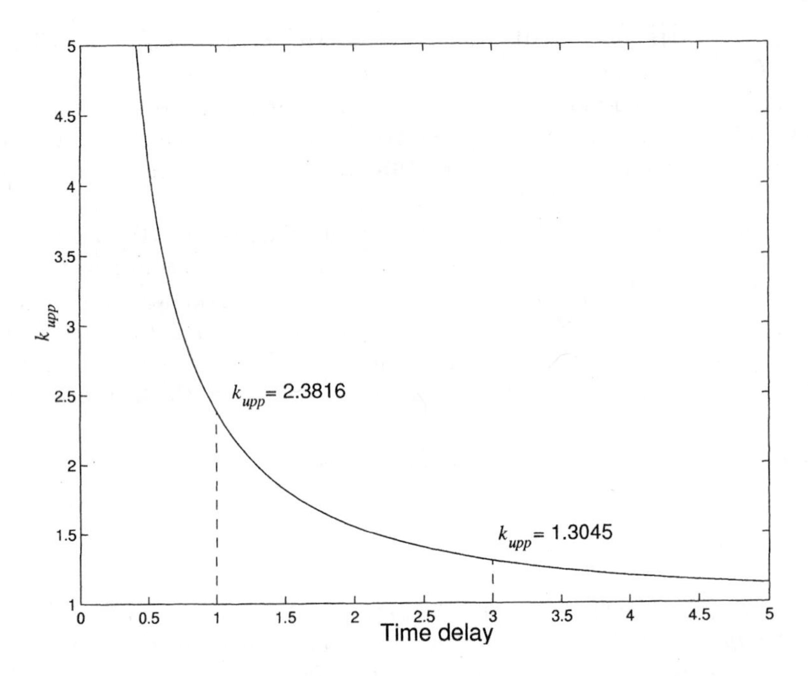

FIGURE 9.6. Plot of the upper bound k_{upp} as a function of the time delay for Example 9.5.

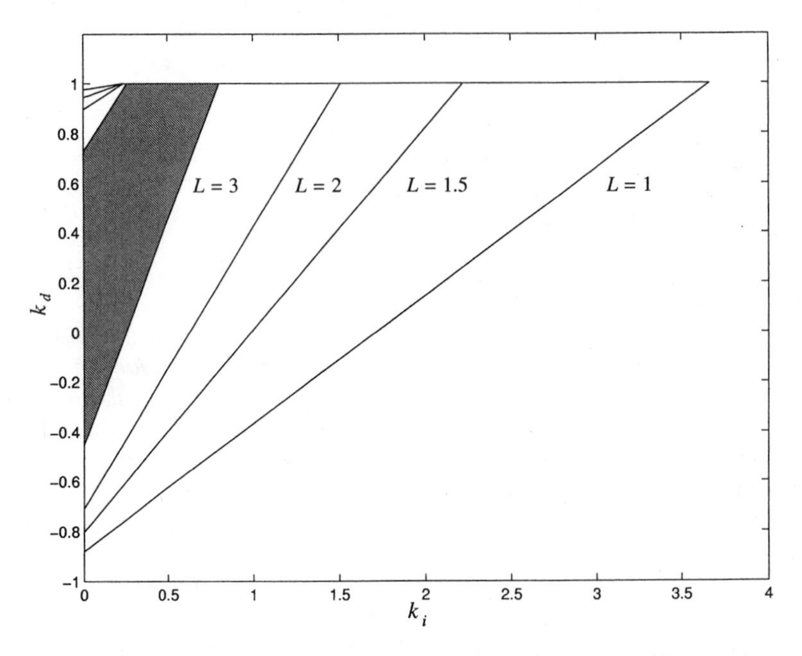

FIGURE 9.7. Sets of stabilizing (k_i, k_d) for Example 9.5 ($k_p = 1.2$).

9.4 Resilient Controller Design

In this section, we study controller fragility and ways to combat it. A controller whose closed-loop system is destabilized by small perturbations in the controller coefficients is said to be *fragile.* Any controller that is to be practically implemented must necessarily be nonfragile or resilient. This is because of two requirements:

1. round-off errors during implementation should not destabilize the closed-loop and
2. tuning of the parameters about the nominal design values should be allowed.

Motivated by this fact, we present in this section a new PID tuning technique that produces a resilient controller. The starting point for this technique is the plant description already introduced in earlier chapters:

$$G(s) = \frac{k}{Ts+1}e^{-Ls} \tag{9.7}$$

where k, T, and L are the three plant parameters. Since most PID tuning rules make use of experimentally measured quantities, let us first see how the experimentally measured quantities can be related to k, T, and L.

9.4.1 Determining k, T, and L from Experimental Data

The parameters T and L in the model (9.7) can be determined experimentally by measuring the ultimate gain and the ultimate period of the plant. These parameters were already discussed in Section 1.4.2 and can be calculated by using the closed-loop configuration shown in Fig. 1.10. The reader will recall that in this figure, the relay is adjusted to induce a self-sustained oscillation in the loop. The ultimate gain (k_u) and the ultimate period (T_u) can now be determined by measuring the amplitude and the period of the oscillations.

We now proceed to derive the relationship between the model parameters T and L and the experimentally observed quantities k_u and T_u. Recall from Section 1.4.2 the following relationships:

$$\begin{aligned} |G(j\omega_u)| &= \frac{1}{N(a)} = \frac{a\pi}{4d} \\ &\triangleq \frac{1}{k_u} \qquad (9.8) \\ \text{and } arg\, G(j\omega_u) &= -\pi \qquad (9.9) \end{aligned}$$

where a is the amplitude of the sinusoidal input signal and d is the relay amplitude (see Fig. 1.10). The quantity k_u defined in (9.8) is called the *ultimate gain* in the sense that if one were to replace the relay nonlinearity in the loop by a static positive gain, setting the gain value equal to the ultimate gain would cause self-sustained oscillations in the loop. The quantity $T_u = \frac{2\pi}{\omega_u}$ is the ultimate period. Substituting $s = j\omega_u$ into (9.7), we obtain the following expressions for the magnitude and phase of $G(j\omega_u)$:

$$|G(j\omega_u)| = \frac{k}{\sqrt{1+{\omega_u}^2T^2}} \quad (9.10)$$

$$\arg\left[G(j\omega_u)\right] = -\text{atan}(\omega_u T) - \omega_u L\,. \quad (9.11)$$

From (9.8), (9.9), (9.10), (9.11), using that $\omega_u = \frac{2\pi}{T_u}$ and solving for T and L, we obtain the following relationships:

$$T = \frac{T_u\sqrt{(kk_u)^2-1}}{2\pi} \quad (9.12)$$

$$L = \frac{T_u\left[\pi - atan(\sqrt{(kk_u)^2-1})\right]}{2\pi}\,. \quad (9.13)$$

Also, the steady-state gain k can be found by applying a unit step input to the plant and observing the amplitude of the steady-state output.

9.4.2 Algorithm for Computing the Largest Ball Inscribed Inside the PID Stabilizing Region

Now that the parameters of the first-order model with time delay (9.7) have been determined, we can proceed to compute the set of stabilizing PID controller parameters for this model. This can be achieved by using Theorem 8.1. We can then choose the PID settings as the *center of the three-dimensional ball of largest radius* inscribed inside the stabilizing region. The radius of this ball represents the maximum l_2 parametric stability margin in the space of the controller parameters. The problem of finding the largest ball inscribed inside the PID stabilizing region for rational plants has already been studied and solved. One of these solutions, developed by Ho, Datta, and Bhattacharyya, is of interest to us. Their method is also applicable here since, for a fixed value of the parameter k_p, the stabilizing regions of (k_i, k_d) values are convex polygons (see Fig. 8.6). Even though the center of the largest ball inscribed inside the stabilizing region cannot be determined in closed form, it can be computed using the following algorithm.

Before presenting the algorithm, we first introduce some concepts. Consider a sphere $\mathcal{B}(x,r)$ in the three-dimensional k_p-k_i-k_d space with radius r and centered at $x \triangleq (x_{k_p}, x_{k_i}, x_{k_d})$. Given any angle $\theta \in \left[-\frac{\pi}{2}, \frac{\pi}{2}\right]$, let $\mathcal{C}(x,r,\theta)$ denote the circle with radius $r\cos(\theta)$, centered at $(x_{k_p} + r\sin(\theta),$

x_{k_i}, x_{k_d}), and parallel to the k_i–k_d plane as illustrated in Fig. 9.8. It is clear that

$$\mathcal{B}(x,r) = \bigcup_{\theta\in[-\frac{\pi}{2},\frac{\pi}{2}]} \mathcal{C}(x,r,\theta).$$

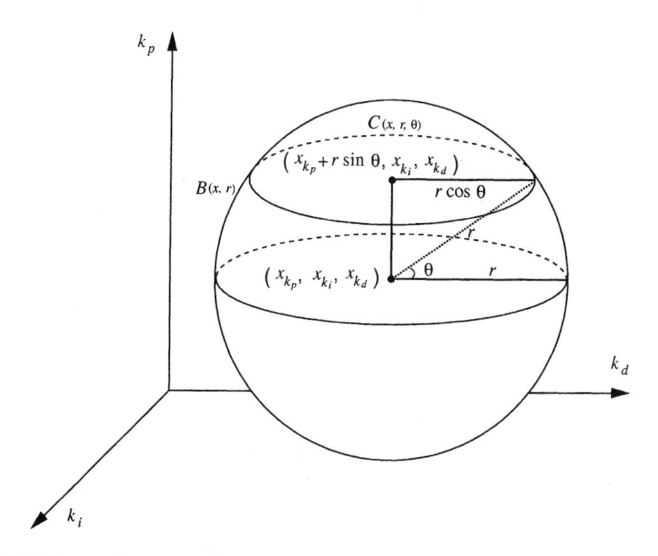

FIGURE 9.8. Sphere $\mathcal{B}(x,r)$ and the definition of the circle $\mathcal{C}(x,r,\theta)$.

Consider $\mathcal{C}(x,r,\theta)$ with fixed x_{k_p}, r, and θ so that $k_p = x_{k_p} + r\sin\theta$ is fixed. The stabilizing set of PID controllers is formed by convex polygons, parallel to the k_i–k_d plane, with either three or four sides (see Fig. 8.6). Let the convex polygon associated with this fixed k_p be given by the set of linear inequalities:

$$\mathcal{P}_\theta = \{x|a_{\theta_j}^T x \leq b_{\theta_j}, j = 1, 2, \ldots, 4\}$$

where $a_{\theta_j} \in R^2$, $b_{\theta_j} \in R$, and each inequality represents the half plane containing one side of the polygon. Define $x_c = (x_{k_i}, x_{k_d})^T$. Then, $\mathcal{C}(x,r,\theta)$ lies inside the stabilizing region $\mathcal{P}_\theta$ if and only if

$$a_{\theta_j}{}^T x_c + r\cos\theta\|a_{\theta_j}\| \quad \leq \quad b_{\theta_j},\ (j = 1, 2, \ldots, 4) \tag{9.14}$$

holds. Let S_θ denote the set of feasible solutions of (9.14). From the geometrical structure, we know that for all $\theta \in [-\frac{\pi}{2},\ \frac{\pi}{2}]$, the centers of the circles $\mathcal{C}(x,r,\theta)$ have the same (k_i, k_d) coordinates. Since S_θ is the set of feasible $(k_i,\ k_d)$ coordinates of the centers associated with $\mathcal{C}(x,r,\theta)$, it follows that $\mathcal{B}(x,r)$ lies inside the stabilizing $(k_p,\ k_i,\ k_d)$ region if and only if

$$\bigcap_{\theta\in[-\frac{\pi}{2},\frac{\pi}{2}]} S_\theta \neq \emptyset.$$

We now present the algorithm.

- **Step 1:** Initialize $k_p = -\frac{1}{k}$ and $step = \frac{1}{N+1}\left(k_{upp} + \frac{1}{k}\right)$, where N is the desired number of points;
- **Step 2:** Increase k_p as follows: $k_p = k_p + step$;
- **Step 3:** If $k_p < k_{upp}$ then go to Step 4. Else, terminate the algorithm;
- **Step 4:** Set $r_L = 0$ and $r_U = min(k_p + \frac{1}{k}, k_{upp} - k_p)$;
- **Step 5:** Set $r = \frac{r_L + r_U}{2}$;
- **Step 6:** Sweep over all $\theta \in \left[-\frac{\pi}{2}, \frac{\pi}{2}\right]$ and determine the set of all feasible solutions S_θ for (9.14) at each stage;
- **Step 7:** If $\bigcap_{\theta \in \left[-\frac{\pi}{2}, \frac{\pi}{2}\right]} S_\theta \neq \emptyset$ then set $r_L = r$; otherwise set $r_U = r$;
- **Step 8:** If $|r_U - r_L| \leq$ specified level, then store r and go to Step 2; otherwise go to Step 5.

The above algorithm determines a family of spheres having different radii and centers, with each sphere corresponding to a particular value of k_p used in Step 2. Among these spheres, we pick the one with the largest radius. Setting the (k_p, k_i, k_d) values at the center of this sphere will yield the maximum l_2 parametric stability margin with respect to the controller parameters. The following example illustrates the steps involved.

Example 9.6 *Consider a first-order plus deadtime model of a system where the parameters are unknown. Using the relay experiment, the ultimate gain and the ultimate period of the plant are determined as* $k_u = 11.44$ *and* $T_u = 0.9582$ *seconds. The steady-state gain of the plant is found by applying a unit step input to the plant. The output value observed is* $k = 1.6667$. *Substituting the* k, k_u, *and* T_u *values in (9.12)–(9.13), the plant parameters were calculated as* $T = 2.9036$ *seconds and* $L = 0.2475$ *seconds. Therefore the first-order plus deadtime model is given by*

$$G(s) = \frac{1.6667}{1 + 2.9036s} e^{-0.2475s} .$$

From Theorem 8.1, the range of k_p *values for which* $G(s)$ *continues to have closed-loop stability with a PID controller in the loop is given by*

$$-0.6 < k_p < 13.0814 .$$

We next sweep over this range and find the largest sphere inscribed inside the stabilizing (k_p, k_i, k_d) *region. Using the algorithm introduced earlier, the sphere with the largest radius was found to be centered at* $k_p^c = 1.9663$, $k_i^c = 1.5195$, $k_d^c = 0.2227$ *and its radius was found to be* $r = 1.5195$.

Figure 9.9 shows the stabilizing set of PID parameters for $G(s)$ and the sphere with the largest radius inscribed inside it. If we consider $k_p = k_p^c + \Delta k_p$, $k_i = k_i^c + \Delta k_i$, $k_d = k_d^c + \Delta k_d$, then all PID controllers with

$$\sqrt{\Delta k_p^2 + \Delta k_i^2 + \Delta k_d^2} < 1.5195$$

stabilize the plant.

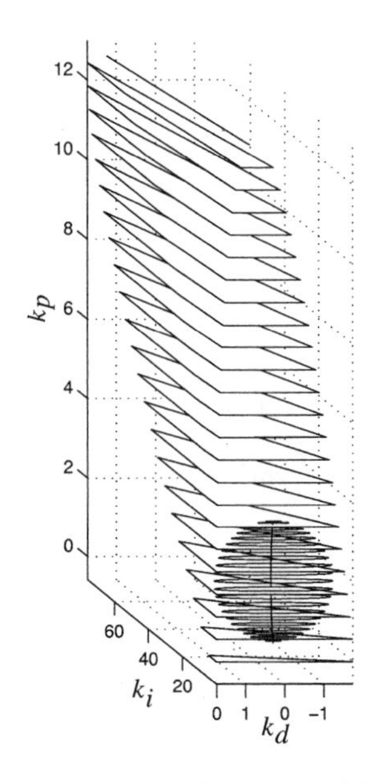

FIGURE 9.9. Largest sphere for the model in Example 9.6.

△

9.5 Time Domain Performance Specifications

Before designing a controller, it is important to understand the primary objective of control. A common type of control objective is to follow a setpoint or a step input. Specifications on setpoint following may include requirements on rise time, settling time, decay ratio, maximum overshoot, and steady-state offset for step changes in the setpoint. These are time domain performance specifications that need to be incorporated into the design of the controller.

As mentioned at the beginning of Chapter 7, the purpose of the integral term in a PI controller is to achieve zero steady-state offset when tracking step inputs. Thus we can employ the time domain specifications mentioned above to quantify the performance of a PI-controlled closed-loop system.

The characterization of all stabilizing (k_p, k_i) values provided in Section 7.3 enables us to graphically display the variation of these performance indices over the entire stabilizing region in the parameter space. Using such a tool, we can select the (k_p, k_i) values that meet the performance specifications. The following examples illustrate the procedure involved.

Example 9.7 *Consider the problem of choosing stabilizing PI gains for the plant in Example 7.1:*

$$G(s) = \frac{1}{4s+1}e^{-s} .$$

Thus the plant parameters are $k = 1$, $L = 1$ *seconds, and* $T = 4$ *seconds. The performance specifications that we are required to meet when designing the PI controller are the following:*

1. *settling time* ≤ 30 *seconds;*
2. *overshoot* $\leq 20\%$.

We now obtain by simulation the transient responses of the closed-loop systems for the (k_p, k_i) *values inside the region depicted in Fig. 7.6. In order to obtain a resilient controller (a concept introduced in the previous section), we only consider the* (k_p, k_i) *values inside the following box defined in the parameter space:*

$$2 \leq k_p \leq 4 \quad and \quad 0.5 \leq k_i \leq 1.5 .$$

In this way we can alleviate the controller fragility problem to some extent. Figure 9.10 displays the variation of the maximum overshoot exhibited by the closed-loop system as a function of the proportional and integral gains. This kind of plot allows us to perform a search for those $(k_p,\ k_i)$ *values that meet the desired performance specifications. Among all these values, we set the controller parameters to:* $k_p = 2.1053$, $k_i = 0.7105$. *Figure 9.11 shows the time response of the corresponding closed-loop system. As we can see from this figure, the closed-loop system with the designed PI controller meets the performance specifications and by design we are also guaranteed to have a nonfragile PI controller.* $\triangle$

The characterization of all stabilizing PID controllers provided in Section 8.3 enables us to graphically display the variation of time domain performance indices over the entire stabilizing region in the parameter space. Using such a tool, we can select $(k_p,\ k_i,\ k_d)$ values that satisfy the given performance specifications. The following example illustrates the steps involved.

Example 9.8 *Consider the PID stabilization problem for a plant described by the differential equation*

$$\frac{dy(t)}{dt} = -0.5y(t) + 0.5u(t-4) .$$

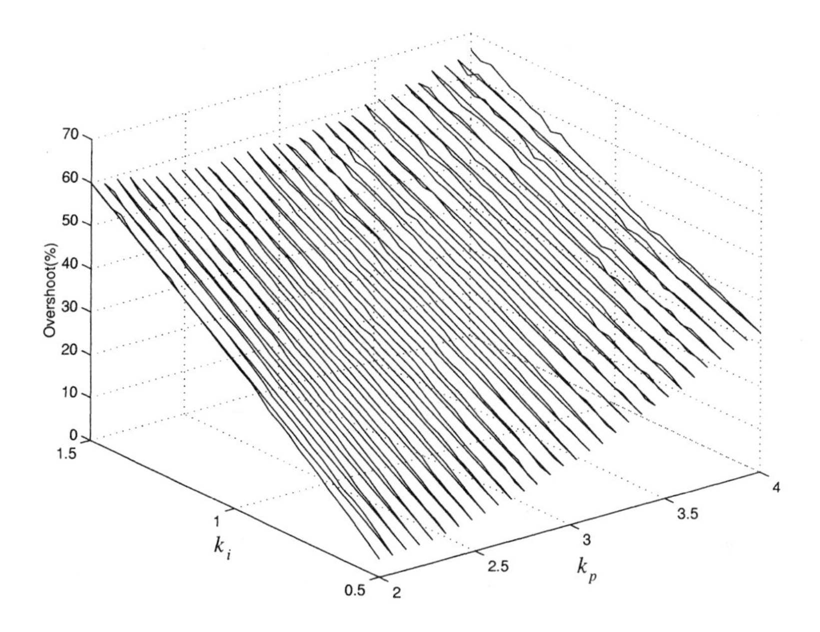

FIGURE 9.10. Plot of maximum overshoot versus (k_p, k_i) for Example 9.7.

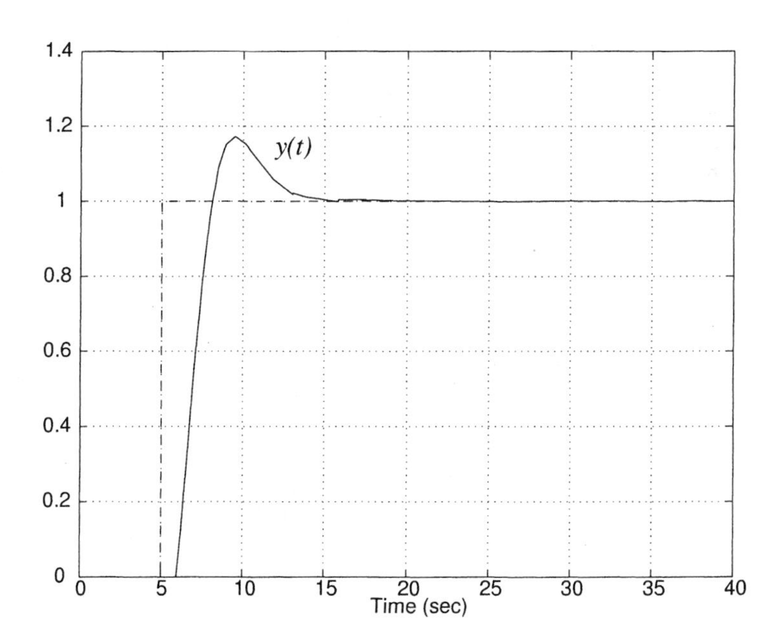

FIGURE 9.11. Time response for the closed-loop system of Example 9.7.

This process can also be described by the transfer function $G(s)$ in (9.7) with the following parameters: $k = 1$, $T = 2$ seconds, and $L = 4$ seconds. Since the system is open-loop stable we use Theorem 8.1 to find the range of k_p values for which a solution to the PID stabilization problem exists. We first compute the parameter $\alpha_1 \in (0, \pi)$ satisfying the following equation:

$$\tan(\alpha) = -0.3333\alpha \,.$$

Solving this equation we obtain $\alpha_1 = 2.4557$. Thus from (8.20) the range of k_p values is given by

$$-1 < k_p < 1.5515 \,.$$

We now sweep over the above range of k_p values and determine the stabilizing set of (k_i, k_d) values at each stage. These regions are sketched in Fig. 9.12.

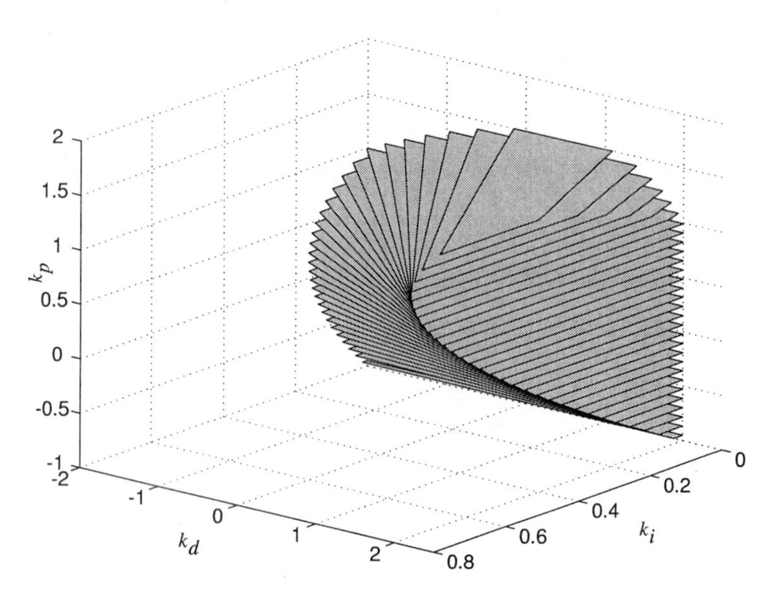

FIGURE 9.12. The stabilizing region of (k_p, k_i, k_d) values for the PID controller in Example 9.8.

Any PID gains selected from these regions will result in closed-loop stability. Now, consider the following performance specifications:

1. *settling time ≤ 60 seconds;*

2. *overshoot $\leq 20\%$.*

We can obtain the transient responses of the closed-loop system for the (k_p, k_i, k_d) values inside the regions depicted in Fig. 9.12. Since we also want *the controller to be resilient, we only consider PID gains lying inside the following box defined in the parameter space:*

$$0.1 \leq k_p \leq 1, \ 0.1 \leq k_i \leq 0.3, \ \textit{and} \ 0.5 \leq k_d \leq 1.5 \, .$$

By searching over this box, several (k_p, k_i, k_d) *values are found to meet the desired performance specifications. We arbitrarily set the controller parameters to* $k_p = 0.3444$, $k_i = 0.1667$, $k_d = 0.8333$. *Figure 9.13 shows the step response of the resulting closed-loop system. It is clear from this figure that the closed-loop system is stable, the output* $y(t)$ *tracks the step input signal, and the performance specifications are met. The figure also shows*

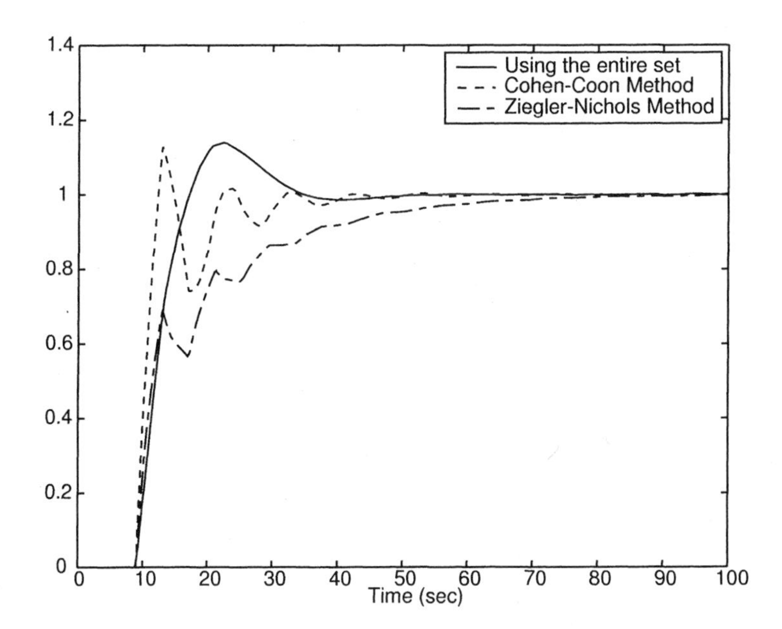

FIGURE 9.13. Time response of the closed-loop system for Example 9.8.

the responses of the closed-loop systems for the case of a PID controller designed using the Cohen-Coon method, and the case of a PID controller designed using the Ziegler-Nichols step response method (these methods will be discussed in detail in the next chapter). Notice that in these cases the system is stable and achieves setpoint following. However, the responses are much more oscillatory. △

Although the designs presented above are essentially brute force optimization searches, nevertheless the fact that the results of Chapters 7 and 8 can be used to confine the search to the stabilizing set makes the design problems orders of magnitude easier.

9.6 Notes and References

For an exhaustive treatment of parametric robust control theory, the reader is referred to [5]. Kharitonov and Zhabko have studied the robust stability of quasi-polynomial families [28]. For a detailed description of the relay experiment the reader is referred to the book by Astrom and Hagglund [2]. The approach for designing nonfragile PID controllers for delay-free systems is due to Ho, Datta, and Bhattacharyya and can be found in [20]. Its application to the case of first-order systems with time delay was derived in [43]. The issue of controller fragility was raised by Keel and Bhattacharyya in [26]. The results in this chapter were first reported in [44].

10 Analysis of Some PID Tuning Techniques

In this chapter we present an analysis of some PID tuning techniques that are based on first-order models with time delays. Using the characterization of all stabilizing PID controllers derived in Chapter 8, each tuning rule is analyzed to first determine if the proportional gain value dictated by that rule lies inside the range of admissible proportional gains. Then the integral and derivative gain values are examined to determine conditions under which the tuning rule exhibits robustness with respect to controller parameter perturbations.

10.1 Introduction

Numerous methods have been developed over the last 40 years for setting the parameters of a PID controller. Some of these methods are based on characterizing the dynamic response of the plant to be controlled with a first-order model with time delay. Traditionally, this model is obtained by applying a step input to the plant and measuring at the output the following three parameters: steady-state gain, time constant, and time delay.

In this chapter, we will analyze several PID tuning techniques that are based on these first-order models with time delay. This analysis will attempt to describe when each tuning technique is appropriate in the sense of providing robust PID controller parameters. As mentioned in Section 9.4, a controller for which the closed-loop system is destabilized by small perturbations in the controller coefficients is said to be fragile. Any controller that is to be practically implemented must necessarily be nonfragile or the system must be controller-robust.

Four tuning techniques will be discussed: (1) the classical Ziegler-Nichols step response method, (2) the Chien, Hrones, and Reswick (CHR) method, (3) the Cohen-Coon method, and (4) the internal model controller (IMC) design technique. In each case, we will first study if the proposed proportional gain value lies inside the allowable range determined in Chapter 8. We then analyze for this fixed proportional gain the possible location of the integral and derivative gains inside the stabilizing region. This procedure will allow us to determine conditions under which each tuning technique provides a good parametric stability margin in the space of the controller parameters. In this way, we will avoid undesirable scenarios such as PID controller parameters that are dangerously close to instability.

10.2 The Ziegler-Nichols Step Response Method

The tuning techniques presented in this chapter are based on characterizing the plant to be controlled by the following transfer function:

$$G(s) = \frac{k}{1+Ts}e^{-Ls} \tag{10.1}$$

where k is the steady-state gain, L is the apparent time delay, and T is the apparent time constant.

A simple way to determine the parameters of a PID controller based on step response data was developed by Ziegler and Nichols in 1942. This method first characterizes the plant by the parameters L and a, where the parameter a is defined as

$$a = k\frac{L}{T}\,.$$

Once these parameters are determined, the PID controller parameters are given in terms of L and a by the following formulas:

$$\overline{k_p} = \frac{1.2}{a} \qquad \overline{k_i} = \frac{0.6}{aL} \qquad \overline{k_d} = \frac{0.6L}{a}\,. \tag{10.2}$$

This tuning rule was developed by empirical simulations of many different systems and is only applicable to open-loop *stable* plants. We now define the parameter τ as the ratio of the apparent time delay to the apparent time constant of the plant, that is,

$$\tau = \frac{L}{T}\,.$$

First we focus on the proportional gain expression given in (10.2) and rewrite it as a function of τ:

$$\overline{k_p} = \frac{1.2}{k\tau}\,. \tag{10.3}$$

Since $k > 0$ and $\tau > 0$ (the plant is open-loop stable), then $\overline{k_p} > 0$. From Theorem 8.1, the range of k_p values for which a given open-loop stable plant, with transfer function $G(s)$ as in (10.1), can be stabilized using a PID controller is given by

$$-\frac{1}{k} < k_p < k_{upp} := \frac{1}{k}\left[\frac{T}{L}\alpha_1 \sin(\alpha_1) - \cos(\alpha_1)\right]$$

where α_1 is the solution of the equation

$$\tan(\alpha) = -\frac{T}{T+L}\alpha$$

in the interval $(0, \pi)$. Let us rewrite the upper bound on k_p as a function of the parameter τ:

$$k_{upp} = \frac{1}{k}\left[\frac{1}{\tau}\alpha_1 \sin(\alpha_1) - \cos(\alpha_1)\right] \tag{10.4}$$

where α_1 is now the solution of the equation

$$\tan(\alpha) = -\frac{1}{1+\tau}\alpha$$

in the interval $(0, \pi)$. We now compare $\overline{k_p}\, k$ and $k_{upp}\, k$ as functions of the parameter τ. As can be seen from Fig. 10.1, the proportional gain value given by the Ziegler-Nichols step response method is always less than the upper bound k_{upp}. Thus this tuning technique always provides a feasible proportional gain value $\overline{k_p}$.

We now set $k_p = \overline{k_p}$ and consider two cases, requiring different treatments according to the results of Chapter 8. Moreover, for clarity of presentation, let us rewrite the parameters $\overline{k_i}$ and $\overline{k_d}$ in (10.2) as

$$\overline{k_i} = \frac{0.6T}{kL^2} \tag{10.5}$$

$$\overline{k_d} = \frac{0.6T}{k}\,. \tag{10.6}$$

Case 1: $\tau \geq 1.2$. In this case, we have $0 < \overline{k_p} \leq \frac{1}{k}$. Then the stabilizing set is given either by Fig. 8.6(a) or by Fig. 8.6(b). Notice from (10.6) that the parameter $\overline{k_d}$ is always less than $\frac{T}{k}$ as illustrated in Fig. 10.2. The derivative gain value provided by the Ziegler-Nichols method is robust in the sense that it is not close to the stability boundary $\frac{T}{k}$. Following the same principle, we would like to guarantee that the integral gain value is also far away from the stability boundary.

Let x_1 be the k_i-coordinate of the point where the line $k_d = \overline{k_d}$ intersects the line $k_d = m_1 k_i + b_1$. From Fig. 10.2, we now find the conditions under

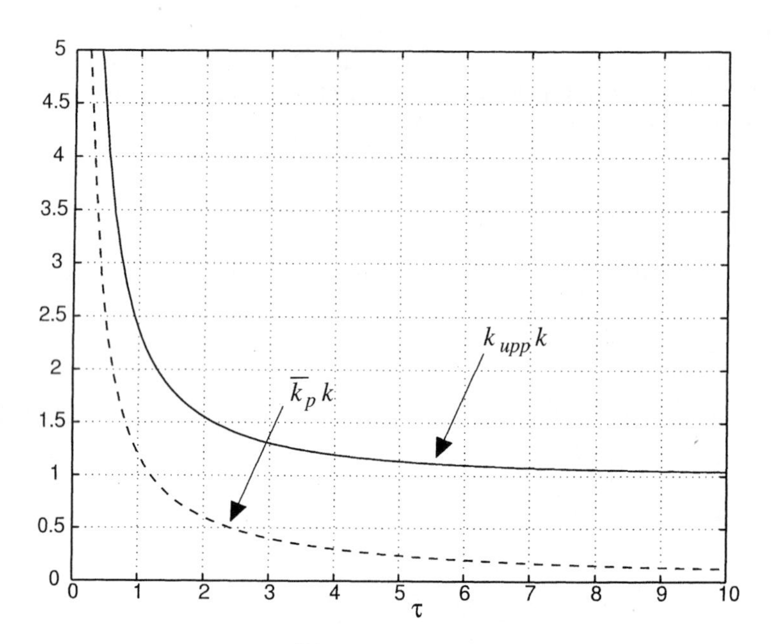

FIGURE 10.1. Comparison of $\overline{k_p}$ given by the Ziegler-Nichols method and the upper bound k_{upp}.

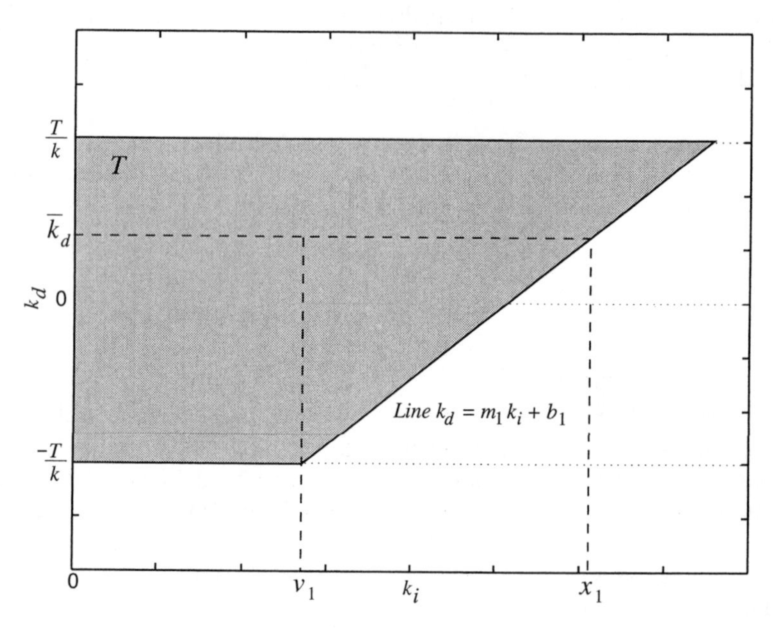

FIGURE 10.2. Location of the parameters $(\overline{k_i},\overline{k_d})$ when $\tau \geq 1.2$.

which the parameter $\overline{k_i}$ lies in the range $(0.2x_1, 0.8x_1)$. Following the same derivation used in (8.18), x_1 can be expressed as follows:

$$x_1 = \frac{T}{kL^2} z_1 \left[\tau \sin(z_1) + z_1 (\cos(z_1) + 0.6)\right] \tag{10.7}$$

where z_1 is the solution of

$$\begin{aligned} k\overline{k_p} + \cos(z) - \frac{T}{L} z \sin(z) &= 0 \\ \Leftrightarrow 1.2 + \tau \cos(z) - z \sin(z) &= 0 \text{ [using (10.3) and the definition of } \tau] \end{aligned}$$

in the interval $(0, \pi)$. From (10.5) and (10.7), we can plot the terms $\lambda \frac{kL^2}{T} x_1$, $\lambda \in (0,1)$, and $\frac{kL^2}{T}\overline{k_i}$ versus τ. This graph is shown in Fig. 10.3 for $\tau \geq 1.2$, and $\lambda = 0.1, 0.2$, and 0.8. As can be seen from this graph, $\overline{k_i}$ does not lie

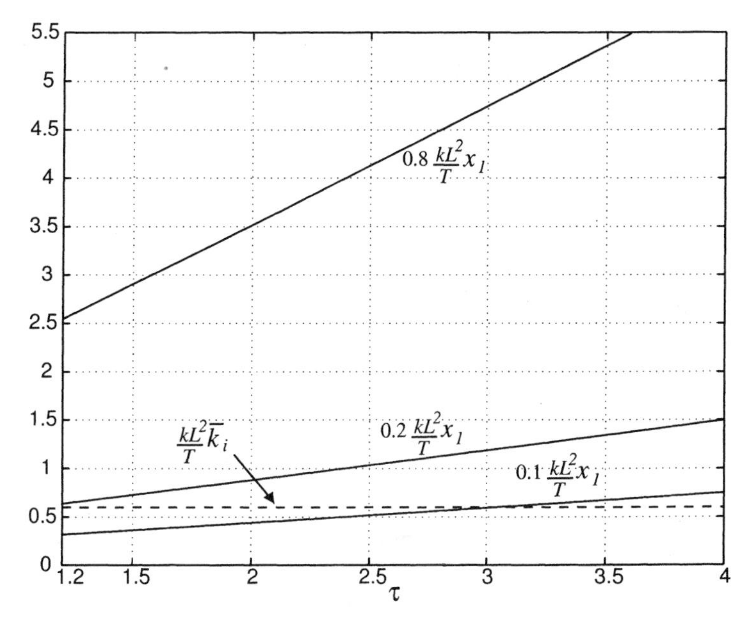

FIGURE 10.3. Comparison of $0.2\frac{kL^2}{T}x_1$, $0.8\frac{kL^2}{T}x_1$, and $\frac{kL^2}{T}\overline{k_i}$ for $\tau \geq 1.2$.

in the range $(0.2x_1, 0.8x_1)$ for any value of τ. If we relax our robustness condition and now make $\overline{k_i}$ lie inside the range $(0.1x_1, 0.8x_1)$, we see from Fig. 10.3 that this occurs for $1.2 \leq \tau < 3$. In this way, for $1.2 \leq \tau < 3$, $\overline{k_i}$ will be located 10% of x_1 away from the k_d axis, which corresponds to a good l_2 parametric stability margin.

Case 2: $0 < \tau < 1.2$. In this case, we have $\frac{1}{k} < \overline{k_p} < k_{upp}$. The stabilizing set is given by Fig. 8.6(c). We now show that the parameter $\overline{k_d}$ is less than

b_2 for all $\tau < 1.2$. From (8.12), b_2 can be rewritten as follows:

$$b_2 = -\frac{T}{k}\left[\tau\frac{\sin(z_2)}{z_2} + \cos(z_2)\right]$$

where $z_2 > z_1 > 0$ is the solution of

$$1.2 + \tau\cos(z) - z\sin(z) = 0$$

in the interval $(0, \pi)$. By sweeping τ in the range $(0, 1.2)$, it can be shown that $\overline{k_d} < b_2$. Figure 10.4 shows the location of $\overline{k_d}$ with respect to the stabilizing set in the space of (k_i, k_d).

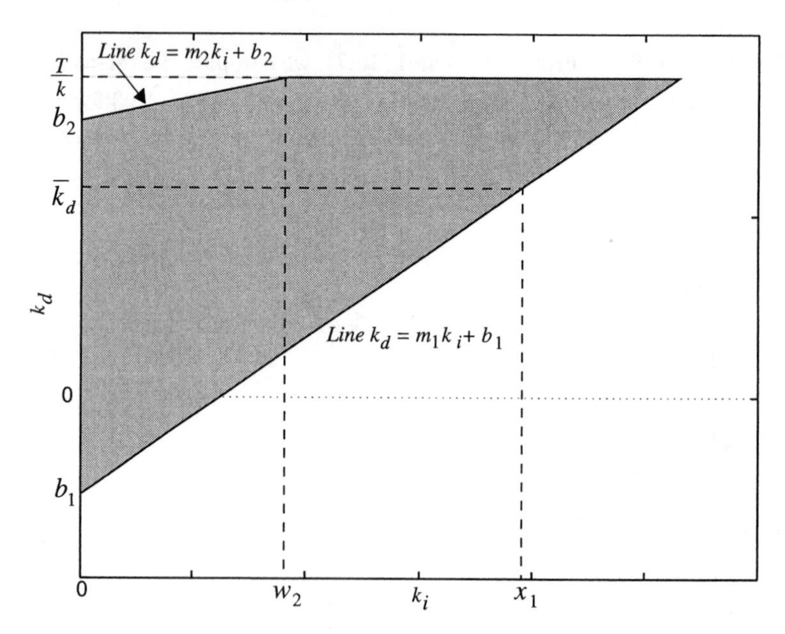

FIGURE 10.4. Location of the parameters $(\overline{k_i}, \overline{k_d})$ when $0 < \tau < 1.2$.

As in the previous case, we will now analyze for which values of τ the parameter $\overline{k_i}$ lies inside the range $(0.2x_1, 0.8x_1)$. As in Case 1, we can plot the terms $0.2\frac{kL^2}{T}x_1$, $0.8\frac{kL^2}{T}x_1$, and $\frac{kL^2}{T}\overline{k_i}$ versus τ. This graph is shown in Fig. 10.5 for $0 < \tau < 1.2$. From this graph we see that $\overline{k_i}$ lies in the range $(0.2x_1, 0.8x_1)$ for $0 < \tau < 1.07$. For the relaxed condition where $\overline{k_i}$ lies in the range $(0.1x_1, 0.8x_1)$, we have $0 < \tau < 1.2$.

From the previous analysis, we conclude that the Ziegler-Nichols step response method gives a controller-robust PID controller for $0 < \tau < 1.07$. Controller-robustness is here understood as a good l_2 parametric stability margin in the space of (k_i, k_d).

Remark 10.1 *It has been determined empirically that the Ziegler-Nichols rule is applicable if* $0.1 < \tau < 0.6$. *In this range, the derivative action often*

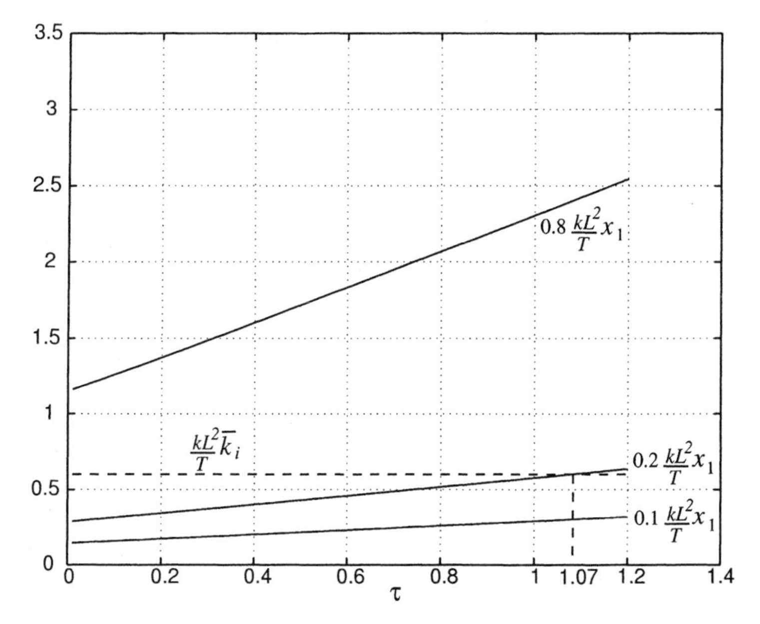

FIGURE 10.5. Comparison of $0.2\frac{kL^2}{T}x_1$, $0.8\frac{kL^2}{T}x_1$, and $\frac{kL^2}{T}\overline{k_i}$ for $0 < \tau < 1.2$.

gives significant improvement of performance. Comparing this range with the one previously obtained for controller-robustness, we see that the former is included in the latter. Thus, for $0.1 < \tau < 0.6$, *the Ziegler-Nichols step response method not only gives good performance but also is robust with respect to controller parameter perturbations.*

10.3 The CHR Method

Since the introduction of the Ziegler-Nichols step response method, there have been many proposed modifications to this tuning rule. One of these modifications was proposed by Chien, Hrones, and Reswick (CHR) in 1952. They made the important observation that tuning for the setpoint response is different from tuning for the load disturbance response and proposed separate tuning formulas for each situation. As in the case of the Ziegler-Nichols method, the parameters a and L of the process model are first determined. Then the controller parameters for setpoint response are given by the following formulas:

$$\overline{k_p} = \frac{0.6}{a} \qquad \overline{k_i} = \frac{0.6}{aT} \qquad \overline{k_d} = \frac{0.3L}{a}\,. \tag{10.8}$$

This tuning rule is based on the 0% overshoot design criterion, also known as the *quickest response without overshoot* criterion.

We now rewrite the controller parameter $\overline{k_p}$ in (10.8) as a function of $\tau = \frac{L}{T}$:

$$\overline{k_p} = \frac{0.6}{k\tau}. \tag{10.9}$$

Clearly $\overline{k_p} > 0$ since $k > 0$ and $\tau > 0$ (open-loop stable plant). We compare $\overline{k_p}$ with the upper bound k_{upp} in (10.4) as the parameter τ is varied. Figure 10.6 shows the plots of both $\overline{k_p}\ k$ and $k_{upp}\ k$ as functions of τ. It is clear from this figure that $\overline{k_p}$ given by the CHR method is less than the upper bound k_{upp} for all values of τ.

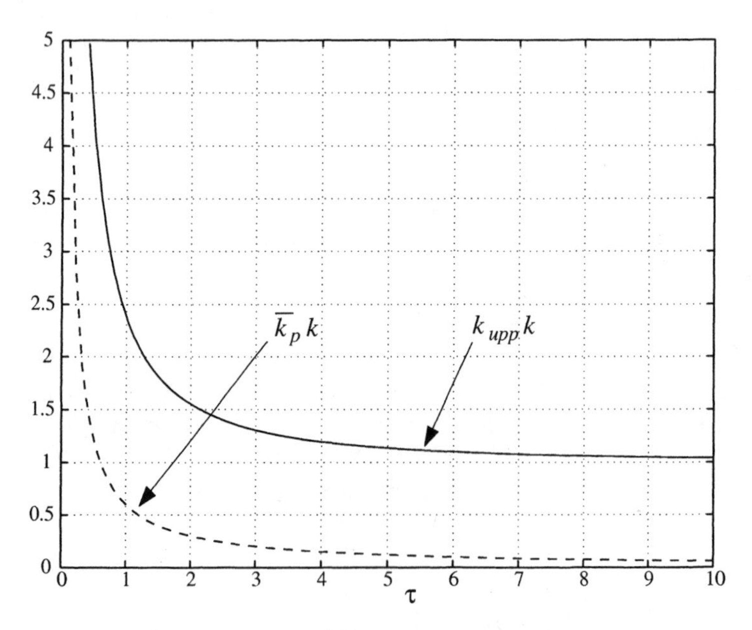

FIGURE 10.6. Comparison of $\overline{k_p}$ given by the CHR method and k_{upp}.

We next analyze the location of the controller parameters $\overline{k_i}$ and $\overline{k_d}$ in the space of (k_i, k_d). From (10.8), we rewrite these parameters as follows:

$$\overline{k_i} = \frac{0.6}{kL} \qquad \overline{k_d} = \frac{0.3T}{k}. \tag{10.10}$$

We now consider two cases.

Case 1: $\tau \geq 0.6$. In this case, we have $0 < \overline{k_p} \leq \frac{1}{k}$. Then the stabilizing set is given by Figs. 8.6(a) or 8.6(b). Notice that the parameter $\overline{k_d}$ given by (10.10) is always less than $\frac{T}{k}$. As in the previous section, we would like to examine the degree of robustness achieved by the CHR method. To analyze the robustness of the controller parameters, we now compare $\overline{k_i}$ given in (10.10) with x_1 (see Fig. 10.2), which is given by

$$x_1 = \frac{1}{kL} z_1 \left[\sin(z_1) + \frac{z_1}{\tau}\left(\cos(z_1) + 0.3\right)\right]$$

where z_1 is the solution of

$$0.6 + \tau \cos(z) - z \sin(z) = 0 \quad \text{[using (10.9) and the definition of } \tau\text{]}$$

in the interval $(0, \pi)$. Figure 10.7 shows the plot of $0.2kLx_1$, $0.8kLx_1$, and $kL\overline{k_i}$ versus τ. From this figure it is clear that $\overline{k_i}$ is inside the range $(0.2x_1, 0.8x_1)$ for all values of $\tau \geq 0.6$.

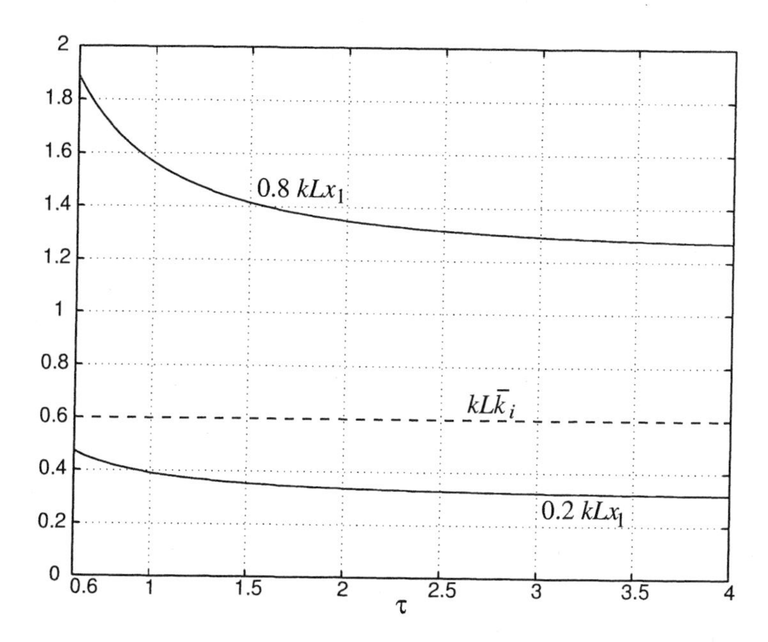

FIGURE 10.7. Comparison of $0.2kLx_1$, $0.8kLx_1$, and $kL\overline{k_i}$ for $\tau \geq 0.6$.

Case 2: $0 < \tau < 0.6$. For this case, $\frac{1}{k} < \overline{k_p} < k_{upp}$ and the stabilizing set is given by Fig. 8.6(c). We now compare the parameter $\overline{k_d}$ given in (10.10) with the parameter b_2 given by

$$b_2 = -\frac{T}{k}\left[\tau \frac{\sin(z_2)}{z_2} + \cos(z_2)\right]$$

where $z_2 > z_1 > 0$ is the solution of

$$0.6 + \tau \cos(z) - z \sin(z) = 0$$

in the interval $(0, \pi)$. Figure 10.8 shows the plot of $\frac{k}{T} b_2$ and $\frac{k}{T}\overline{k_d}$ as a function of the parameter τ. It is clear from this plot that $\overline{k_d} < b_2$ for $0 < \tau < 0.6$.

As in Case 1, we now find the values of τ for which the parameter $\overline{k_i}$ lies inside the range $(0.2x_1, 0.8x_1)$. From Fig. 10.9 we see that this occurs when $0.37 < \tau < 0.6$.

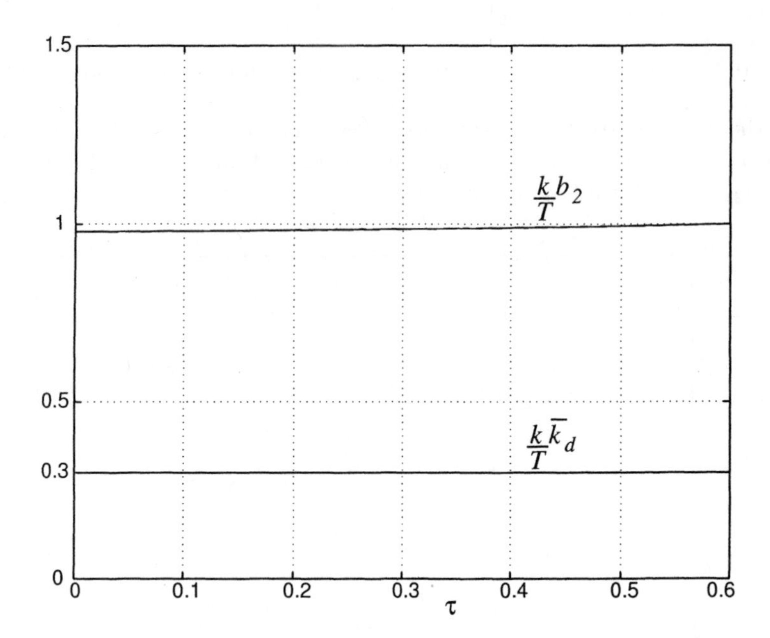

FIGURE 10.8. Comparison of $\frac{k}{T}b_2$ and $\frac{k}{T}\overline{k_d}$ for $0 < \tau < 0.6$.

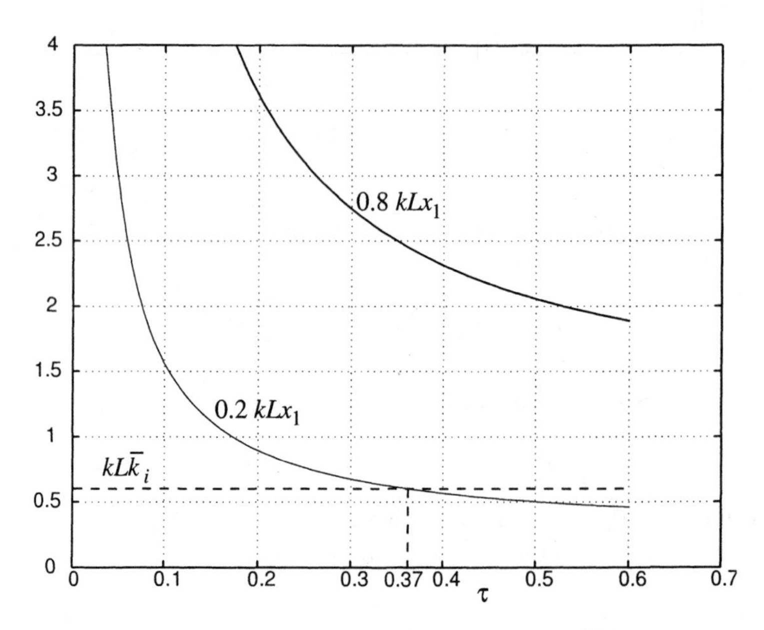

FIGURE 10.9. Comparison of $0.2kLx_1$, $0.8kLx_1$, and $kL\overline{k_i}$ for $0 < \tau < 0.6$.

Thus we conclude from the previous analysis that the CHR method gives a controller-robust PID controller for $\tau \in (0.37, 0.6)$, i.e., for these values of τ, we obtain a good parametric stability margin with respect to the controller coefficients.

10.4 The Cohen-Coon Method

Early papers on PID control introduced a tuning technique called *dominant pole design.* This technique attempts to position a few closed-loop poles to achieve certain control performance specifications. The Cohen-Coon method is a dominant pole design method based on the first-order model with time delay (10.1). It attempts to locate three closed-loop poles, a pair of complex poles, and one real pole, such that the amplitude decay ratio for load disturbance response is $\frac{1}{4}$ and the integral error $\int_0^\infty e(t)dt$ is minimized. Based on analytical and numerical computations, Cohen and Coon derived some formulas for the PID controller parameters in terms of the plant parameters k, T, and L. These formulas are

$$\overline{k_p} = \frac{1.35}{a}\left(1+\frac{0.18b}{1-b}\right) \tag{10.11}$$

$$\overline{k_i} = \frac{1.35}{aL}\left(1+\frac{0.18b}{1-b}\right)\left(\frac{1-0.39b}{2.5-2b}\right) \tag{10.12}$$

$$\overline{k_d} = \frac{1.35L}{a}\left(1+\frac{0.18b}{1-b}\right)\left(\frac{0.37-0.37b}{1-0.81b}\right) \tag{10.13}$$

where

$$a = \frac{kL}{T} \qquad b = \frac{L}{L+T}\,.$$

We start our robustness analysis by rewriting the parameter $\overline{k_p}$ as a function of $\tau = \frac{L}{T}$:

$$\overline{k_p} = \frac{1.35}{k}\left(\frac{1}{\tau}+0.18\right)\,. \tag{10.14}$$

If we now consider $k > 0$ and an open-loop stable plant ($\tau > 0$), then $\overline{k_p} > 0$. As in the previous section, we can plot $\overline{k_p}$ and the upper bound k_{upp} in (10.4) as a function of τ. These plots (scaled by k) are shown in Fig. 10.10. It is clear from this figure that $\overline{k_p} < k_{upp}$ for all values of τ. Thus, the Cohen-Coon method always provides a proportional gain value inside the stabilizing range dictated by Theorem 8.1.

We now analyze the location of the controller parameters $\overline{k_i}$ and $\overline{k_d}$ in the space of (k_i, k_d). From (10.12)– (10.13), we can express these parameters in terms of τ as follows:

$$\overline{k_i} = \frac{T}{kL^2}\left[0.54\frac{(1+0.18\tau)(1+0.61\tau)}{(1+0.2\tau)}\right] \tag{10.15}$$

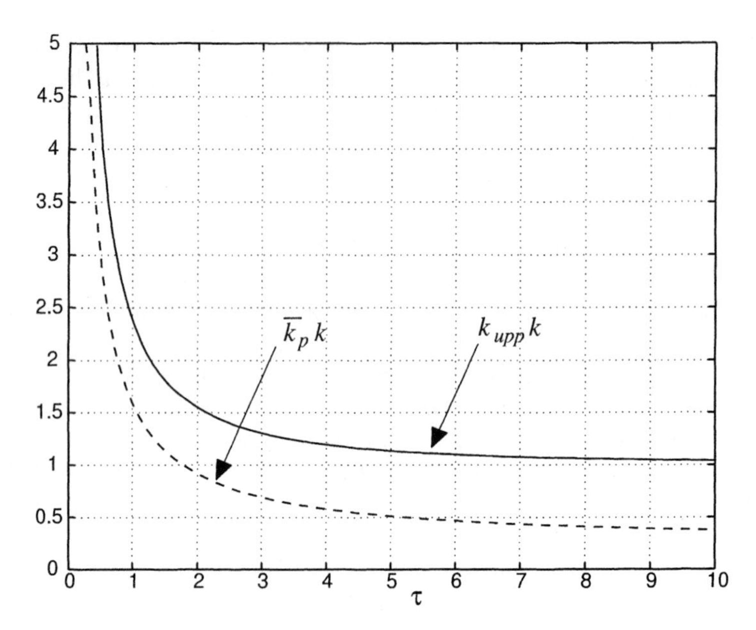

FIGURE 10.10. Comparison of $\overline{k_p}$ given by the Cohen-Coon method and the upper bound k_{upp}.

$$\overline{k_d} \quad = \quad \frac{T}{k}\left[0.4995\frac{(1+0.18\tau)}{(1+0.19\tau)}\right] \quad . \tag{10.16}$$

We now consider two different scenarios. These scenarios arise from the different geometries of the stabilizing sets in the space of $(k_i,\, k_d)$ as presented in Chapter 8.

Case 1: $\tau \geq 1.7834$. In this case, we have $0 < \overline{k_p} \leq \frac{1}{k}$, so the stabilizing set in the space of (k_i, k_d) is given by Figs. 8.6(a) or 8.6(b). Since $\tau > 0$ then $1+0.18\tau < 1+0.19\tau$. Thus from (10.16) it is not difficult to see that $0 < \overline{k_d} < 0.4995\frac{T}{k}$. This implies that the parameter $\overline{k_d}$ given by the Cohen-Coon method lies inside the stabilizing range of derivative gain values for $\tau \geq 1.7834$ (see Figs. 8.6(a) or 8.6(b)).

To study the robustness of the tuning technique, we now compare the parameter $\overline{k_i}$ in (10.15) with x_1 defined in Fig. 10.2. This parameter is given in this case by

$$x_1 = \frac{T}{kL^2}z_1\left[\tau\sin(z_1) + z_1\left(\cos(z_1) + 0.4995\frac{(1+0.18\tau)}{(1+0.19\tau)}\right)\right]$$

where z_1 is the solution of

$$1.35(1+0.18\tau) + \tau\cos(z) - z\sin(z) = 0$$

in the interval $(0, \pi)$. We now find the values of the parameter τ for which $\overline{k_i}$ lies inside the range $(0.2x_1, 0.8x_1)$. Figure 10.11 shows the plot of $0.2\frac{kL^2}{T}x_1$,

$0.8\frac{kL^2}{T}x_1$, and $\frac{kL^2}{T}\overline{k_i}$ as a function of τ. From this figure we conclude that for $1.7834 \le \tau < 8.53$ the parameter $\overline{k_i}$ lies inside the interval $(0.2x_1, 0.8x_1)$.

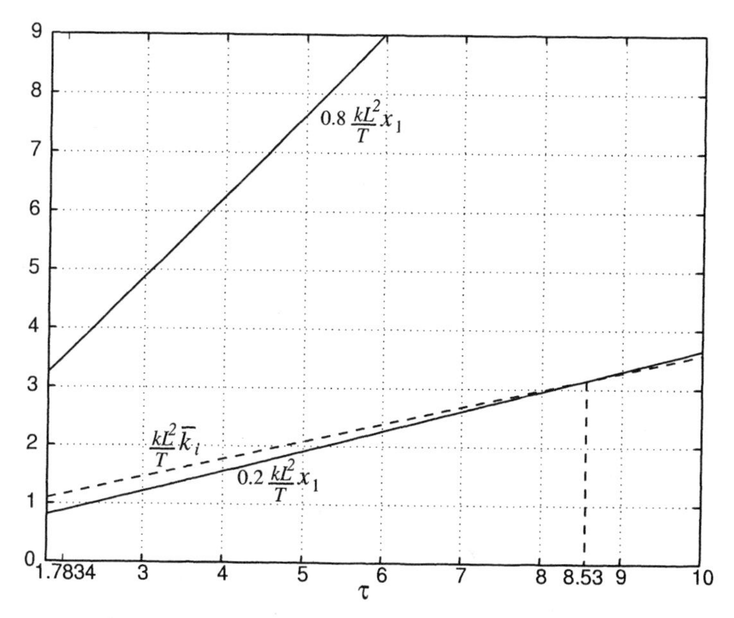

FIGURE 10.11. Comparison of $0.2\frac{kL^2}{T}x_1$, $0.8\frac{kL^2}{T}x_1$, and $\frac{kL^2}{T}\overline{k_i}$ for $\tau \ge 1.7834$.

Case 2: $\tau < 1.7834$. In this case $\frac{1}{k} < \overline{k_p} < k_{upp}$ and the stabilizing set is given by Fig. 8.6(c). First we compare the parameter $\overline{k_d}$ given by (10.16) with the parameter b_2 given by

$$b_2 = -\frac{T}{k}\left[\tau\frac{\sin(z_2)}{z_2} + \cos(z_2)\right]$$

where $z_2 > z_1 > 0$ is the solution of

$$1.35(1 + 0.18\tau) + \tau\cos(z) - z\sin(z) = 0$$

in the interval $(0, \pi)$. The plot of $\frac{k}{T}\overline{k_d}$ and $\frac{k}{T}b_2$ versus τ is shown in Fig. 10.12. From this figure it is clear that $\overline{k_d} < b_2$ for all $0 < \tau < 1.7834$. We next study the location of the parameter $\overline{k_i}$ to achieve some robustness. Figure 10.13 shows the plot of $0.2\frac{kL^2}{T}x_1$, $0.8\frac{kL^2}{T}x_1$, and $\frac{kL^2}{T}\overline{k_i}$ versus τ. It can be shown using this figure that the parameter $\overline{k_i}$ lies inside the interval $(0.2x_1, 0.8x_1)$ for all $0 < \tau < 1.7834$.

Thus we conclude from Cases 1 and 2 that the Cohen-Coon method gives controller-robust PID parameters in the sense of the parametric stability margin when the plant under study satisfies the condition $0 < \frac{L}{T} < 8.53$.

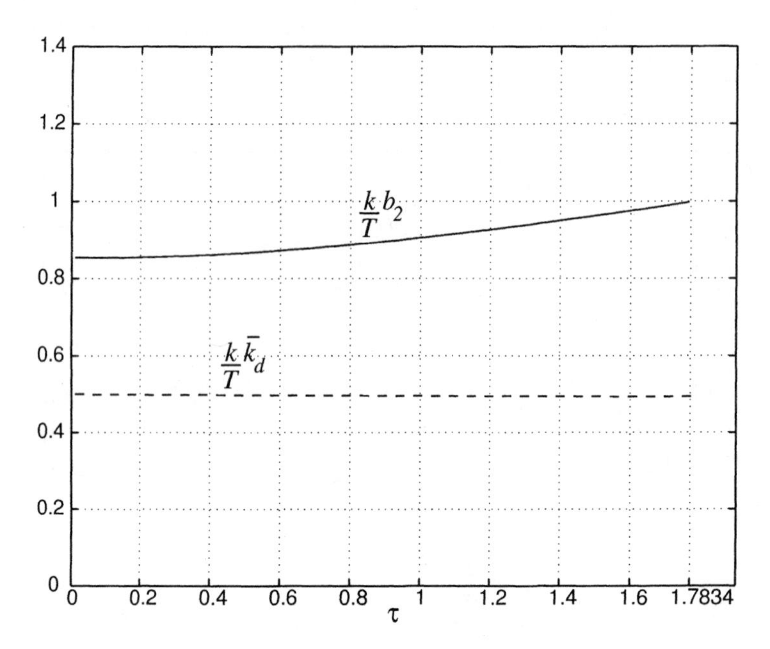

FIGURE 10.12. Comparison of $\frac{k}{T}b_2$ and $\frac{k}{T}\overline{k_d}$ for $0 < \tau < 1.7834$.

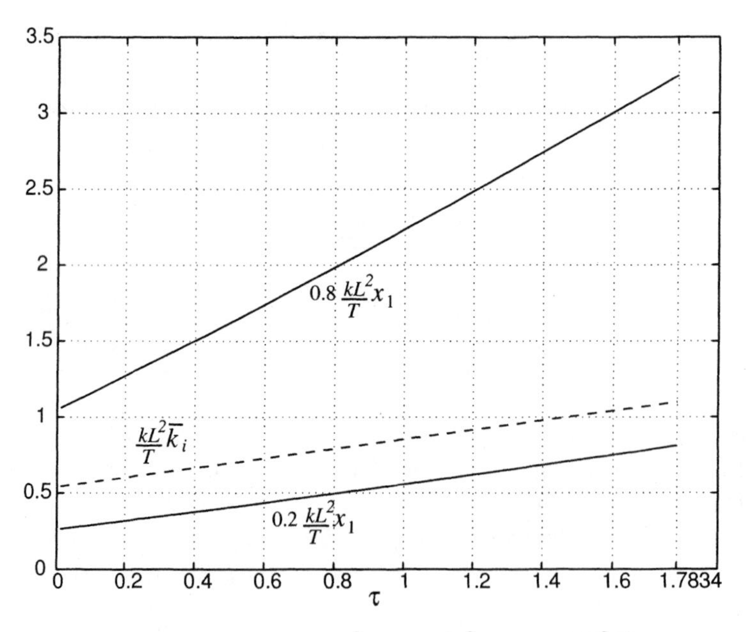

FIGURE 10.13. Comparison of $0.2\frac{kL^2}{T}x_1$, $0.8\frac{kL^2}{T}x_1$, and $\frac{kL^2}{T}\overline{k_i}$ for $0 < \tau < 1.7834$.

10.5 The IMC Design Technique

The IMC structure has become popular in process control applications. In this structure the controller includes an explicit model of the plant. This structure is particularly appropriate for the design and implementation of controllers for open-loop stable systems. The fact that many of the plants encountered in process control happen to be open-loop stable possibly accounts for the popularity of IMC among practicing engineers.

The IMC principle is a general method that can be applied to the design of PID controllers. The plant to be controlled is modeled as a first-order system with time delay as in (10.1), and the deadtime is approximated by a first-order Padé approximation. Following the standard IMC procedure, the following parameters are obtained for a PID controller:

$$\overline{k_p} = \frac{2T + L}{2k(L + \lambda)} \tag{10.17}$$

$$\overline{k_i} = \frac{1}{k(L + \lambda)} \tag{10.18}$$

$$\overline{k_d} = \frac{TL}{2k(L + \lambda)} \tag{10.19}$$

where $\lambda > 0$ is a small number. By properly selecting the design variable λ, the resulting PID controller can achieve a good compromise between performance and robustness. It has been suggested in the literature that a suitable choice for λ should satisfy $\lambda > 0.2T$ and $\lambda > 0.25L$.

We start our robustness analysis by studying the parameter $\overline{k_p}$. From (10.17), $\overline{k_p}$ can be rewriten in terms of $\tau = \frac{L}{T}$ as follows:

$$\overline{k_p} = \frac{1}{k}\left[\frac{2+\tau}{2\tau(1+\lambda/L)}\right]. \tag{10.20}$$

Since the plant is open-loop stable, then $\tau > 0$. Moreover, we know that $\lambda/L \geq 0$. Thus $\overline{k_p} > 0$. For different values of the parameter λ/L, we plot $\overline{k_p}$ and k_{upp} given by (10.4) versus τ. Figure 10.14 shows these plots (scaled by k) for $\lambda/L = 0, 0.25, 0.5, 1$. It is clear from this figure that for these values of λ/L, the parameter $\overline{k_p}$ given by the IMC design technique is inside the allowable range of stabilizing proportional gain values. For a fixed value of τ, we see from (10.20) that $\overline{k_p}$ is a monotonically decreasing function of λ/L. We have also seen from Fig. 10.14 that for $\lambda/L = 0$, the corresponding $\overline{k_p}$ is less than k_{upp} for all $\tau > 0$. Thus for any value of $\lambda/L > 0$, the corresponding $\overline{k_p}$ is less than k_{upp} for all $\tau > 0$. We conclude that for any $\lambda/L > 0$, the IMC design technique provides a stabilizing proportional gain value.

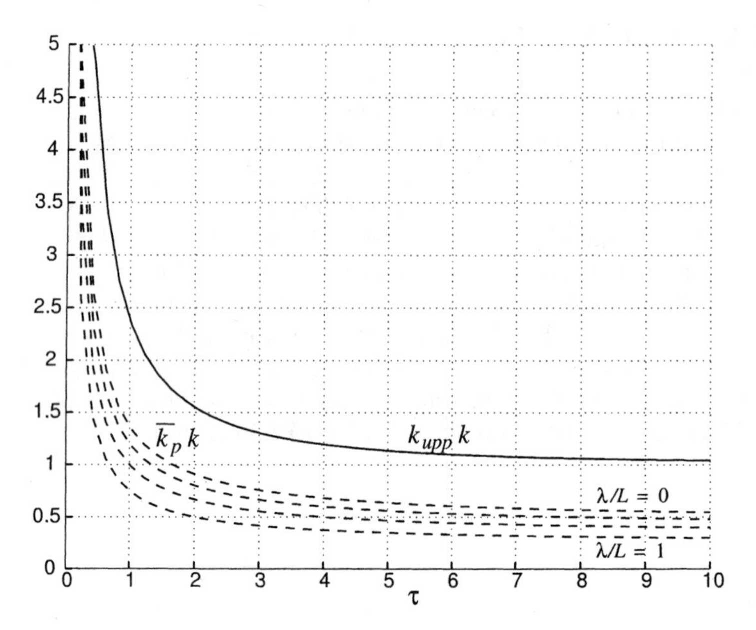

FIGURE 10.14. Comparison of $\overline{k_p}$ given by the IMC design technique and the upper bound k_{upp}.

Next we express the parameters $\overline{k_i}$ and $\overline{k_d}$ given by (10.18) and (10.19), respectively, as functions of λ/L:

$$\overline{k_i} = \frac{1}{kL}\left(\frac{1}{1+\lambda/L}\right) \tag{10.21}$$

$$\overline{k_d} = \frac{T}{k}\left(\frac{0.5}{1+\lambda/L}\right). \tag{10.22}$$

As in the previous sections, we now consider two different cases.

Case 1: $\tau \geq \frac{2}{1+2\lambda/L}$. Thus $0 < \overline{k_p} \leq \frac{1}{k}$ and the stabilizing region in the (k_i, k_d) space is given by Figs. 8.6(a) or 8.6(b). Since $\lambda/L > 0$, it is not difficult to show from (10.22) that

$$0 < \overline{k_d} < 0.5\frac{T}{k} \quad \text{for any } \lambda/L > 0.$$

Thus the parameter $\overline{k_d}$ provided by the IMC design technique lies, in this case, inside the stabilizing set of derivative gain values as illustrated in Fig. 10.2.

We next turn our attention to the parameter $\overline{k_i}$. We want to find the values of τ for which $\overline{k_i}$ lies inside the range $(0.2x_1, 0.8x_1)$, where x_1 is given by

$$x_1 = \frac{1}{kL}z_1\left[\sin(z_1) + \frac{z_1}{\tau}\left(\cos(z_1) + \frac{0.5}{1+\lambda/L}\right)\right]$$

and z_1 is the solution of

$$\frac{2+\tau}{2(1+\lambda/L)} + \tau\cos(z) - z\sin(z) = 0$$

in the interval $(0,\pi)$. By fixing the parameter λ/L at the values 0.1, 0.25, 0.5, 1, we can plot $kL\overline{k_i}$, $0.2kLx_1$, and $0.8kLx_1$ versus τ. Figure 10.15 shows these plots. From these figures it is clear that $\overline{k_i}$ lies inside the suggested range for all $\tau \geq \frac{2}{1+2\lambda/L}$, where $\lambda/L \in [0.1, 1]$.

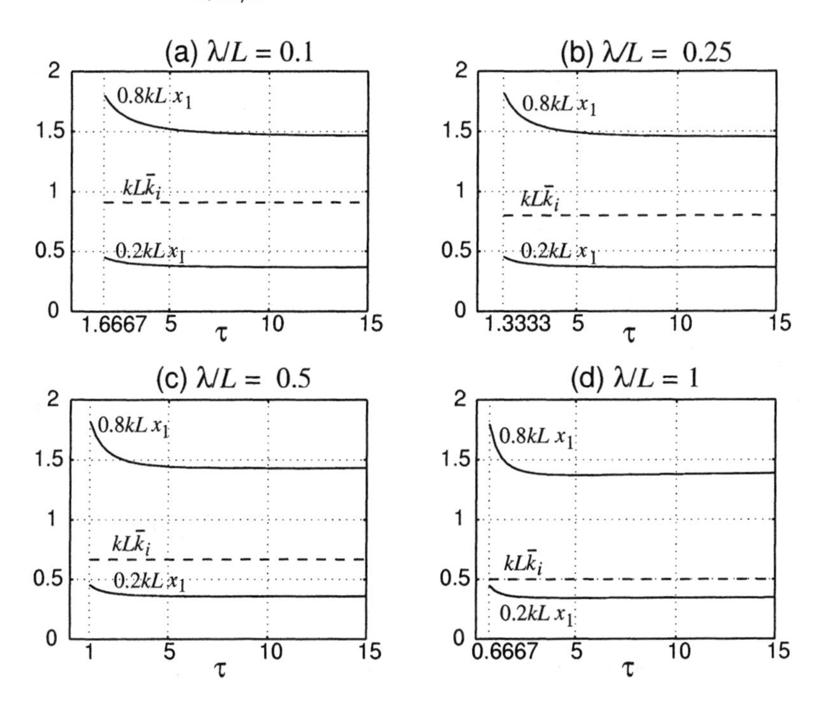

FIGURE 10.15. Comparison of $0.2kLx_1$, $0.8kLx_1$, and $kL\overline{k_i}$ for $\tau \geq \frac{2}{1+2\lambda/L}$.

Case 2: $0 < \tau < \frac{2}{1+2\lambda/L}$. In this case, $\frac{1}{k} < \overline{k_p} < k_{upp}$ and the stabilizing region in the (k_i, k_d) space is given by Fig. 8.6(c). We first analyze the parameter $\overline{k_d}$ and compare it with b_2 given by

$$b_2 = -\frac{T}{k}\left[\tau\frac{\sin(z_2)}{z_2} + \cos(z_2)\right]$$

where $z_2 > z_1$ is the solution of

$$\frac{2+\tau}{2(1+\lambda/L)} + \tau\cos(z) - z\sin(z) = 0$$

in the interval $(0,\pi)$. By ploting $\frac{k}{T}\overline{k_d}$ and $\frac{k}{T}b_2$ versus τ for different values of the parameter λ/L, it can be shown that $\overline{k_d} < b_2$ for all $0 < \tau < \frac{2}{1+2\lambda/L}$ and all $\lambda/L > 0$.

We now examine the parameter $\overline{k_i}$ and find the values of τ for which $\overline{k_i}$ lies inside the range $(0.2x_1, 0.8x_1)$. Figure 10.16 shows the plots of $0.2kLx_1$, $0.8kLx_1$, and $kL\overline{k_i}$ versus τ for $\lambda/L = 0.1, 0.25, 0.5, 1$. From these plots

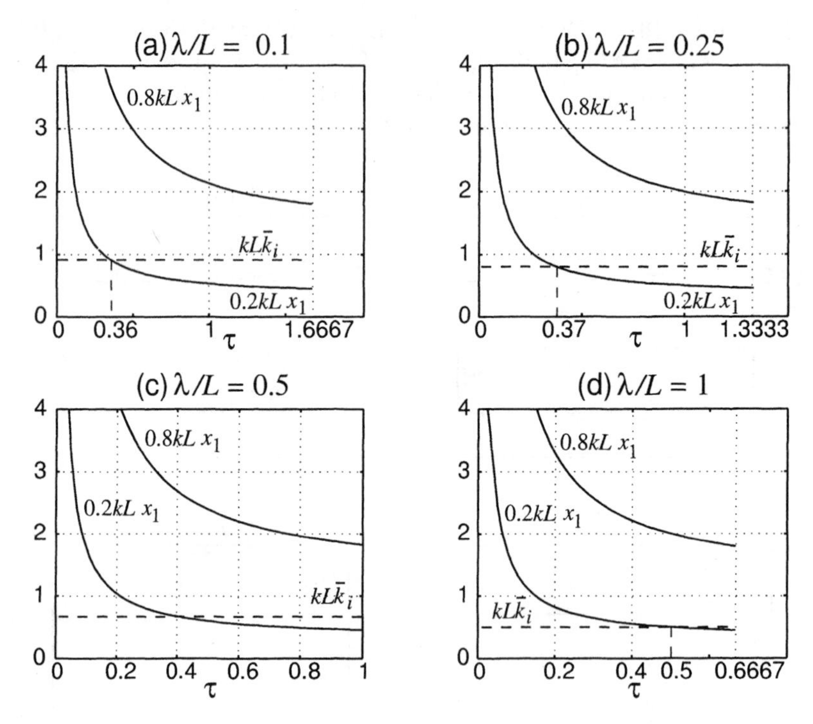

FIGURE 10.16. Comparison of $0.2kLx_1$, $0.8kLx_1$, and $kL\overline{k_i}$ for $0 < \tau < \frac{2}{1+2\lambda/L}$.

we can find the range of τ values for which the parameter $\overline{k_i}$ lies inside the interval $(0.2x_1, 0.8x_1)$. For example, for $\lambda/L = 0.25$, the range of τ is $(0.37, \infty)$.

Thus from the previous analysis we conclude that the robustness with respect to the controller parameters depends not only on the ratio $\frac{L}{T}$ but also on the parameter λ/L. The following values of τ guarantee a good parametric stability margin with respect to the controller:

λ/L	τ
0.1	$(0.36, \infty)$
0.25	$(0.37, \infty)$
0.5	$(0.41, \infty)$
1	$(0.50, \infty)$

Remark 10.2 *It is commonly recommended that λ/L should be fixed at 0.25. A ratio of $\lambda/L = 0.25$ offers a good compromise between performance and robustness. However, from the previous table, it can be seen that as λ/L*

increases, the lower bound on τ for controller-robustness increases. Thus if the plant under analysis has a small τ, a smaller λ/L should be selected.

10.6 Summary

In this chapter we have presented an analysis of the robustness of some common PID tuning techniques. This analysis was motivated by the fact that a good PID controller design should exhibit robustness with respect to small perturbations in the controller coefficients. Our criterion was to ensure first that the controller parameters k_p and k_d were inside the stabilizing set of gain values. Then the parameter k_i was forced to lie inside a box located 20% of x_1 from the boundaries of the stabilizing set in the (k_i, k_d) space. Here, x_1 represents the maximum stabilizing integral gain value for the fixed proportional and derivative gains provided by the particular tuning rule. As a result of this criterion, the range of $\frac{L}{T}$ values that ensures robustness was determined for each tuning technique. These values are summarized below:

$$\begin{aligned} \text{Ziegler-Nichols step response method}: &\quad 0 < \frac{L}{T} < 1.07 \\ \text{CHR method}: &\quad 0.37 < \frac{L}{T} \\ \text{Cohen-Coon method}: &\quad 0 < \frac{L}{T} < 8.53 \\ \text{IMC design technique}: &\quad 0.37 < \frac{L}{T} \quad (\text{for } \lambda/L = 0.25) \end{aligned}$$

10.7 Notes and References

The classical step respond method developed by Ziegler and Nichols first appeared in [54]. A detailed treatment of this method and interesting variations of it can be found in the book by Astrom and Hagglund [2]. The CHR method developed by Chien, Hrones, and Reswick can be found in [7]. The dominant pole design method developed by Cohen and Coon was introduced in [9]. For an excellent description of the IMC structure and its applications to process control, the reader is referred to the book by Morari and Zafiriou [31]. The results presented in this chapter are based on [43] and [45].

11

PID Stabilization of Arbitrary Linear Time-Invariant Systems with Time Delay

In this chapter we present an approach for solving the problem of finding the set of all PID controllers that stabilize an *arbitrary*-order plant with time delay. The results presented in Chapters 6 through 8 do not readily extend to the case of higher-order plants with time delay, and an alternative procedure is presented here. This procedure is based on a connection linking Pontryagin's results on quasi-polynomials to the Nyquist criterion.

11.1 Introduction

In some of the previous chapters, we have concentrated on developing tools for analyzing the stability of feedback systems containing first-order plants with time delay. These tools were later used to design PID controllers that satisfy performance specifications while maintaining some degree of robustness. Throughout the book we have emphasized the importance of knowing *a priori* the set of PID controllers that stabilize a given first-order plant with time delay. This set was characterized using the results presented in Chapter 5, where we introduced the concept of a quasi-polynomial. While first-order systems with time delays are widely used to model the behavior of industrial plants, it is also of interest to study the more general case of nth order systems with time delay.

This chapter deals with the problem of stabilizing an arbitrary-order plant with time delay. Although the results from Chapters 6 through 8 are important in their own right, they do not readily extend to the case of higher-order plants. On the other hand, a generalized Nyquist criterion developed by Tsypkin has often been used to *analyze* arbitrary-order plants with time delay. Its graphical simplicity provides a promising tool for attacking the *synthesis* problem of PID controllers. However, unlike Pontryagin's theory, the generalization of the Nyquist criterion presented in the literature lacks a solid theoretical justification. This is because the proof of Tsypkin's generalization may be inappropriate if the closed-loop system has an unbounded number of RHP poles. In this chapter, the conditions under which one can use the generalized Nyquist criterion are derived based on the material presented in Chapter 5. Based on this, a method to compute the complete set of PID controllers to stabilize a given arbitrary-order plant with time delay is developed.

The chapter is organized as follows. Section 11.2 is devoted to a study of the generalized Nyquist criterion and the conditions under which it is applicable. Section 11.3 introduces the control problem that will be treated in this chapter. In Sections 11.4 through 11.6 we apply the generalized Nyquist criterion to solve the constant gain, PI, and PID stabilization problems for arbitrary-order plants with time delay.

11.2 A Study of the Generalized Nyquist Criterion

The generalized Nyquist criterion developed by Tsypkin extends the well-known Nyquist criterion to systems with time delay. However, as the following example shows, this criterion can lead to misleading conclusions when applied to such a system.

Example 11.1 *Given a system with nominal open-loop transfer function*

$$G(s) = \frac{2s+1}{s+2}, \tag{11.1}$$

we can draw its Nyquist plot, as shown in Fig. 11.1.

The closed-loop system is stable with unity negative feedback and the plot intersects the unit circle at $\omega_0 = 1$. Thus from the graph, using Tsypkin's result, the closed-loop system can tolerate a time delay up to

$$L_0 = \frac{\pi + \arg[G(j\omega_0)]}{\omega_0} = 3.7851\ .$$

However, when we add a 1-second delay to the nominal transfer function, the closed-loop system becomes unstable, as shown in Fig. 11.2.

△

In this section, we use theorems from Chapter 5 to derive conditions under which a modified generalized Nyquist criterion can be used to correctly analyze the stability of a system with time delay.

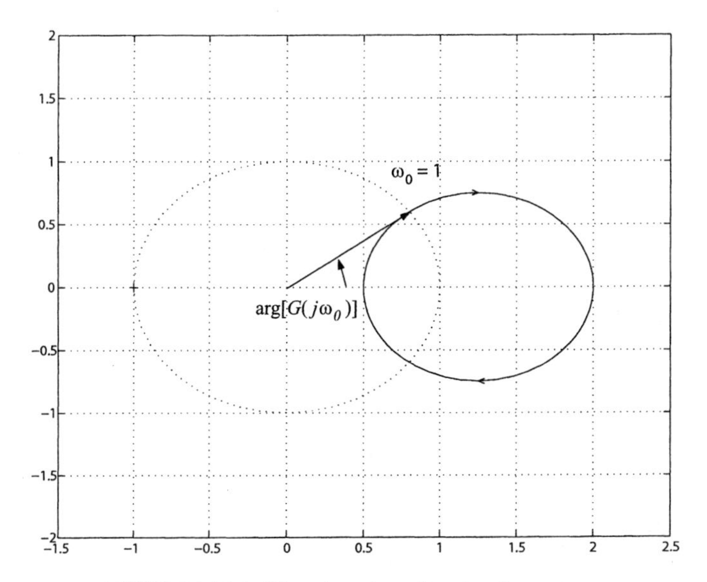

FIGURE 11.1. Nyquist plot of a simple system.

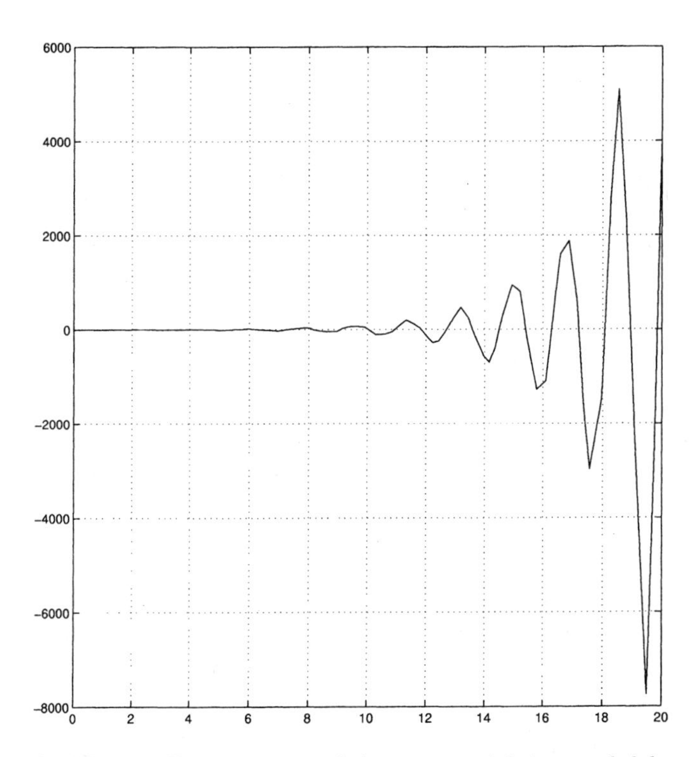

FIGURE 11.2. Time response of the system with 1-second delay.

As in Section 5.4, let $f(s,t)$ be a polynomial in the two variables s and t with constant coefficients, i.e.,

$$f(s,t) = \sum_{h=0}^{M} \sum_{k=0}^{N} a_{hk} s^h t^k . \tag{11.2}$$

Moreover, as before, define $F(s) = f(s, e^s)$. As in Definition 5.1, we denote the principal term of $f(s,t)$ by $a_{MN} s^M t^N$. Following (5.22), we can write (11.2) as

$$f(s,t) = X_M^{(N)}(t)s^M + X_{M-1}^{(N)}(t)s^{M-1} + \cdots + X_1^{(N)}(t)s + X_0^{(N)}(t) \tag{11.3}$$

where $X_h^{(N)}(t), h = 0, 1, 2, \ldots, M$ are polynomials in t with degree at most equal to N.

The reader should recall that Theorem 5.4 provided more information about the existence of roots of the function $F(s)$ in the open right half of the complex plane. We note that in Theorem 5.4, the situation when $X_M^{(N)}(e^s)$ has zero(s) on the imaginary axis is not mentioned. We will look into this more deeply. Let us look at the distribution of the zeros of $F(s)$ when $|s| \to \infty$. As $|s| \to \infty$, $F(s) = 0$ can be approximated as $X_M^{(N)}(e^s) = 0$. That means the roots of $X_M^{(N)}(e^s) = 0$ determine the zeros of $F(s)$ at infinity. We have seen in Section 5.5 that the roots of $F(s)$ form certain chains and they go deep into the LHP, the RHP, or go to infinity within strips with finite real parts. Thus if $X_M^{(N)}(e^s)$ has zeros on the imaginary axis, $F(s)$ has root chains that approach the imaginary axis at infinity.

The following theorem gives us the conditions that should be satisfied when using the Nyquist criterion with the conventional Nyquist contour (the contour consisting of the imaginary axis and a semicircle of arbitrarily large radius in the RHP).

Theorem 11.1 *Suppose we are given a unity feedback system with an open-loop transfer function*

$$G(s) = G_0(s)e^{-Ls} = \frac{N(s)}{D(s)} e^{-Ls}$$

where $N(s)$ and $D(s)$ are real polynomials of degree p and q, respectively, and L is a fixed delay. Then we have the following conclusions:

1. *If $q < p$, or $q = p$ and $|\frac{b_q}{a_q}| \geq 1$, where a_q, b_p are the leading coefficients of $D(s)$ and $N(s)$, respectively, the Nyquist criterion is not applicable and the system is unstable according to Pontryagin's theorems.*

2. *If $q > p$, or $q = p$ and $|\frac{b_q}{a_q}| < 1$, the Nyquist criterion is applicable and we can use it to check the stability of the closed-loop system.*

Proof. The characteristic equation of the closed-loop system is

$$\delta(s) = D(s) + N(s)e^{-Ls} . \tag{11.4}$$

Multiply (11.4) by e^{Ls} to obtain

$$\delta^*(s) = \delta(s)e^{Ls} = D(s)e^{Ls} + N(s) . \tag{11.5}$$

Let $z = Ls$ so that

$$\delta^*(z) = D_z(z)e^z + N_z(z) , \tag{11.6}$$

with $N_z(z)$ and $D_z(z)$ appropriately defined. Note that for $L > 0$, both the above operations do not affect the number of RHP roots of the original equation. With

$$\begin{aligned} D(s) &= a_q s^q + a_{q-1}s^{q-1} + \cdots + a_1 s + a_0 \\ N(s) &= b_p s^p + b_{p-1}s^{p-1} + \cdots + b_1 s + b_0 , \end{aligned}$$

we have

$$\begin{aligned} D_z(z) &= a_q L^{-q} z^q + a_{q-1}L^{-q+1}z^{q-1} + \cdots + a_1 L^{-1} z + a_0 \\ N_z(z) &= b_p L^{-p} z^p + b_{p-1}L^{-p+1}z^{p-1} + \cdots + b_1 L^{-1} z + b_0 . \end{aligned}$$

Now we will discuss the possible stability of (11.6) in the following three cases.

1. $\deg[D_z(z)] < \deg[N_z(z)]$, i.e., $q < p$.
   In this case, $\delta^*(z)$ does not have a principal term. According to Theorem 5.1, it has an unbounded number of RHP roots. The Nyquist criterion is inapplicable but we already know that $\delta^*(z)$ is unstable.

2. $\deg[D_z(z)] > \deg[N_z(z)]$, i.e., $q > p$.
   Here $\delta^*(z)$ has the principal term $a_q L^{-q} z^q e^z$. The coefficient of z^q is

   $$X_q^{(1)}(e^z) = \frac{a_q}{L^q}e^z ,$$

   and the solution for $X_q^{(1)}(e^z) = 0$ is $z = -\infty$, which lies in the LHP. So by Theorem 5.4, $\delta^*(z)$ can only have a bounded set of RHP zeros. This bounded set is also a finite set, and the Nyquist criterion can be used for stability analysis.

3. $\deg[D_z(z)] = \deg[N_z(z)]$, i.e., $q = p$.
   $\delta^*(z)$ has the principal term $a_q L^{-q} z^q e^z$ in this case too. However, the coefficient of z^q is

   $$X_q^{(1)}(e^z) = \frac{a_q}{L^q}e^z + \frac{b_q}{L^q} .$$

   To make $X_q^{(1)}(e^z) = 0$, we must have $e^z = -\frac{b_q}{a_q}$. Let $z = x + jy$ and $x, y \in \mathcal{R}$, then we have $e^x e^{jy} = -\frac{b_q}{a_q}$. The solutions are

- Case 1: $\frac{b_q}{a_q} > 0$. Then $e^x = |\frac{b_q}{a_q}|, e^{jy} = -1$ so that

$$x = \ln |\frac{b_q}{a_q}|, y = 2k\pi + \pi, k \in \mathbf{Z} .$$

- Case 2: $\frac{b_q}{a_q} < 0$. Then $e^x = |\frac{b_q}{a_q}|, e^{jy} = 1$ so that

$$x = \ln |\frac{b_q}{a_q}|, y = 2k\pi, k \in \mathbf{Z} .$$

Depending on the value of $|\frac{b_q}{a_q}|$, we will arrive at different conclusions:

(3a) If $|\frac{b_q}{a_q}| > 1$, then $x > 0$, i.e., $X_q^{(1)}$ has RHP zeros. So $\delta^*(z)$ has an unbounded set of RHP zeros. Again, the Nyquist criterion is inapplicable but the closed-loop system is unstable.

(3b) If $|\frac{b_q}{a_q}| < 1$, then $x < 0$, i.e., $X_q^{(1)}$ has only LHP zeros. So $\delta^*(z)$ has no more than a bounded and finite set of RHP zeros and the closed-loop stability can be ascertained using the Nyquist criterion.

(3c) If $|\frac{b_q}{a_q}| = 1$, then $x = 0$, i.e., $X_q^{(1)}$ has zeros on the imaginary axis. So $\delta^*(z)$ has root chains approaching the imaginary axis, and is, therefore, unstable. The Nyquist criterion is inapplicable in this case also.

Since for a fixed $L > 0$, $\delta^*(z)$ has the same number of RHP zeros as $\delta(s)$, from the above analysis, we can see that in cases (1), (3a), and (3c), $\delta(s)$ is unstable, while in cases (2) and (3b), $\delta(s)$ has no more than a bounded set of zeros in the RHP; hence it is possibly stable.

So only in cases (2) and (3b), the Nyquist criterion can be used to ascertain possible stability. Thus Tsypkin's results and the proof of his generalized Nyquist criterion are valid only for these two cases. ■

Remark 11.1 *In all fairness, it is appropriate to point out that most likely Tsypkin assumed the plant to be strictly proper, though he did not state it explicitly in the literature. For our purposes, attaching a PID controller to a proper or strictly proper plant opens up the very real possibility of ending up with an improper or a proper open-loop transfer function. This is the reason that the above study had to be undertaken.*

11.3 Problem Formulation and Solution Approach

Theorem 11.1 sets the stage for determining all stabilizing P, PI, and PID controllers for plants with time delay. As in previous chapters, we will

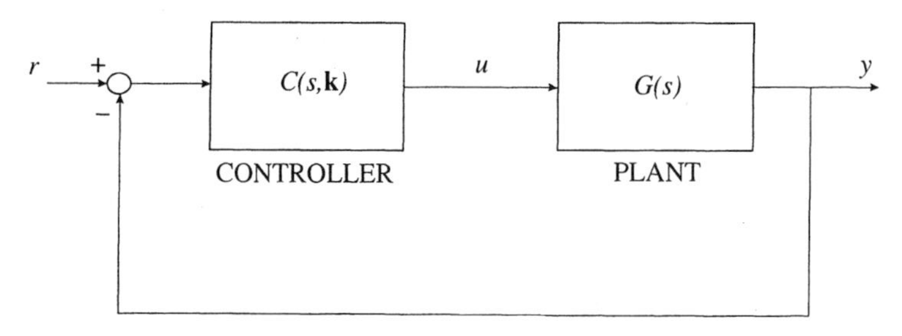

FIGURE 11.3. Feedback control system.

study the feedback control system illustrated in Fig. 11.3. In this figure, $G(s)$ represents a given linear time-invariant plant with time delay L,

$$G(s) = G_0(s)e^{-Ls} = \frac{N(s)}{D(s)}e^{-Ls} \tag{11.7}$$

and $C(s, \mathbf{k})$ represents a fixed-structure controller in the unity feedback configuration, with $\mathbf{k}$ being the vector of adjustable parameters. The problem of interest is to find the complete set of $\mathbf{k}$s that can stabilize the system for any $L \in [0, L_0]$.

The approach developed in this chapter to solve this problem involves the following steps:

1. Find the complete set of $\mathbf{k}$s that stabilize the delay-free plant $G_0(s)$ and denote this set by $\mathcal{S}_0$.

2. Define the set $\mathcal{S}_N$, which is the set of $\mathbf{k}$s such that $C(s, \mathbf{k})G_0(s)$ is an improper transfer function or

$$\mathcal{S}_N = \left\{\mathbf{k} \mid \lim_{s\to\infty} |C(s, \mathbf{k})G_0(s)| \geq 1\right\} . \tag{11.8}$$

   Note that the elements in $\mathcal{S}_N$ make the closed-loop system unstable after the delay is introduced (Theorem 11.1). Next we exclude $\mathcal{S}_N$ from $\mathcal{S}_0$ and denote the new set by $\mathcal{S}_1$, i.e., $\mathcal{S}_1 = \mathcal{S}_0 \backslash \mathcal{S}_N$.

3. Compute the set $\mathcal{S}_L$:

$$\begin{aligned} \mathcal{S}_L \;=\; & \{\mathbf{k} | \mathbf{k} \notin \mathcal{S}_N \text{ and } \exists\, L \in [0, L_0], \omega \in \mathbf{R}, \text{s.t.} \\ & C(j\omega, \mathbf{k})G_0(j\omega)e^{-jL\omega} = -1\} . \end{aligned} \tag{11.9}$$

   From this definition, $\mathcal{S}_L$ is the set of $\mathbf{k}$s that make $C(s, \mathbf{k})G(s)$ have a minimal destabilizing delay that is less than or equal to L_0.

4. The set $\mathcal{S}_R \triangleq \mathcal{S}_1 \backslash \mathcal{S}_L$ is the solution to our problem.

This is formalized in the following theorem.

Theorem 11.2 *The set $\mathcal{S}_R$ defined above is the complete set of controllers in the unity feedback configuration that stabilize the plant $G(s)$ with delay L going from 0 up to L_0.*

Proof. For any $\mathbf{k}_0 \in \mathcal{S}_R$, since $\mathcal{S}_R \subseteq \mathcal{S}_1 \subseteq \mathcal{S}_0$, we have $\mathbf{k}_0 \in \mathcal{S}_0$, i.e., there is no RHP pole when the controller $C(s, \mathbf{k}_0)$ is applied to the plant $G(s)$ with $L = 0$. Since $\mathbf{k}_0 \notin \mathcal{S}_N$, with the increase of L, there is no unbounded RHP pole (Theorem 11.1) and the possible RHP poles are the poles that come from the LHP by crossing the imaginary axis. However, from $\mathbf{k}_0 \notin \mathcal{S}_L$, we know that there are no boundary crossing poles. So, the closed-loop system does not have RHP poles for L going from 0 to L_0 and it is, therefore, stable for those Ls.

We next consider $\mathbf{k}_1 \notin \mathcal{S}_R$. Such a $\mathbf{k}_1$ must fall into one or more of the following categories.

1. $\mathbf{k}_1 \notin \mathcal{S}_0$, which means the controller cannot even stabilize the delay-free plant ($L = 0$).

2. $\mathbf{k}_1 \in \mathcal{S}_N$, which means that the closed-loop system is unstable regardless of the size of the delay (Theorem 11.1).

3. $\mathbf{k}_1 \in \mathcal{S}_L$, which means that some of the poles are on the imaginary axis for certain $L_1 \leq L_0$. These poles will either go into the RHP or return to the LHP. However, the stability at that L_1 has already been destroyed.

We can see from the above analysis that $\mathcal{S}_R$ is exactly the complete set of stabilizing controller parameters that is of interest to us. ■

In the rest of this chapter, this general method will be applied to the special case of P, PI, and PID controllers to find all the P, PI, and PID controllers that can stabilize a given plant with time delay up to a certain value.

11.4 Stabilization Using a Constant Gain Controller

In this section we consider the problem of determining the complete set of constant gain controllers that stabilize a general linear time-invariant plant with time delay. For a proportional controller, we have

$$C(s) = k_p$$

and the plant $G(s)$ is given by (11.7). Our objective is to find all the k_p values that stabilize $G(s)$ with time delay $L \in [0, L_0]$.

To implement the method proposed in Section 11.3, the key step is to find $\mathcal{S}_L$. The Nyquist curve of the system crossing $(-1, 0)$ is equivalent to

$k_p G_0(j\omega)e^{-jL\omega} = -1$ for certain L and ω. This, in turn, is equivalent to the following two conditions:

$$\arg[k_p G_0(j\omega)] - L\omega = 2h\pi - \pi, h \in \mathbf{Z} \tag{11.10}$$

$$|k_p G_0(j\omega)| = 1\,. \tag{11.11}$$

Here by convention, we restrict the range of the argument function to the interval $[-\pi, \pi)$. Also we only need to explicitly consider $\omega > 0$ since the Nyquist plot for $\omega < 0$ can be obtained from symmetry. Since we are interested in the minimal positive L which satisfies (11.10), (11.11), the phase condition (11.10) can be rewritten as

$$\arg[k_p G_0(j\omega)] - L\omega = -\pi\,.$$

Note that such a reasoning also applies to the PI and PID cases, to be considered later.

The two conditions above yield

$$L(\omega, k_p) = \frac{\arg[k_p G_0(j\omega)] + \pi}{\omega} \tag{11.12}$$

$$k_p(\omega) = \pm\frac{1}{|G_0(j\omega)|}. \tag{11.13}$$

For $k_p > 0$, we have

$$L(\omega, k_p) = L(\omega) = \frac{\arg[G_0(j\omega)] + \pi}{\omega}\,.$$

We next solve $L(\omega) \leq L_0$ to get the set Ω^+ defined as

$$\Omega^+ = \left\{\omega \in \mathbf{R}^+ \mid L(\omega) \leq L_0\right\}\,.$$

From the magnitude condition (11.11), we can get a set of positive k_p gain values corresponding to Ω^+; let us call this set $\mathcal{S}_L^+$. This set consists of all the positive k_p gain values that make the system have poles on the imaginary axis for certain $L \leq L_0$. Similarly, for $k_p < 0$, we will have a set Ω^- and a corresponding set $\mathcal{S}_L^-$.

Now the union of $\mathcal{S}_L^+$ and $\mathcal{S}_L^-$ is the complete set $\mathcal{S}_L$, i.e., $\mathcal{S}_L = \mathcal{S}_L^+ \cup \mathcal{S}_L^-$. The above discussion leads to the following steps for computing $\mathcal{S}_R$.

1. Compute the delay-free stabilizing k_p set, $\mathcal{S}_0$, either by the Routh-Hurwitz criterion or the method proposed in Chapter 3.

2. Find $\mathcal{S}_N$ defined as

$$\mathcal{S}_N = \left\{k_p \in \mathbf{R} \mid \lim_{s\to\infty}\left|k_p\frac{N(s)}{D(s)}\right| \geq 1\right\}\,.$$

- If $\deg[N(s)] > \deg[D(s)]$, $\mathcal{S}_N = \mathbf{R}$, which means $\mathcal{S}_R = \emptyset$.
- If $\deg[N(s)] < \deg[D(s)]$, $\mathcal{S}_N = \emptyset$.
- If $\deg[N(s)] = \deg[D(s)]$, $\mathcal{S}_N = \{k_p | \, |k_p| \geq |\frac{a_q}{b_q}|\}$, where a_q, b_q are the leading coeffients of $D(s)$ and $N(s)$, respectively.

3. Compute $\mathcal{S}_1 = \mathcal{S}_0 \backslash \mathcal{S}_N$.

4. Compute $\mathcal{S}_L$ according to the analysis in this section.

5. Compute $\mathcal{S}_R = \mathcal{S}_1 \backslash \mathcal{S}_L$.

We next present a numerical example to illustrate the above steps.

Example 11.2 *Consider the problem of finding all proportional controllers that stabilize the plant*

$$G(s) = \frac{s^2 + 3s - 2}{s^3 + 2s^2 + 3s + 2} e^{-Ls} \tag{11.14}$$

with delay up to $L_0 = 1.8$ *seconds.*

For the delay-free plant, the stabilizing k_p *range is* $\mathcal{S}_0 = (-0.4093, 1)$. *Since* $\deg[N(s)] = 2 < 3 = \deg[D(s)]$, $\mathcal{S}_N = \emptyset$ *and* $\mathcal{S}_1 = \mathcal{S}_0$.

Next, we obtain the plot of $L(\omega)$ *versus* ω *for* $k_p > 0$, *as shown in Fig. 11.4. From this figure we see that for* $k_p > 0$, $\Omega^+ = [1.5129, +\infty)$. *To determine the corresponding* $\mathcal{S}_L^+$ *set, we plot* $|k_p(\omega)|$ *as a function of* ω. *This plot is shown in Fig. 11.5, and from this figure we see that* $\mathcal{S}_L^+ = [0.4473, +\infty)$.

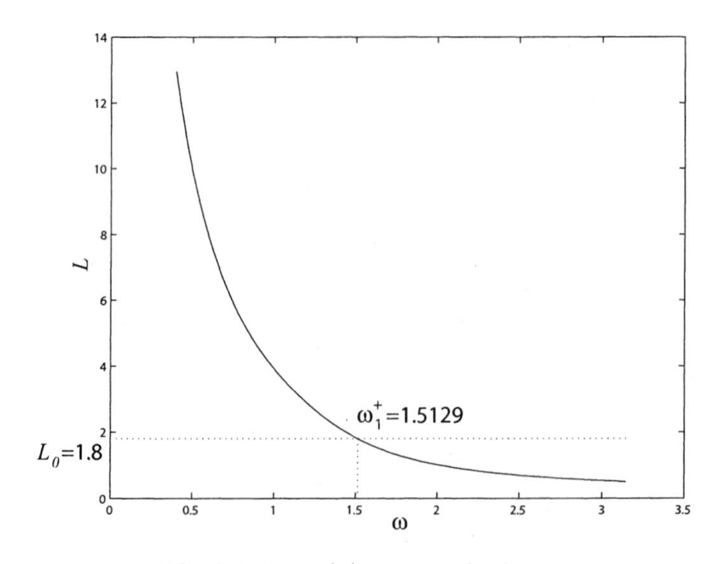

FIGURE 11.4. $L(\omega)$ versus ω for $k_p > 0$.

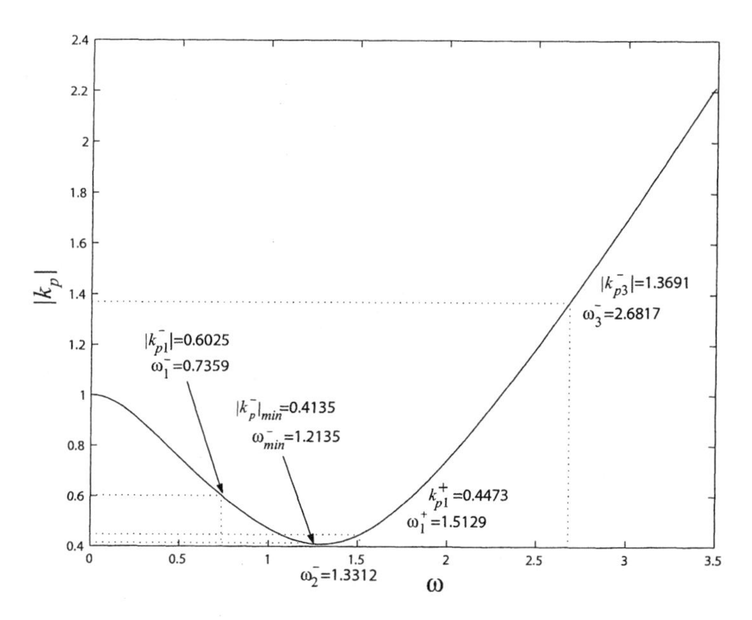

FIGURE 11.5. $|k_p(\omega)|$ versus ω.

Similarly, for $k_p < 0$, we plot $L(\omega)$ as a function of ω and obtain $\Omega^- = [0.7359, 1.3312] \cup [2.6817, +\infty)$ *(see Fig. 11.6). The corresponding* $\mathcal{S}_L^- = (-\infty, -1.3691] \cup [-0.6025, -0.4135]$ *(see Fig. 11.5, which is the same figure shared by the case $k_p > 0$).*

The set of all stabilizing k_ps for the plant with time delay up to 1.8 *is*

$$\begin{aligned} \mathcal{S}_R &= \mathcal{S}_1 \backslash \mathcal{S}_L \\ &= (-0.4093, 1) \backslash \{(-\infty, -1.3691] \cup [-0.6025, -0.4135] \cup [0.4473, +\infty)\} \\ &= (-0.4093, 0.4473). \end{aligned}$$

△

11.5 Stabilization Using a PI Controller

The general approach presented in Section 11.3 is also applicable to the case of a PI controller. In this section we discuss the details for this specific case. For a PI controller, we have

$$C(s) = k_p + \frac{k_i}{s} = \frac{k_p s + k_i}{s}$$

and the open-loop transfer function becomes

$$H(s) = C(s)G(s) = C(s)G_0(s)e^{-Ls} = H_0(s)e^{-Ls}$$

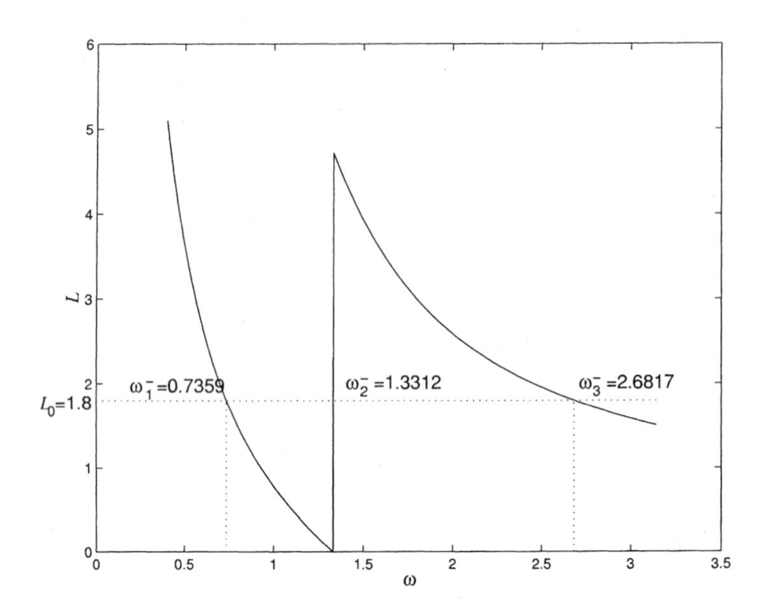

FIGURE 11.6. $L(\omega)$ versus ω for $k_p < 0$.

where

$$\begin{aligned} H_0(s) &= C(s)G_0(s) \\ &= \frac{k_p s + k_i}{s} \cdot \frac{N(s)}{D(s)} \\ &= (k_p s + k_i) \cdot \frac{N(s)}{sD(s)} \\ &= (k_p s + k_i) \cdot R_0(s), \end{aligned}$$

with $R_0(s) \stackrel{\Delta}{=} \frac{N(s)}{sD(s)}$.

The magnitude and phase conditions

$$\begin{aligned} \arg[(k_i + jk_p\omega)R_0(j\omega)] - L\omega &= -\pi \\ |(k_i + jk_p\omega)R_0(j\omega)| &= 1 \end{aligned}$$

can be written as

$$L(\omega, k_p, k_i) = \frac{\arg[(k_i + jk_p\omega)R_0(j\omega)] + \pi}{\omega} \tag{11.15}$$

$$k_i = \pm\sqrt{\frac{1}{|R_0(j\omega)|^2} - k_p^2\omega^2}\,. \tag{11.16}$$

We can first fix k_p and define

$$M(\omega) = \frac{1}{|R_0(j\omega)|^2} - k_p^2\omega^2\,.$$

Thus

$$k_i = \pm\sqrt{M(\omega)}\,. \tag{11.17}$$

Note that since $k_i \in \mathbf{R}$, only those ω values for which $M(\omega) \geq 0$ need to be considered when we compute $\mathcal{S}_L$. Substituting (11.17) into (11.15), we have

$$L(\omega) = \frac{\arg\{[\pm\sqrt{M(\omega)} + jk_p\omega]R_0(j\omega)\} + \pi}{\omega}\,.$$

Before proceeding further, we need to introduce some notation. For a given set in the controller parameter space, if one of the controller parameters appears as a subscript, then the new set represents the subset of the original one with that parameter fixed at some value. For example, $\mathcal{S}_{R,k_p}$ is a subset of $\mathcal{S}_R$ with k_p fixed at some value.

Based on the above discussion, the following steps can be used for computing $\mathcal{S}_R$:

1. Compute $\mathcal{S}_0$ using the results of Chapter 3.
2. Find $\mathcal{S}_N$, defined as
$$\mathcal{S}_N = \left\{(k_p, k_i) \in \mathbf{R}^2 \mid \lim_{s\to\infty}\left|\frac{(k_p s + k_i)N(s)}{sD(s)}\right| \geq 1\right\}\,.$$
   - If $\deg[N(s)] > \deg[D(s)]$, $\mathcal{S}_N = \mathbf{R}^2$, which means $\mathcal{S}_R = \emptyset$.
   - If $\deg[N(s)] < \deg[D(s)]$, $\mathcal{S}_N = \emptyset$.
   - If $\deg[N(s)] = \deg[D(s)]$, $\mathcal{S}_N = \{(k_p, k_i) | k_p, k_i \in \mathbf{R}$ and $|k_p| \geq |\frac{a_q}{b_q}|\}$, where a_q, b_q are the leading coeffients of $D(s)$ and $N(s)$, respectively.
3. Compute $\mathcal{S}_1 = \mathcal{S}_0 \backslash \mathcal{S}_N$.
4. For a fixed k_p, find $\mathcal{S}_{R,k_p}$.
   - First determine the sets Ω^+ and S^+_{L,k_p}:
$$\begin{aligned}\Omega^+ &= \Big\{\omega|\omega > 0 \text{ and } M(\omega) \geq 0 \text{ and}\\ &\qquad L(\omega) = \frac{\arg\{[\sqrt{M(\omega)} + jk_p\omega]R_0(j\omega)\} + \pi}{\omega} \leq L_0\Big\}\\ \mathcal{S}^+_{L,k_p} &= \{k_i | k_i \notin \mathcal{S}_{N,k_p} \text{ and } \exists\, \omega \in \Omega^+ \text{ s.t. } k_i = \sqrt{M(\omega)}\}.\end{aligned}$$
   - Next determine the sets Ω^- and S^-_{L,k_p}:
$$\Omega^- = \Big\{\omega|\omega > 0 \text{ and } M(\omega) \geq 0 \text{ and}$$

$$L(\omega) = \frac{\arg\{[-\sqrt{M(\omega)} + jk_p\omega]R_0(j\omega)\} + \pi}{\omega} \leq L_0 \Bigg\}$$

$$\mathcal{S}^-_{L,k_p} = \{k_i | k_i \notin \mathcal{S}_{N,k_p} \text{ and } \exists\, \omega \in \Omega^- \text{ s.t. } k_i = -\sqrt{M(\omega)}\}.$$

Compute $\mathcal{S}_{L,k_p} = \mathcal{S}^+_{L,k_p} \cup \mathcal{S}^-_{L,k_p}$ and $\mathcal{S}_{R,k_p} = \mathcal{S}_{1,k_p} \backslash \mathcal{S}_{L,k_p}$.

5. By sweeping over k_p, we will have the complete set of PI controllers that stabilize all plants with delay up to L_0:

$$\mathcal{S}_R = \bigcup_{k_p} \mathcal{S}_{R,k_p} \ .$$

11.6 Stabilization Using a PID Controller

In this section, we complete the chapter by extending the results presented earlier to the case of a PID controller. As before, the feedback system is as shown in Fig. 11.3 with the system $G(s)$ given by

$$G(s) = G_0(s)e^{-Ls} = \frac{N(s)}{D(s)}e^{-Ls}$$

where $N(s)$, $D(s)$ are polynomials with real coefficients. The controller, which in this case is a PID, is given by the following transfer function

$$C(s) = k_p + \frac{k_i}{s} + k_d s = \frac{k_d s^2 + k_p s + k_i}{s} \tag{11.18}$$

where k_p, k_i, and k_d represent the proportional, integral, and derivate gains, respectively. As in previous sections, our control objective is to determine the set of (k_p, k_i, k_d) values that stabilize the closed-loop system.

The open-loop transfer function becomes

$$H(s) = C(s)G(s) = C(s)G_0(s)e^{-Ls} = H_0(s)e^{-Ls}$$

where

$$\begin{aligned} H_0(s) &= C(s)G_0(s) \\ &= \frac{k_d s^2 + k_p s + k_i}{s} \cdot \frac{N(s)}{D(s)} \\ &= (k_d s^2 + k_p s + k_i) \cdot \frac{N(s)}{sD(s)} \\ &= (k_d s^2 + k_p s + k_i) \cdot R_0(s) \ , \end{aligned}$$

with $R_0(s) \triangleq \frac{N(s)}{sD(s)}$. The phase and magnitude conditions are given by

$$\begin{aligned} \arg[(k_i - k_d\omega^2 + jk_p\omega)R_0(j\omega)] - L\omega &= -\pi \\ \text{and } |(k_i - k_d\omega^2 + jk_p\omega)R_0(j\omega)| &= 1\,. \end{aligned}$$

These conditions can be further reduced to

$$L(\omega, k_p, k_i, k_d) = \frac{\pi + \arg\{[(k_i - k_d\omega^2) + jk_p\omega] \cdot R_0(j\omega)\}}{\omega} \tag{11.19}$$

$$k_i - k_d\omega^2 = \pm\sqrt{\frac{1}{|R_0(j\omega)|^2} - (k_p\omega)^2}\,. \tag{11.20}$$

Similar to the PI case presented in the previous section, for fixed k_p, we define

$$M(\omega) = \frac{1}{|R_0(j\omega)|^2} - (k_p\omega)^2\,.$$

Then (11.20) can be rewritten as follows:

$$k_i - k_d\omega^2 = \pm\sqrt{M(\omega)}\,. \tag{11.21}$$

Like the PI case, we only need to consider ω values with $M(\omega) \geq 0$ when we compute $\mathcal{S}_L$.

Substituting (11.21) into (11.19), we have

$$L(\omega, k_p, k_i, k_d) = L(\omega) = \frac{\pi + \arg\{[\pm\sqrt{M(\omega)} + jk_p\omega] \cdot R_0(j\omega)\}}{\omega}\,.$$

The following steps can be used for computing $\mathcal{S}_R$:

1. Compute $\mathcal{S}_0$ using the results of Chapter 4.
2. Find $\mathcal{S}_N$, defined as

$$\mathcal{S}_N = \left\{(k_p, k_i, k_d) \in \mathbf{R}^3 \mid \lim_{s\to\infty}\left|\frac{(k_d s^2 + k_p s + k_i)N(s)}{sD(s)}\right| \geq 1\right\}\,.$$

   - If $\deg[N(s)] > \deg[D(s)] - 1$, $\mathcal{S}_N = \mathbf{R}^3$, which means $\mathcal{S}_R = \emptyset$.
   - If $\deg[N(s)] < \deg[D(s)] - 1$, $\mathcal{S}_N = \emptyset$.
   - If $\deg[N(s)] = \deg[D(s)] - 1$, then

$$\mathcal{S}_N = \left\{(k_p, k_i, k_d) \in \mathbf{R}^3 \mid |k_d| \geq \left|\frac{a_q}{b_{q-1}}\right|\right\}$$

   where a_q, b_{q-1} are the leading coefficients of $D(s)$ and $N(s)$, respectively.

3. Compute $\mathcal{S}_1 = \mathcal{S}_0 \backslash \mathcal{S}_N$.

4. For a fixed k_p, determine the set $\mathcal{S}_{R,k_p}$ as follows:
   - First determine the sets Ω^+ and $\mathcal{S}^+_{L,k_p}$:

$$\begin{aligned}\Omega^+ &= \Bigg\{\omega|\omega > 0 \text{ and } M(\omega) \geq 0 \text{ and } L(\omega) = \\ &\frac{\pi + \arg\{[\sqrt{M(\omega)} + jk_p\omega] \cdot R_0(j\omega)\}}{\omega} \leq L_0\Bigg\} \qquad (11.22)\\ \mathcal{S}^+_{L,k_p} &= \{(k_i, k_d)|(k_i, k_d) \notin \mathcal{S}_{N,k_p} \text{ and } \exists\, \omega \in \Omega^+ \text{ such that} \\ &k_i - k_d\omega^2 = \sqrt{M(\omega)}\}. \qquad (11.23)\end{aligned}$$

Note that $\mathcal{S}^+_{L,k_p}$ is a set of straight lines in the (k_i, k_d) space.

   - Next determine the sets Ω^- and $\mathcal{S}^-_{L,k_p}$:

$$\begin{aligned}\Omega^- &= \Bigg\{\omega|\omega > 0 \text{ and } M(\omega) \geq 0 \text{ and } L(\omega) = \\ &\frac{\pi + \arg\{[-\sqrt{M(\omega)} + jk_p\omega] \cdot R_0(j\omega)\}}{\omega} \leq L_0\Bigg\} \qquad (11.24)\\ \mathcal{S}^-_{L,k_p} &= \{(k_i, k_d)|(k_i, k_d) \notin \mathcal{S}_{N,k_p} \text{ and } \exists\, \omega \in \Omega^- \text{ such that} \\ &k_i - k_d\omega^2 = -\sqrt{M(\omega)}\}. \qquad (11.25)\end{aligned}$$

Compute $\mathcal{S}_{L,k_p} = \mathcal{S}^+_{L,k_p} \cup \mathcal{S}^-_{L,k_p}$ and $\mathcal{S}_{R,k_p} = \mathcal{S}_{1,k_p} \backslash \mathcal{S}_{L,k_p}$.

5. By sweeping over k_p, we will have the complete set of PID controllers that stabilize all plants with delay up to L_0:

$$\mathcal{S}_R = \bigcup_{k_p} \mathcal{S}_{R,k_p}. \qquad (11.26)$$

We next present a numerical example to illustrate the above steps.

Example 11.3 *Use a PID controller to stabilize the time-delay plant*

$$G(s) = \frac{s^3 - 4s^2 + s + 2}{s^5 + 8s^4 + 32s^3 + 46s^2 + 46s + 17} e^{-Ls} \qquad (11.27)$$

with L up to $L_0 = 1$ second, i.e., for all $L \in [0, 1]$.

We start by fixing $k_p = 1$. We can use the method proposed in Chapter 4 to get the stabilizing k_i, k_d values for the delay-free plant, $\mathcal{S}_{0,k_p}$. This set is shown in Fig. 11.7.

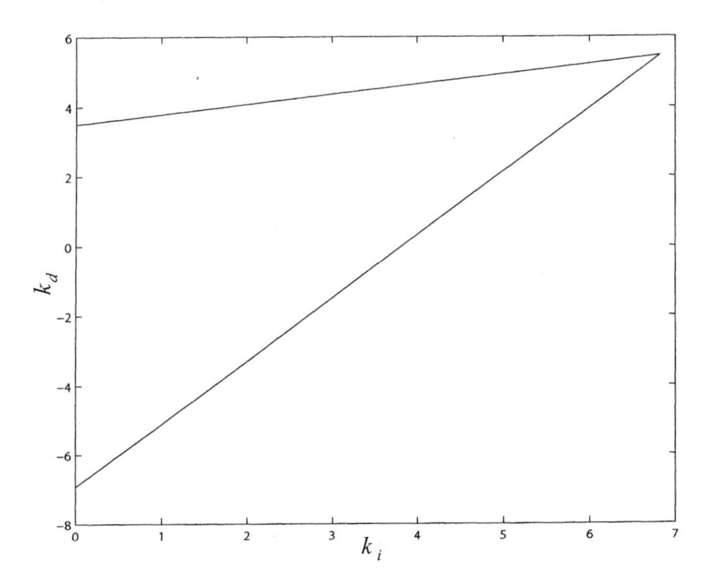

FIGURE 11.7. Stabilizing region in the (k_i, k_d) space with $k_p = 1$ for delay-free plant in Example 11.3.

Since $\deg[D(s)] - \deg[N(s)] > 1$, $\mathcal{S}_N = \emptyset$ *and* $\mathcal{S}_1 = \mathcal{S}_0$.

Next, for $k_i - k_d\omega^2 = \sqrt{M(\omega)} > 0$, *we plot* $L(\omega)$ *as a function of* ω. *This plot is shown in Fig. 11.8. From this figure we determine that the set of* ω *where* $L(\omega) \leq L_0$ *is* $\Omega^+ = [0.524825, 0.742302] \cup [2.57318, +\infty)$. *Also, we plot* $\sqrt{M(\omega)}$ *as a function of* ω *(see Fig. 11.9). From this figure we can find the corresponding values of* $\sqrt{M(\omega)}$ *and* $\mathcal{S}^+_{L,k_p}$, *i.e., the straight lines defined by* $k_i - k_d\omega^2 = \sqrt{M(\omega)}$ *for* $\omega \in \Omega^+$.

In a similar fashion, we proceed to plot $L(\omega)$ *as a function of* ω *for* $k_i - k_d\omega^2 = -\sqrt{M(\omega)} < 0$. *This plot is depicted in Fig. 11.10 and allows us to determine the set* $\Omega^- = [1.35894, 1.8659] \cup [4.37326, +\infty)$. *From this set we can then get* $\mathcal{S}^-_{L,k_p}$.

Finally, we can exclude $\mathcal{S}^+_{L,k_p}$ *and* $\mathcal{S}^-_{L,k_p}$ *from* $\mathcal{S}_{1,k_p}$ *to get* $\mathcal{S}_{R,k_p}$. *This procedure is depicted in Fig. 11.11 and allows us to determine the complete set of* (k_i, k_d) *gain values that stabilize the plant with* $k_p = 1$ *and a delay up to 1 second.*

$\triangle$

The following example shows how the general approach of this chapter can be used to recover the results presented in Chapter 8 for first-order plants with time delay.

Example 11.4 *Determine all PID controllers that stabilize a first-order plant with time delay up to* L_0. *To this end, consider the first-order plant*

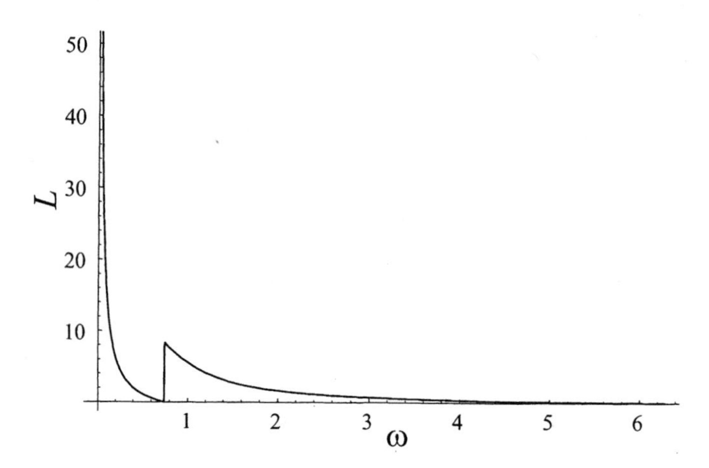

FIGURE 11.8. $L(\omega)$ versus ω with $k_i - k_d\omega^2 = \sqrt{M(\omega)} > 0$.

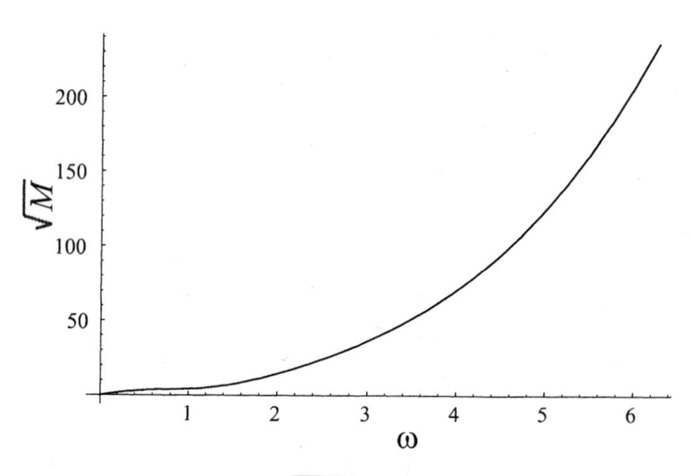

FIGURE 11.9. $\sqrt{M(\omega)}$ versus ω with $k_p = 1$.

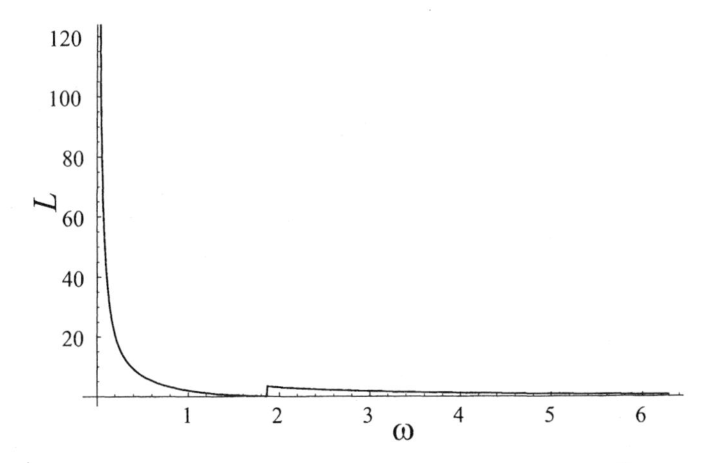

FIGURE 11.10. $L(\omega)$ versus ω with $k_i - k_d\omega^2 = -\sqrt{M(\omega)} < 0$.

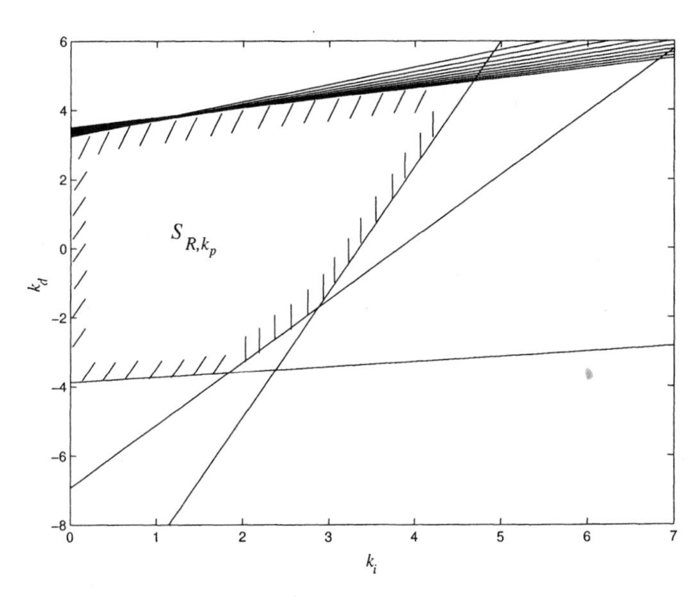

FIGURE 11.11. Stabilizing region of (k_i, k_d) with $k_p = 1$ for plant with delay up to 1 second.

with time delay:

$$G(s) = \frac{k}{Ts+1}e^{-Ls}, \ \ L \in [0, L_0] \, .$$

The stabilizing PID parameters for the delay-free plant are

$$\begin{aligned} \mathcal{S}_0 &= \left\{ (k_p, k_i, k_d) | k_p > -\frac{1}{k}, k_i > 0, k_d > -\frac{T}{k} \right. \\ & \quad \left. \text{or } k_p < -\frac{1}{k}, k_i < 0, k_d < -\frac{T}{k} \right\} . \end{aligned}$$

Since $\deg[D(s)] - \deg[N(s)] = 1$, *we have*

$$\mathcal{S}_N = \left\{ (k_p, k_i, k_d) \in \mathbf{R}^3 \ s.t. \ |k_d| \geq \left| \frac{T}{k} \right| \right\} .$$

Without loss of generality, let us assume that $k > 0$. *Then*

$$\mathcal{S}_1 = \mathcal{S}_0 \backslash \mathcal{S}_N = \left\{ (k_p, k_i, k_d) | k_p > -\frac{1}{k}, k_i > 0, \frac{T}{k} > k_d > -\frac{T}{k} \right\}$$

for $T > 0$, *and*

$$\mathcal{S}_1 = \left\{ (k_p, k_i, k_d) | k_p < -\frac{1}{k}, k_i < 0, \frac{T}{k} < k_d < -\frac{T}{k} \right\}$$

for $T < 0$.

A detailed analysis of this example using the techniques of this chapter is given in Appendix C. The results derived in this appendix agree with those presented in Chapter 8 and are as follows.

For $T > 0$, *with different* k_p *values, the stabilizing regions in the* (k_i, k_d) *space take on different but simple shapes:*

- *For* $-\frac{1}{k} < k_p \leq \frac{1}{k}$, $\mathcal{S}_{R,k_p}$ *is a trapezoid as shown in Fig. 11.12(a).*
- *For* $k_p > \frac{1}{k}$, $\mathcal{S}_{R,k_p}$ *is a quadrilateral as shown in Fig. 11.12(b), (c).*

Similar results can also be obtained for $T < 0$. △

Remark 11.2 *In comparing Fig. 11.12 with Fig. 8.6, note that Fig. 11.12(a) corresponds to Figs. 8.6(a) and (b). The triangle in Fig. 8.6(b) is really a degenerate case of the trapezoid T in Fig. 8.6(a). Furthermore, Figs. 11.12(b) and (c) correspond to Fig. 8.6(c). From the analysis presented in Appendix C, it is clear that for* $k_p > \frac{1}{k}$, *two different cases have to be considered. However, in each case, the stabilizing regions in the* (k_i, k_d) *space are qualitatively similar and this is clear from Figs. 11.12(b) and (c) and also from Fig. 8.6(c), where we have only one figure encompassing these two cases.*

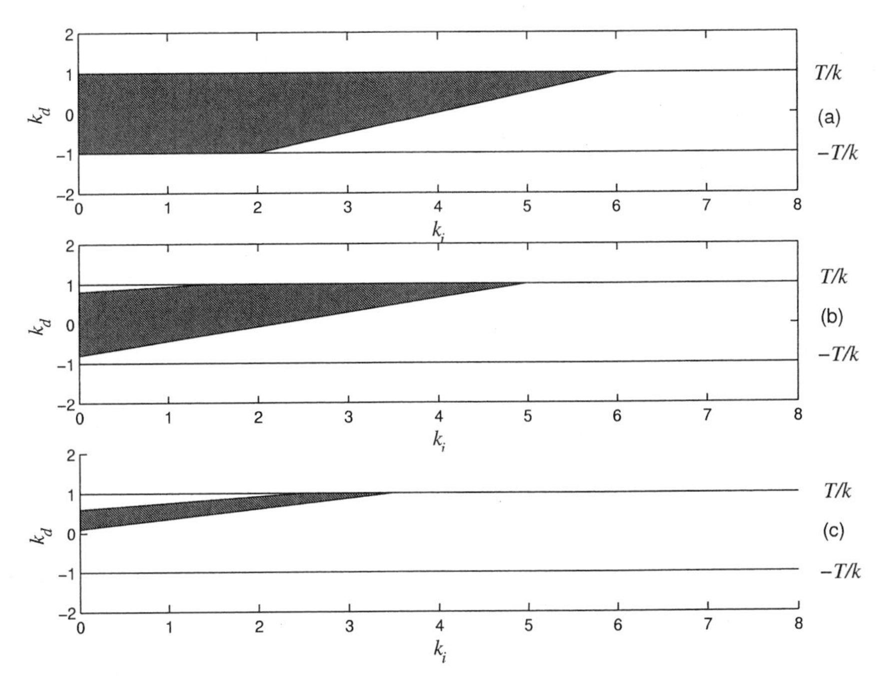

FIGURE 11.12. First-order plant: stabilizing region of (k_i, k_d) with different k_p.

11.7 Notes and References

The details of the generalized Nyquist criterion developed by Tsypkin can be found in the English translation of his original paper [47]. The results presented in this chapter are due to Xu, Datta, and Bhattacharyya [51].

12 Algorithms for Real and Complex PID Stabilization

This final chapter presents a summary of algorithms that can be used to generate the entire set of stabilizing PID controllers for single-input single-output (1) continuous-time rational plants of arbitrary order, (2) discrete-time rational plants of arbitrary order, and (3) continuous-time first-order plants with time delay. These algorithms follow from the material presented throughout the book. They display the rich mathematical structure underlying the topology of PID stabilizing sets. By presenting these algorithms without the highly technical details of the underlying theory, we seek to make the results accessible to as many engineers as possible. We have incorporated the bare minimum mathematical background required to make it self-contained.

12.1 Introduction

As pointed out earlier in this book, the design of PID controllers is in most cases carried out using *ad hoc* tuning rules. These rules have been developed over the years based primarily on empirical observations and industrial experience. In part this state of affairs occurs because the state feedback observer-based theory of modern and post modern control theory including H_2, H_∞, μ, and l_1 optimal control cannot be applied to PID control. Indeed, until recently it was not known how to even determine whether stabilization of a nominal system was possible using PID controllers.

The results presented throughout this book represent a collection of significant results on PID stabilization developed in recent years. We believe that these results could assist the industrial practitioner to carry out computer-aided designs of PID controllers with guaranteed stability and performance. Given that PID controllers are used in applications as diverse as process control, rolling mills, aerospace, motion control, pneumatic, hydraulic, electrical and mechanical systems, disc drives and digital cameras, the impact of these results could be enormous.

The theoretical development of these results is quite involved and technical, which could make the results inaccessible to practicing engineers. However, the algorithms that result are straightforward and can be easily programmed on a computer. The objective of this final chapter is to present these PID stabilization and design algorithms, devoid of detailed mathematical proofs, and show via examples how these algorithms can be used by the industrial practitioner to carry out computer-aided designs. In particular, the graphical displays of feasible design regions using two-dimensional and three-dimensional graphics should appeal to control designers and are very suitable for computer-aided design where several performance objectives have to be overlaid and intersected. Specific design problems in which these algorithms can be profitably used are discussed.

The chapter is organized as follows. In Section 12.2, we present an algorithm for determining the set of all stabilizing PID controllers for a continuous-time delay-free plant of arbitrary order. An example is included to illustrate the detailed calculations involved. In Section 12.3, we show how all PID stabilizers for a discrete-time plant of arbitrary order can be determined by suitably modifying the algorithm of Section 12.2. Once again, an illustrative example is included. In Section 12.4, we present an algorithm to determine the set of all stabilizing PID controllers for a continuous-time first-order plant with time delay. An example is included to demonstrate the use of this algorithm. Section 12.5 discusses some PID controller *design* problems involving frequency domain performance specifications, which can be solved by using a complex version of the algorithm of Section 12.2. The modifications required are indicated.

12.2 Algorithm for Linear Time-Invariant Continuous-Time Systems

Consider the general feedback system shown in Fig. 12.1. Here r is the command signal, y is the output, $G(s)$ is the plant to be controlled, and $C(s)$ is the controller to be designed. The controller $C(s)$ will be assumed to be of the PID type so that

$$C(s) = k_p + \frac{k_i}{s} + k_d s$$

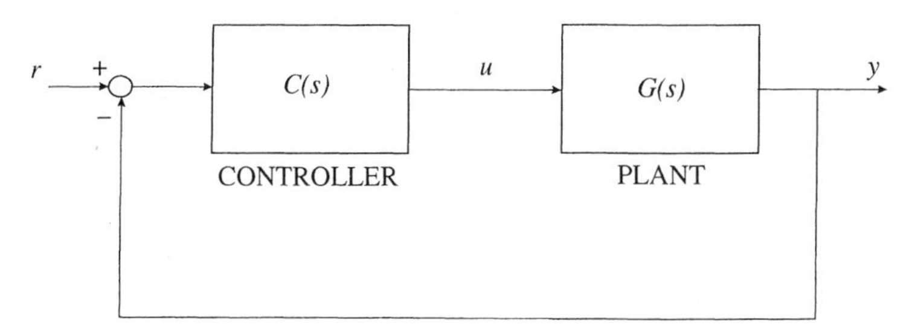

FIGURE 12.1. Feedback control system.

where k_p, k_i, and k_d are the proportional, integral, and derivative gains, respectively. For this section, the plant transfer function $G(s)$ will be assumed to be rational so that

$$G(s) = \frac{N(s)}{D(s)}$$

where $N(s)$, $D(s)$ are polynomials in the Laplace variable s. With this plant description, the closed-loop characteristic polynomial becomes

$$\delta(s, k_p, k_i, k_d) = sD(s) + (k_i + k_d s^2)N(s) + k_p sN(s). \tag{12.1}$$

The problem of stabilization using a PID controller should be clear to the reader by now: to determine the values of k_p, k_i, and k_d for which the closed-loop characteristic polynomial $\delta(s, k_p, k_i, k_d)$ is Hurwitz, that is, has all its roots in the open LHP. Since plants with a zero at the origin cannot be stabilized by PID controllers, we exclude such plants at the outset. In this section, we simply present an algorithm for computationally characterizing all stabilizing PID controllers for a given plant with $N(0) \neq 0$. For a proof of the derivation, the reader is referred to Chapter 4.

To make this chapter self-contained, we recall here some definitions and notations introduced earlier in Chapter 2.

Definition 12.1 *The standard signum function* sgn : $\mathcal{R} \to \{-1, 0, 1\}$ *is defined by*

$$\text{sgn}[x] = \begin{cases} -1 & \text{if } x < 0 \\ 0 & \text{if } x = 0 \\ 1 & \text{if } x > 0. \end{cases}$$

Definition 12.2 *Let* $a(s) = a_0 + a_1 s + \cdots + a_n s^n$ *be a given real polynomial of degree* n. *Let* C^- *denote the LHP and* C^+ *the open RHP. Then* $l(a(s))$ *and* $r(a(s))$ *denote the numbers of roots of* $a(s)$ *in* C^- *and* C^+, *respectively.*

Definition 12.3 *Given a real polynomial* $a(s)$ *of degree* n, *the even-odd decomposition of* $a(s)$ *is defined as*

$$a(s) = a_e(s^2) + s a_o(s^2)$$

where $a_e(s^2)$ and $sa_o(s^2)$ are the components of $a(s)$ made up of even and odd powers of s, respectively.

To motivate the manipulations to follow we observe first that for a given real polynomial $a(s)$, the real and imaginary parts of $a(j\omega)$ are given by $a_e(-\omega^2)$ and $\omega a_o(-\omega^2)$, respectively. It will turn out that the root distribution (numbers of LHP and RHP roots) of $a(s)$ can be determined from the zeros of its imaginary part and the signs of the real parts at these zeros. Finally, if $a(s)$ has unknown design parameters, this approach to determining the root distribution is most conveniently applied when the unknown parameter sets appearing in the real and imaginary parts are separated, that is, have no common elements. These ideas were used in Chapter 4 to develop an algorithm to determine the complete set of parameters k_p, k_i, k_d resulting in the Hurwitz stability of (12.1). In the following we describe the essentials of the algorithm without mathematical proofs.

Using the even-odd decomposition of $N(s)$, define

$$N^*(s) = N(-s) = N_e(s^2) - sN_o(s^2)\,.$$

Also let n, m be the degrees of $\delta(s, k_p, k_i, k_d)$ and $N(s)$, respectively. To achieve the parameter separation mentioned earlier, multiply $\delta(s, k_p, k_i, k_d)$ by $N^*(s)$ and rewrite $N(s)$, $D(s)$ in terms of their even-odd decompositions to obtain

$$\begin{aligned} \nu(s) \;&:=\; \delta(s, k_p, k_i, k_d)N^*(s) \\ &= [s^2(N_e(s^2)D_o(s^2) - D_e(s^2)N_o(s^2)) \\ &\quad +(k_i + k_d s^2)(N_e(s^2)N_e(s^2) - s^2 N_o(s^2)N_o(s^2))] \\ &\quad +s[D_e(s^2)N_e(s^2) - s^2 D_o(s^2)N_o(s^2) \\ &\quad +k_p(N_e(s^2)N_e(s^2) - s^2 N_o(s^2)N_o(s^2))]. \end{aligned} \tag{12.2}$$

Note that the polynomial $\nu(s)$ has degree $n+m$ and $\delta(s, k_p, k_i, k_d)$ is Hurwitz if and only if $\nu(s)$ has exactly the same number of closed RHP zeros as $N^*(s)$; this is the condition we will use for stability. Note also that while the characteristic polynomial $\delta(s, k_p, k_i, k_d)$ has all three parameters appearing in both the even and odd parts, the test polynomial $\nu(s)$ exhibits parameter separation; that is, k_p appears in the odd part only while k_i and k_d appear in the even part only. This will facilitate the application of root-counting formulas to $\nu(s)$.

To proceed, substitute $s = j\omega$ into (12.2) to obtain

$$\delta(j\omega, k_p, k_i, k_d)N^*(j\omega) \;=\; p(\omega,\, k_i,\, k_d) + jq(\omega,\, k_p)$$

where

$$\begin{aligned} p(\omega,\, k_i,\, k_d) &= p_1(\omega) + (k_i - k_d\omega^2)p_2(\omega) \\ q(\omega,\, k_p) &= q_1(\omega) + k_p q_2(\omega) \end{aligned}$$

$$\begin{aligned}
p_1(\omega) &= -\omega^2(N_e(-\omega^2)D_o(-\omega^2) - D_e(-\omega^2)N_o(-\omega^2)) && (12.3)\\
p_2(\omega) &= N_e(-\omega^2)N_e(-\omega^2) + \omega^2 N_o(-\omega^2)N_o(-\omega^2)) && (12.4)\\
q_1(\omega) &= \omega(D_e(-\omega^2)N_e(-\omega^2) + \omega^2 D_o(-\omega^2)N_o(-\omega^2)) && (12.5)\\
q_2(\omega) &= \omega(N_e(-\omega^2)N_e(-\omega^2) + \omega^2 N_o(-\omega^2)N_o(-\omega^2)). && (12.6)
\end{aligned}$$

The PID stabilization algorithm to be presented below is based on a fundamental result generalizing the classical Hermite-Biehler Theorem to the case of root distribution determination of real polynomials that are not necessarily Hurwitz. This generalization, introduced earlier in Chapter 2, provides an analytical expression for the difference between the numbers of roots of a real polynomial in the open LHP and open RHP. In our case we will exploit these results to impose the stability condition that $\nu(s)$ has exactly the same number of RHP roots as $N(-s)$. For this to happen a necessary condition is that $q(\omega, k_p)$ has at least

$$\begin{cases} \frac{|n-(l(N(s))-r(N(s)))|}{2} & \text{for } m+n \text{ even} \\ \frac{|n-(l(N(s))-r(N(s)))|+1}{2} & \text{for } m+n \text{ odd} \end{cases} \tag{12.7}$$

real, nonnegative, distinct roots of odd multiplicity. The ranges of k_p satisfying (12.7) are called *allowable.* Let

$$0 = \omega_0 \ < \ \omega_1 \ < \ ... \ < \ \omega_{l-1} \tag{12.8}$$

denote the real nonnegative distinct roots of $q(\omega, k_p)$ of odd multiplicity, and with $\omega_l := \infty$ write

$$\text{sgn}[p(\omega_j)] = i_j, \ j = 0, 1, ..., l. \tag{12.9}$$

It can be shown, using the root-counting results mentioned earlier, that the stability condition reduces to

$$\begin{aligned}
n - (l(N(s)) - r(N(s))) \ &= \ \{i_0 - 2i_1 + 2i_2 + \cdots + (-1)^{l-1}2i_{l-1} \\
&\quad +(-1)^l i_l\} \cdot (-1)^{l-1}\text{sgn}[q(\infty, k_p)] \quad (12.10)
\end{aligned}$$

and therefore the string of integers $\{i_0, i_1, ..., i_l\}$ will be called *admissible* if it satisfies (12.10).

Using the above we can present the following algorithm for determining all stabilizing (k_p, k_i, k_d) values for the given plant. The reader is referred to Chapter 4 for a more complete development.

PID Stabilization Algorithm For LTI Plants:

Step 1: For the given $N(s)$ and $D(s)$, compute the corresponding $p_1(\omega)$, $p_2(\omega)$, $q_1(\omega)$, and $q_2(\omega)$ from (12.3)–(12.6);

Step 2: Determine the allowable ranges $P_i, i = 1, 2, \ldots, d$ of k_p from (12.7). The resulting ranges of k_p are the only ranges of k_p for which stabilizing $(k_i,\ k_d)$ values may exist;

Step 3: If there is no k_p satisfying **Step 2** then output **NO SOLUTION** and **EXIT**;

Step 4: Initialize $j = 1$ and $P = P_j$;

Step 5: Pick a range $[k_{low}, k_{upp}]$ in P and initialize $k_p = k_{low}$;

Step 6: Pick the number of grid points N and set $step = \frac{1}{N+1}[k_{upp} - k_{low}]$;

Step 7: Increase k_p as follows: $k_p = k_p + step$. If $k_p > k_{upp}$ then **GOTO Step 14**;

Step 8: For fixed k_p in **Step 7**, solve for the real, non-negative, distinct finite zeros of $q(\omega, k_p)$ with odd multiplicities and denote them by $0 = \omega_0 < \omega_1 < \omega_2 < \cdots < \omega_{l-1}$. Also define $\omega_l = \infty$;

Step 9: Construct sequences of numbers $i_0, i_1, i_2, \ldots, i_l$ as follows:
(i)

$$i_0 = \begin{cases} \text{sgn}[p_{1_f}^{(k_n)}(0)] & \text{if } N^*(s) \text{ has a zero of} \\ & \text{multiplicity } k_n \text{ at the origin} \\ \alpha & \text{otherwise} \end{cases}$$

where $\alpha \in \{-1, 1\}$ and

$$p_{1_f}(\omega) := \frac{p_1(\omega)}{(1+\omega^2)^{\frac{(m+n)}{2}}};$$

(ii) For $t = 1, 2, \ldots, l-1$:

$$i_t = \begin{cases} 0 & \text{if } N^*(j\omega_t) = 0 \\ \alpha & \text{otherwise} \end{cases}$$

(iii)

$$i_l = \begin{cases} \alpha & \text{if } n+m \text{ is even} \\ 0 & \text{if } n+m \text{ is odd} \end{cases}.$$

With $i_0, i_1, \ldots$ defined in this way, we define the string $\mathcal{I} : N \rightarrow R$ as the following sequence of numbers:

$$\mathcal{I} := \{i_0, i_1, \ldots, i_l\}.$$

Define A_{k_p} to be the set of all possible strings $\mathcal{I}$ that can be generated to satisfy the preceding requirements.

Step 10: Determine the admissible strings $\mathcal{I}$ in A_{k_p} from (12.10). If there is no admissible string then **GOTO Step 7**;

Step 11: For an admissible string $\mathcal{I}$, determine the set of $(k_i,\ k_d)$ values that simultaneously satisfy the following string of linear inequalities:

$$[p_1(\omega_t) + (k_i - k_d\omega_t^2)p_2(\omega_t)]i_t \ > \ 0, \forall\, t = 0, 1, 2, \ldots$$

for which $N^*(j\omega_t) \neq 0$;

Step 12: Repeat **Step 11** for all admissible strings $\mathcal{I}_1, \mathcal{I}_2, \ldots, \mathcal{I}_v$ to obtain the corresponding admissible (k_i, k_d) sets $\mathcal{S}_1, \mathcal{S}_2, \ldots, \mathcal{S}_v$. The set of all stabilizing (k_i, k_d) values corresponding to the fixed k_p is then given by

$$\mathcal{S}_{(k_p)} \ = \ \cup_{x=1,\,2,\cdots,\,v}\, \mathcal{S}_x;$$

Step 13: GOTO Step 7;

Step 14: Set $j = j + 1$ and $P = P_j$. If $j \leq d$ **GOTO STEP 5**; else, terminate the algorithm.

We now present an example to illustrate the detailed calculations involved in determining the stabilizing (k_p, k_i, k_d) gain values.

Example 12.1 *Consider the problem of determining stabilizing PID gains for the plant* $G(s) = \frac{N(s)}{D(s)}$ *where*

$$\begin{aligned} N(s) &= s^3 - 2s^2 - s - 1 \\ D(s) &= s^6 + 2s^5 + 32s^4 + 26s^3 + 65s^2 - 8s + 1. \end{aligned}$$

The closed-loop characteristic polynomial is

$$\delta(s, k_p, k_i, k_d) = sD(s) + (k_i + k_d s^2)N(s) + k_p s N(s).$$

Thus $n = 7$ *and* $m = 3$. *Also*

$$\begin{aligned} N_e(s^2) &= -2s^2 - 1, \qquad & N_o(s^2) &= s^2 - 1, \\ D_e(s^2) &= s^6 + 32s^4 + 65s^2 + 1, \qquad & D_o(s^2) &= 2s^4 + 26s^2 - 8, \end{aligned}$$

and

$$N^*(s) = (-2s^2 - 1) - s(s^2 - 1).$$

Therefore, from (12.2) we obtain

$$\begin{aligned}\delta(s, k_p, k_i, k_d)N^*(s) &= [s^2(-s^8 - 35s^6 - 87s^4 + 54s^2 + 9) + (k_i + k_d s^2) \\ &\quad (-s^6 + 6s^4 + 3s^2 + 1)] + s[(-4s^8 - 89s^6 - 128s^4 \\ &\quad -75s^2 - 1) + k_p(-s^6 + 6s^4 + 3s^2 + 1)]\end{aligned}$$

so that

$$\begin{aligned}\delta(j\omega, k_p, k_i, k_d)N^*(j\omega) &= [p_1(\omega) + (k_i - k_d\omega^2)p_2(\omega)] \\ &\quad +j[q_1(\omega) + k_p q_2(\omega)]\end{aligned}$$

where

$$\begin{aligned}p_1(\omega) &= \omega^{10} - 35\omega^8 + 87\omega^6 + 54\omega^4 - 9\omega^2 \\ p_2(\omega) &= \omega^6 + 6\omega^4 - 3\omega^2 + 1 \\ q_1(\omega) &= -4\omega^9 + 89\omega^7 - 128\omega^5 + 75\omega^3 - \omega \\ q_2(\omega) &= \omega^7 + 6\omega^5 - 3\omega^3 + \omega.\end{aligned}$$

In **Step 2**, *the range of k_p such that $q_f(\omega, \ k_p)$ has at least three real, non-negative, distinct, finite zeros with odd multiplicities was determined to be $(-24.7513, \ 1)$ which is the allowable range. Now for a fixed $k_p \in (-24.7513, 1)$, for instance, $k_p = -18$, we have*

$$\begin{aligned}q(\omega, -18) &= q_1(\omega) - 18q_2(\omega) \\ &= -4\omega^9 + 71\omega^7 - 236\omega^5 + 129\omega^3 - 19\omega.\end{aligned}$$

Then the real, non-negative, distinct finite zeros of $q_f(\omega, \ -18)$ with odd multiplicities are

$$\omega_0 = 0, \ \omega_1 = 0.5195, \ \omega_2 = 0.6055, \ \omega_3 = 1.8804, \ \omega_4 = 3.6848.$$

Also define $\omega_5 = \infty$. Since $m + n = 10$, which is even, and $N^*(s)$ has no $j\omega$-axis roots, from **Step 9**, *the set $A_{(-18)}$ becomes*

$$\left\{\begin{array}{ccc}
\{-1,-1,-1,-1,-1,-1\} & \{1,-1,-1,-1,-1,-1\} & \{-1,1,-1,-1,-1,-1\} \\
\{1,1,-1,-1,-1,-1\} & \{-1,-1,1,-1,-1,-1\} & \{1,-1,1,-1,-1,-1\} \\
\{-1,1,1,-1,-1-1\} & \{1,1,1,-1,-1,-1\} & \{-1,-1,-1,1,-1,-1\} \\
\{1,-1,-1,1,-1,-1\} & \{-1,1,-1,1,-1,-1\} & \{1,1,-1,1,-1,-1\} \\
\{-1,-1,1,1,-1,-1\} & \{1,-1,1,1,-1,-1\} & \{-1,1,1,1,-1,-1\} \\
\{1,1,1,1,-1,-1\} & \{-1,-1,-1,-1,1,-1\} & \{1,-1,-1,-1,1,-1\} \\
\{-1,1,-1,-1,1,-1\} & \{1,1,-1,-1,1,-1\} & \{-1,-1,1,-1,1,-1\} \\
\{1,-1,1,-1,1,-1\} & \{-1,1,1,-1,1,-1\} & \{1,1,1,-1,1,-1\} \\
\{-1,-1,-1,1,1,-1\} & \{1,-1,-1,1,1,-1\} & \{-1,1,-1,1,1,-1\} \\
\{1,1,-1,1,1,-1\} & \{-1,-1,1,1,1,-1\} & \{1,-1,1,1,1,-1\} \\
\{-1,1,1,1,1,-1\} & \{1,1,1,1,1,-1\} & \{-1,-1,-1,-1,-1,1\} \\
\{1,-1,-1,-1,-1,1\} & \{-1,1,-1,-1,-1,1\} & \{1,1,-1,-1,-1,1\} \\
\{-1,-1,1,-1,-1,1\} & \{1,-1,1,-1,-1,1\} & \{-1,1,1,-1,-1,1\} \\
\{1,1,1,-1,-1,1\} & \{-1,-1,-1,1,-1,1\} & \{1,-1,-1,1,-1,1\} \\
\{-1,1,-1,1,-1,1\} & \{1,1,-1,1,-1,1\} & \{-1,-1,1,1,-1,1\} \\
\{1,-1,1,1,-1,1\} & \{-1,1,1,1,-1,1\} & \{1,1,1,1,-1,1\} \\
\{-1,-1,-1,-1,1,1\} & \{1,-1,-1,-1,1,1\} & \{-1,1,-1,-1,1,1\} \\
\{1,1,-1,-1,1,1\} & \{-1,-1,1,-1,1,1\} & \{1,-1,1,-1,1,1\} \\
\{-1,1,1,-1,1,1\} & \{1,1,1,-1,1,1\} & \{-1,-1,-1,1,1,1\} \\
\{1,-1,-1,1,1,1\} & \{-1,1,-1,1,1,1\} & \{1,1,-1,1,1,1\} \\
\{-1,-1,1,1,1,1\} & \{1,-1,1,1,1,1\} & \{-1,1,1,1,1,1\} \\
\{1,1,1,1,1,1\} & &
\end{array}\right\}$$

Since $l(N(s)) = 2$ and $r(N(s)) = 1$,

$$l(N(s)) - r(N(s)) = 1$$

and

$$(-1)^{l-1}\text{sgn}[q(\infty,\ -18)] = -1,$$

it follows from **Step 10** *that every admissible string*

$$\mathcal{I} = \{i_0,\ i_1,\ i_2,\ i_3,\ i_4,\ i_5\}$$

must satisfy

$$\{i_0 - 2i_1 + 2i_2 - 2i_3 + 2i_4 - i_5\} \cdot (-1) = 6.$$

Hence the admissible strings are

$$\begin{array}{rcl}
\mathcal{I}_1 & = & \{-1,\ -1,\ -1,\ 1,\ -1,\ 1\} \\
\mathcal{I}_2 & = & \{-1,\ 1,\ 1,\ 1,\ -1,\ 1\} \\
\mathcal{I}_3 & = & \{-1,\ 1,\ -1,\ -1,\ -1,\ 1\} \\
\mathcal{I}_4 & = & \{-1,\ 1,\ -1,\ 1,\ 1,\ 1\} \\
\mathcal{I}_5 & = & \{1,\ 1,\ -1,\ 1,\ -1,\ -1\}.
\end{array}$$

From **Step 11**, *for $\mathcal{I}_1$ it follows that the stabilizing (k_i, k_d) values corresponding to $k_p = -18$ must satisfy the string of inequalities:*

$$\begin{cases} p_1(\omega_0) + (k_i - k_d\omega_0^2)p_2(\omega_0) < 0 \\ p_1(\omega_1) + (k_i - k_d\omega_1^2)p_2(\omega_1) < 0 \\ p_1(\omega_2) + (k_i - k_d\omega_2^2)p_2(\omega_2) < 0 \\ p_1(\omega_3) + (k_i - k_d\omega_3^2)p_2(\omega_3) > 0 \\ p_1(\omega_4) + (k_i - k_d\omega_4^2)p_2(\omega_4) < 0 \\ p_1(\omega_5) + (k_i - k_d\omega_5^2)p_2(\omega_5) > 0\,. \end{cases}$$

Substituting for ω_0, ω_1, ω_2, ω_3, ω_4, and ω_5 in the above expressions, we obtain

$$\begin{cases} k_i < 0 \\ k_i - 0.2699k_d < -4.6836 \\ k_i - 0.3666k_d < -10.0797 \\ k_i - 3.5358k_d > 3.912 \\ k_i - 13.5777k_d < 140.2055\,. \end{cases} \tag{12.11}$$

The set of values of $(k_i,\ k_d)$ for which (12.11) holds can be solved by linear programming and is denoted by $\mathcal{S}_1$. For $\mathcal{I}_2$, we have

$$\begin{cases} k_i < 0 \\ k_i - 0.2699k_d > -4.6836 \\ k_i - 0.3666k_d > -10.0797 \\ k_i - 3.5358k_d > 3.912 \\ k_i - 13.5777k_d < 140.2055\,. \end{cases} \tag{12.12}$$

The set of values of $(k_i,\ k_d)$ for which (12.12) holds can also be solved by linear programming and is denoted by $\mathcal{S}_2$. Similarly, we obtain

$$\begin{cases} \mathcal{S}_3 = \emptyset \text{ for } \mathcal{I}_3 \\ \mathcal{S}_4 = \emptyset \text{ for } \mathcal{I}_4 \\ \mathcal{S}_5 = \emptyset \text{ for } \mathcal{I}_5\,. \end{cases}$$

Then the stabilizing set of $(k_i,\ k_d)$ values when $k_p = -18$ is given by

$$\begin{aligned} \mathcal{S}_{(-18)} &= \cup_{x=1,\,2,\cdots,\,5}\mathcal{S}_x \\ &= \mathcal{S}_1 \cup \mathcal{S}_2\,. \end{aligned}$$

The set $\mathcal{S}_{(-18)}$ and the corresponding $\mathcal{S}_1$ and $\mathcal{S}_2$ are shown in Fig. 12.2. By sweeping over different k_p values within the interval $(-24.7513,\ 1)$ and repeating the above procedure at each stage, we can generate the set of stabilizing $(k_p,\ k_i,\ k_d)$ values. This set is shown in Fig. 12.3. △

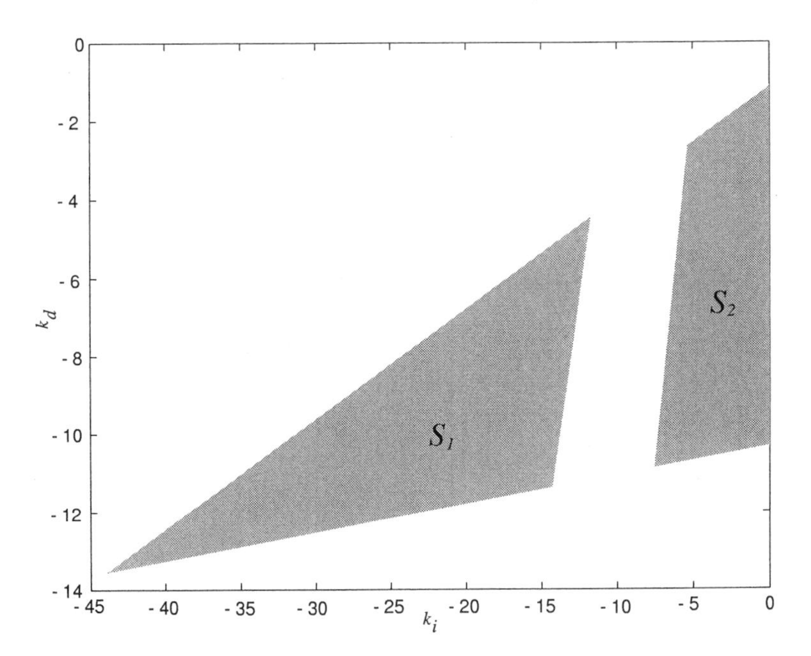

FIGURE 12.2. The stabilizing set of (k_i, k_d) values when $k_p = -18$.

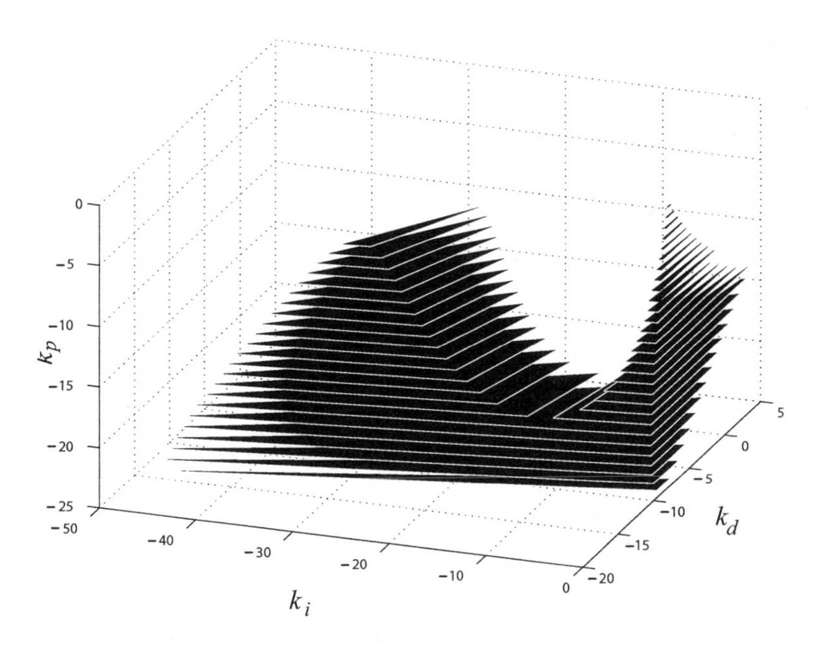

FIGURE 12.3. The stabilizing set of (k_p, k_i, k_d) values.

12.3 Discrete-Time Systems

In the case of a discrete-time system, the plant is given by

$$G_z(z) = \frac{N_z(z)}{D_z(z)}$$

where $N_z(z)$ and $D_z(z)$ are polynomials in the forward shift operator z. The discrete-time PID controller is given by

$$\begin{aligned} C_z(z) &= k_p + k_i \frac{1}{1 - z^{-1}} + k_d \frac{1 - 2z^{-1} + z^{-2}}{1 - z^{-1}} \\ &= \frac{(k_p + k_i + k_d)z^2 - (k_p + 2k_d)z + k_d}{z^2 - z} \end{aligned}$$

where k_p, k_i, and k_d are the proportional, integral, and derivative gains, respectively. Plants with a zero at $z = 1$ cannot be stabilized by PID controllers because of the unstable pole-zero cancellation implied and are excluded at the outset. Using the bilinear transformation $z = \frac{w+1}{w-1}$, we obtain the w-domain plant:

$$\frac{N(w)}{D(w)} = G_z(z)|_{z=\frac{w+1}{w-1}}$$

and the w-domain PID controller

$$\frac{B(w)}{A(w)} = \frac{k_i w^2 + 2(k_p + k_i)w + 2k_p + k_i + 4k_d}{2w + 2}.$$

The corresponding w-domain closed-loop characteristic polynomial becomes

$$\begin{aligned} \delta(w, k_p, k_i, k_d) = & (2w + 2)D(w) + [k_i w^2 + 2(k_p + k_i)w \\ & +2k_p + k_i + 4k_d]N(w) . \end{aligned} \quad (12.13)$$

The Hurwitz stability of this polynomial is equivalent to the stability of the original discrete-time system.

Following Chapter 4 we proceed as in the last section and multiply (12.13) by the factor $N(-w)$ to obtain

$$\delta^*(w, k_p, k_i, k_d) = N(-w)\delta(w, k_p, k_i, k_d).$$

By using the substitution

$$k_i = k_s - k_p \quad (12.14)$$

we can write

$$\begin{aligned} \delta^*(w, k_p, k_d, k_s) &= \delta'_e(w^2, k_p, k_d, k_s) + w\delta'_o(w^2, k_s) \\ &= [k_p\delta'_{ep}(w^2) + k_s\delta'_{es}(w^2) + k_d\delta'_{ed}(w^2) + \delta'_{ec}(w^2)] \\ &\quad +w[k_s\delta'_{os}(w^2) + \delta'_{oc}(w^2)] \end{aligned} \quad (12.15)$$

where

$$
\begin{aligned}
\delta'_{ep}(w^2) &= (1-w^2)(N_e^2 - w^2 N_o^2) \\
\delta'_{es}(w^2) &= (1+w^2)(N_e^2 - w^2 N_o^2) \\
\delta'_{ed}(w^2) &= 4(N_e^2 - w^2 N_o^2) \\
\delta'_{ec}(w^2) &= 2(N_e D_e + w^2 N_e D_o - w^2 N_o D_e - w^2 N_o D_o) \\
\delta'_{os}(w^2) &= 2(N_e^2 - w^2 N_o^2) \\
\delta'_{oc}(w^2) &= 2(N_e D_e + N_e D_o - N_o D_e - w^2 N_o D_o)\,.
\end{aligned}
$$

It is clear from (12.15) that we can now proceed as in the previous section, i.e., fix k_s, then use linear programming to solve for the stabilizing values of k_p and k_d. In other words, the entire development in the last section can be repeated by replacing $\delta(s, k_p, k_i, k_d)N^*(s)$ in (12.2) by $\delta^*(w, k_p, k_d, k_s)$ and proceeding as before. However, this procedure will yield the stabilizing parameters only in the space of (k_p, k_d, k_s). In order to recover the stabilizing parameters in the original (k_p, k_i, k_d) space, we need to go through the inverse linear transformation.

Example 12.2 *Consider a PID controller to stabilize the discrete-time system* $\frac{N_z(z)}{D_z(z)}$ *where*

$$
\begin{aligned}
N_z(z) &= z+1 \\
D_z(z) &= z^2 - 1.5z + 0.5.
\end{aligned}
$$

Using the bilinear transformation, we obtain the w-domain plant $\frac{N(w)}{D(w)}$ *where*

$$
\begin{aligned}
N(w) &= 2w^2 - 2w \\
D(w) &= w + 3.
\end{aligned}
$$

Figure 12.4 shows the stabilizing regions in the space of (k_p, k_d, k_s) determined using the procedure outlined above. After going through the inverse linear transformation we obtained the stabilizing regions in the space of (k_p, k_i, k_d). This region is shown in Fig. 12.5. △

12.4 Algorithm for Continuous-Time First-Order Systems with Time Delay

In this section, we consider the feedback system of Fig. 12.1 where the plant $G(s)$ is described by

$$
G(s) = \frac{k}{1+Ts} e^{-Ls}. \tag{12.16}
$$

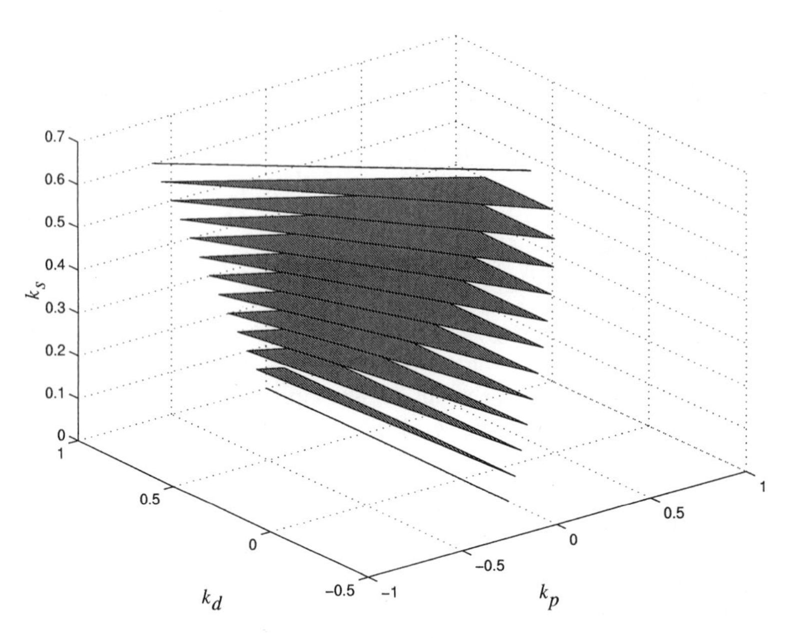

FIGURE 12.4. The stabilizing region in the space of $(k_p,\ k_d,\ k_s)$.

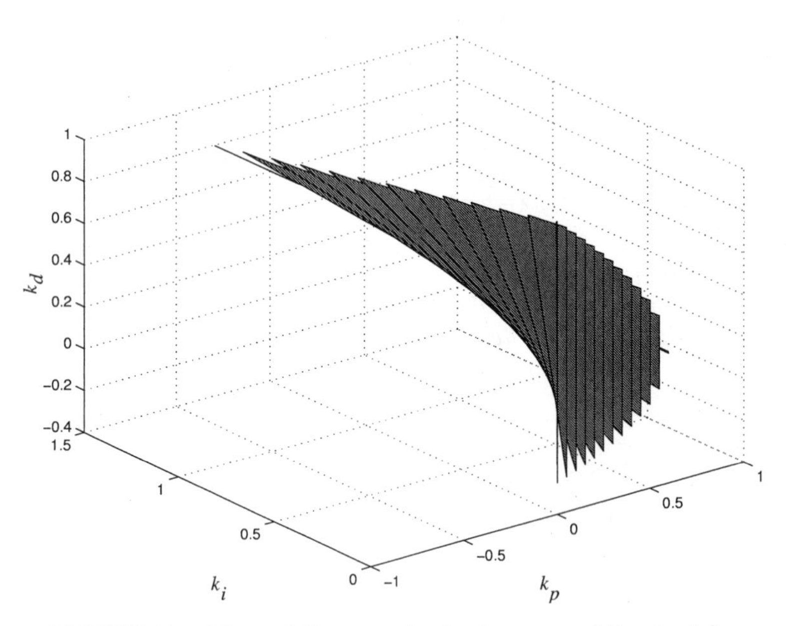

FIGURE 12.5. The stabilizing region in the space of $(k_p,\ k_i,\ k_d)$.

Here k represents the steady-state gain of the plant, L the time delay, and T the time constant of the plant. As before, the controller is of the PID type, i.e.,

$$C(s) = k_p + \frac{k_i}{s} + k_d s \, .$$

The objective is to determine the set of controller parameters (k_p,k_i,k_d) for which the closed-loop system is stable. A complete solution to this problem has already been presented in Chapter 8. We provide a brief summary of these results.

12.4.1 Open-Loop Stable Plant

In this case $T > 0$. We make the standing assumption that $k > 0$ and $L > 0$. The next theorem presents the complete set of stabilizing PID controllers for an open-loop stable plant described by (12.16).

Theorem 12.1 *The range of k_p values for which a given open-loop stable plant with transfer function $G(s)$ as in (12.16) continues to have closed-loop stability with a PID controller in the loop is given by*

$$-\frac{1}{k} < k_p < \frac{1}{k}\left[\frac{T}{L}\alpha_1 \sin(\alpha_1) - \cos(\alpha_1)\right] \tag{12.17}$$

where α_1 is the solution of the equation

$$\tan(\alpha) = -\frac{T}{T+L}\alpha$$

in the interval $(0, \pi)$. For k_p values outside this range, there are no stabilizing PID controllers. The complete stabilizing region is given by (see Fig. 12.6):

1. *For each $k_p \in (-\frac{1}{k}, \frac{1}{k})$, the cross-section of the stabilizing region in the (k_i, k_d) space is the trapezoid T;*
2. *For $k_p = \frac{1}{k}$, the cross-section of the stabilizing region in the (k_i, k_d) space is the triangle Δ;*
3. *For each $k_p \in \left(\frac{1}{k}, k_u := \frac{1}{k}\left[\frac{T}{L}\alpha_1 \sin(\alpha_1) - \cos(\alpha_1)\right]\right)$, the cross-section of the stabilizing region in the (k_i, k_d) space is the quadrilateral Q.*

The parameters m_j, b_j, w_j, $j = 1, 2$ necessary for determining the boundaries of T, Δ, and Q can be determined using the following equations:

$$m_j = \frac{L^2}{z_j^2} \tag{12.18}$$

$$b_j = -\frac{L}{k z_j}\left[\sin(z_j) + \frac{T}{L} z_j \cos(z_j)\right] \tag{12.19}$$

$$w_j = \frac{z_j}{kL}\left[\sin(z_j) + \frac{T}{L} z_j (\cos(z_j) + 1)\right] \tag{12.20}$$

where z_j, $j = 1, 2, \ldots$, are the real, positive solutions of

$$kk_p + \cos(z) - \frac{T}{L} z \sin(z) = 0 \tag{12.21}$$

arranged in ascending order of magnitude.

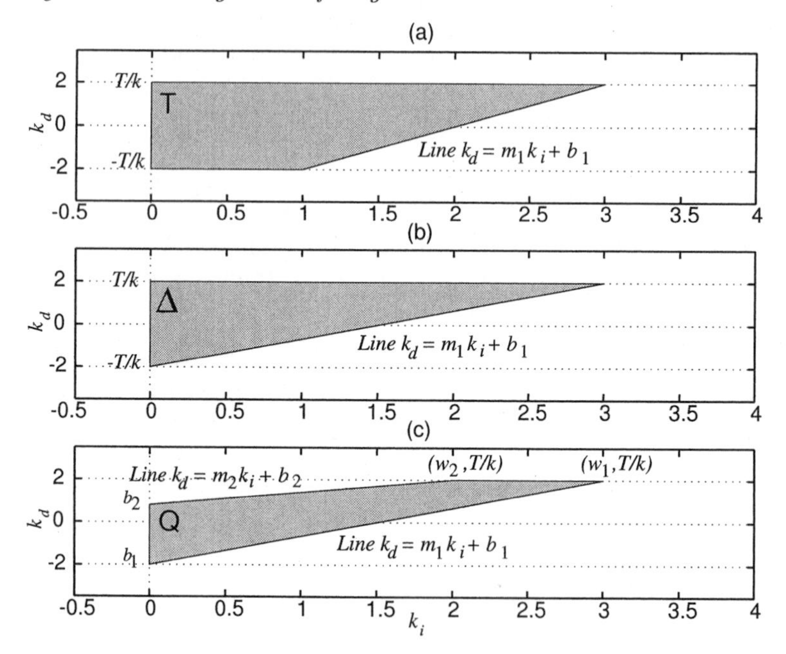

FIGURE 12.6. The stabilizing region of (k_i, k_d) for (a) $-\frac{1}{k} < k_p < \frac{1}{k}$, (b) $k_p = \frac{1}{k}$, (c) $\frac{1}{k} < k_p < k_u$.

12.4.2 Open-Loop Unstable Plant

In this case $T < 0$ in (12.16). Let us assume that $k > 0$ and $L > 0$.

Theorem 12.2 *A necessary and sufficient condition for the existence of a stabilizing PID controller for the open-loop unstable plant (12.16) is $|\frac{T}{L}| >$ 0.5. If this condition is satisfied, then the range of k_p values for which a given open-loop unstable plant with transfer function $G(s)$ as in (12.16) can be stabilized using a PID controller is given by*

$$\frac{1}{k}\left[\frac{T}{L}\alpha_1 \sin(\alpha_1) - \cos(\alpha_1)\right] < k_p < -\frac{1}{k}$$

where α_1 is the solution of the equation

$$\tan(\alpha) = -\frac{T}{T+L}\alpha$$

in the interval $(0, \pi)$. *In the special case of* $|\frac{T}{L}| = 1$, *we have* $\alpha_1 = \frac{\pi}{2}$. *For* k_p *values outside this range, there are no stabilizing PID controllers. Moreover, the complete stabilizing region is characterized by (see Fig. 12.7):*

> *For each* $k_p \in \left(k_l := \frac{1}{k}\left[\frac{T}{L}\alpha_1 \sin(\alpha_1) - \cos(\alpha_1)\right], -\frac{1}{k}\right)$, *the cross-section of the stabilizing region in the* (k_i, k_d) *space is the quadrilateral* Q.

The parameters m_j, b_j *and* w_j, $j = 1, 2$ *necessary for determining the boundary of* Q *are as defined in the statement of Theorem 12.1.*

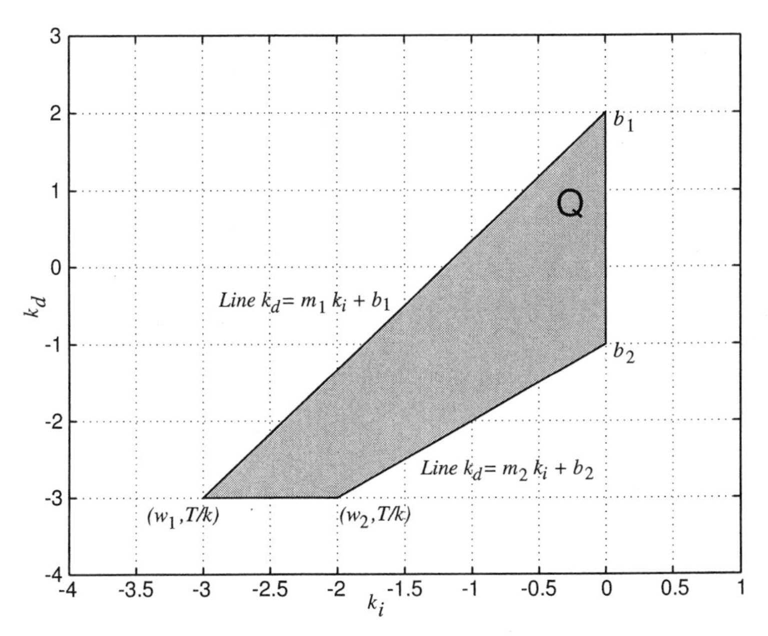

FIGURE 12.7. The stabilizing region of (k_i, k_d) for $k_l < k_p < -\frac{1}{k}$.

In view of Theorem 12.1, we now propose an algorithm to determine the set of stabilizing parameters for the plant (12.16) with $T > 0$.

PID Stabilization Algorithm for Time-Delay Plants:

Step 1: Initialize $k_p = -\frac{1}{k}$ and $step = \frac{1}{N+1}\left(k_u + \frac{1}{k}\right)$, where N is the desired number of grid points;

Step 2: Increase k_p as follows: $k_p = k_p + step$;

Step 3: If $k_p < k_u$ then **GOTO Step 4**. Else, terminate the algorithm and **EXIT**;

Step 4: Find the roots z_1 and z_2 of (12.21);

Step 5: Compute the parameters m_j and b_j, $j = 1, 2$ associated with the previously found z_j by using (12.18) and (12.19);

Step 6: Determine the stabilizing region in the k_i-k_d space using Fig. 12.6;

Step 7: GOTO Step 2.

A similar algorithm can be written for the case of an open-loop unstable plant by using Theorem 12.2.

We next present an example that illustrates the use of the above algorithm to determine stabilizing PID parameters.

Example 12.3 *Consider the PID stabilization problem for a plant described by the transfer function*

$$G(s) = \frac{1}{3s+1}e^{-2.8s} .$$

In this case the plant parameters are $k = 1$, $T = 3$ seconds, and $L = 2.8$ seconds. Since the system is open-loop stable we use Theorem 12.1 to find the range of k_p values for which a solution to the PID stabilization problem exists. We first compute the parameter $\alpha_1 \in (0, \pi)$ satisfying the following equation:

$$\tan(\alpha) = -0.5172\alpha .$$

Solving this equation we obtain $\alpha_1 = 2.2752$. From (12.17) the range of k_p values is given by

$$-1 < k_p < 2.5051 .$$

We now sweep over the above range of k_p values and determine the stabilizing set of (k_i, k_d) values at each stage using the previous algorithm. These regions are sketched in Fig. 12.8.

Any PID gains selected from these regions will result in closed-loop stability and any gains outside will result in instability. Consider the following performance specifications:

1. *settling time ≤ 40 seconds;*

2. *overshoot $\leq 20\%$.*

We can obtain the transient responses of the closed-loop system for the $(k_p,\ k_i,\ k_d)$ values inside the regions depicted in Fig. 12.8. By searching over these regions, several $(k_p,\ k_i,\ k_d)$ values are found to meet the desired performance specifications. We arbitrarily set the controller parameters to $k_p = 1.2$, $k_i = 0.3$, $k_d = 1.6667$. Figure 12.9 shows the step response of the resulting closed-loop system. The step signal is applied at $t = 5$ seconds. It is clear from the figure that the closed-loop system is stable, the output $y(t)$ tracks the step input signal, and the performance specifications are met. $\triangle$

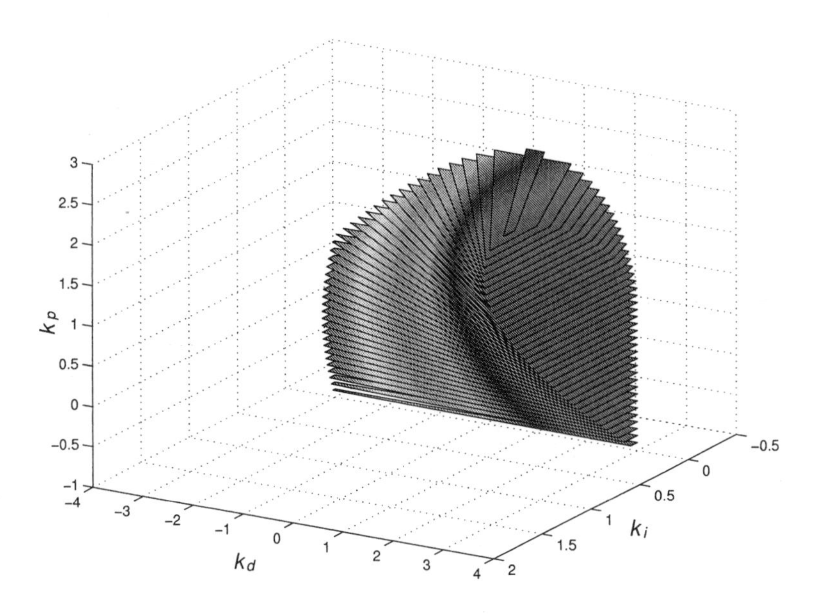

FIGURE 12.8. The stabilizing region of (k_p, k_i, k_d) values for the PID controller in Example 12.3.

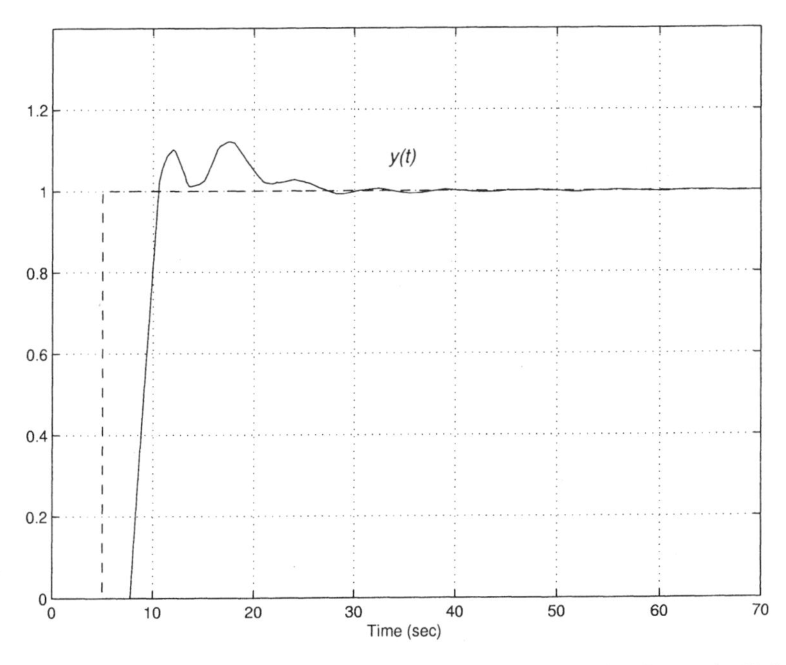

FIGURE 12.9. Time response of the closed-loop system for Example 12.3.

Although the design presented above is essentially an optimization search by gridding, the fact that the algorithm of this section can be used to confine the search to the stabilizing set makes the design problem orders of magnitude easier.

12.5 Algorithms for PID Controller Design

The PID stabilization algorithms presented in the last three sections can be used to determine the entire set of stabilizing PID controllers. Hence, in principle, they can be used to facilitate PID design. Indeed, by confining the search for the PID parameters to the stabilizing regions, it is possible to optimize different performance indices while ensuring that the stability constraint is always satisfied. This, however, constitutes a numerical design approach; Example 12.3 illustrates this point. In certain situations, nevertheless, it is possible to do better than mere numerical optimization. This section is devoted to a discussion of such situations that have arisen so far in our research.

In many situations control system performance can be specified by a frequency domain inequality or, equivalently, an H_∞ norm constraint on a closed-loop transfer function $G(s) = \frac{N(s)}{D(s)}$:

$$\|G(s)\|_\infty < \gamma.$$

It has been shown that the above condition is equivalent to Hurwitz stability of the complex polynomial family:

$$\gamma D(s) + e^{j\theta} N(s), \theta \in [0, 2\pi].$$

In our PID design problem the polynomials $D(s), N(s)$ will have the PID gains embedded in them, and the set of parameters achieving specifications is given by those achieving simultaneously the stabilization of the complex polynomial family as well as the real closed-loop characteristic polynomial. It turns out that the set of PID gains achieving stabilization of a complex polynomial family and therefore attaining the specifications can be found by an extension of the algorithm given for the real case. Toward this end, consider a complex polynomial of the form:

$$c(s,\ k_p,\ k_i,\ k_d) \quad = \quad L(s) + (k_d s^2 + k_p s + k_i) M(s) \qquad (12.22)$$

where $L(s)$ and $M(s)$ are given complex polynomials. The results on PID stabilization presented in Section 12.2 have been extended to the stabilization of (12.22). The algorithm, described below, is similar to the stabilization algorithm given for the real case. We will therefore not write the algorithm in detail but only point out the differences in the formulas and steps from that of the real case. We then show through examples how many PID *performance or design* problems can be converted into stabilization problems of complex polynomial families of the form of (12.22) and solved using this algorithm.

12.5.1 Complex PID Stabilization Algorithm

The complex PID stabilization algorithm is similar to the algorithm given for the real case in Section 12.2 and we need only to point out the differences in some formulas and steps. To do that we first introduce some definitions and notation.

Definition 12.4 *Let $a(s)$ be a given complex polynomial of degree n:*

$$\begin{aligned} a(s) &= (a_0 + jb_0) + (a_1 + jb_1)s + \cdots + (a_{n-1} + jb_{n-1})s^{n-1} \\ &\quad + (a_n + jb_n)s^n, \ a_n + jb_n \neq 0. \end{aligned}$$

The real-imaginary decomposition of $a(s)$ is defined as

$$a(s) = a_R(s) + a_I(s)$$

where

$$\begin{aligned} a_R(s) &= a_0 + jb_1 s + a_2 s^2 + jb_3 s^3 + \cdots \\ a_I(s) &= jb_0 + a_1 s + jb_2 s^2 + a_3 s^3 + \cdots. \end{aligned}$$

Now we consider a complex polynomial of the form

$$c(s, k_p, k_i, k_d) = L(s) + (k_d s^2 + k_p s + k_i)M(s) \qquad (12.23)$$

where $L(s)$ and $M(s)$ are two given complex polynomials. Write $L(s)$ and $M(s)$ in terms of their real-imaginary decompositions:

$$\begin{aligned} L(s) &= L_R(s) + L_I(s) \\ M(s) &= M_R(s) + M_I(s) \end{aligned}$$

and define

$$M^*(s) = M_R(s) - M_I(s)$$

and

$$\nu(s) = c(s, k_p, k_i, k_d)M^*(s) .$$

Also let n, m be the degrees of $c(s, \ k_p, \ k_i, \ k_d)$ and $M(s)$, respectively. Evaluating the polynomial $\nu(s)$ at $s = j\omega$, we obtain

$$\nu(j\omega) = c(j\omega, \ k_p, \ k_i, \ k_d)M^*(j\omega) = p(\omega, \ k_i, \ k_d) + jq(\omega, \ k_p)$$

where

$$\begin{aligned}
p(\omega,\ k_i,\ k_d) &= p_1(\omega) + (k_i - k_d\,\omega^2)p_2(\omega) && (12.24)\\
q(\omega,\ k_p) &= q_1(\omega) + k_p\, q_2(\omega) && (12.25)\\
p_1(\omega) &= L_R(j\omega)M_R(j\omega) - L_I(j\omega)M_I(j\omega) && (12.26)\\
p_2(\omega) &= M_R^2(j\omega) - M_I^2(j\omega) && (12.27)\\
q_1(\omega) &= \frac{1}{j}[L_I(j\omega)M_R(j\omega) - L_R(j\omega)M_I(j\omega)] && (12.28)\\
q_2(\omega) &= \omega[M_R^2(j\omega) - M_I^2(j\omega)]. && (12.29)
\end{aligned}$$

Let ξ denote the leading coefficient of $c(s,\ k_p,\ k_i,\ k_d)M^*(s)$. The procedure for determining all stabilizing $(k_p,\ k_i,\ k_d)$ for which $c(s,\ k_p,\ k_i,\ k_d)$ is Hurwitz for the given $L(s)$ and $M(s)$ is identical to the stabilization algorithm of Section 12.2 except for the following steps below, labeled Step lc, in the computation of the *allowable* range and *admissible* strings.

Differences Between Real and Complex PID Stabilization Algorithms:

Step 1c: Compute $p_1(\omega), p_2(\omega), q_1(\omega), q_2(\omega)$ from (12.26)–(12.29);

Step 2c: The *allowable* ranges of k_p are such that $q(\omega,\ k_p)$ has at least

$$\begin{cases} |n - (l(M(s)) - r(M(s)))| - 1 & \text{if } m+n \text{ is even and } \xi \text{ is purely real, or } m+n \text{ is odd and } \xi \text{ is purely imaginary} \\ |n - (l(M(s)) - r(M(s)))| & \text{if } m+n \text{ is even and } \xi \text{ is not purely real, or } m+n \text{ is odd and } \xi \text{ is not purely imaginary} \end{cases}$$

real, distinct finite zeros with odd multiplicities. The resulting ranges of k_p are the only ranges of k_p for which stabilizing $(k_i,\ k_d)$ values may exist;

Step 8c: For fixed k_p solve for the real, distinct finite zeros of $q(\omega, k_p)$ with odd multiplicities and denote them by $\omega_1 < \omega_2 < \cdots < \omega_{l-1}$ and let $\omega_0 = -\infty$ and $\omega_l = \infty$;

Step 9c: The construction of the sequences of numbers $i_0, i_i, i_2, \ldots, i_l$ is as follows:
If $M^*(j\omega_t) = 0$ for some $t = 0, 1, \ldots, l$, then define

$$i_t = 0;$$

else

$$i_t \in \{-1, 1\}, \text{ for all other } t = 0, 1, \ldots, l.$$

With $i_0, i_1, \ldots$ defined in this way, define the set A_{k_p} as

$$A_{k_p} = \begin{cases} \{\{i_0, i_1, \ldots, i_l\}\}, & \text{if } m+n \text{ is even and } \xi \text{ is purely real, or } m+n \text{ is odd and } \xi \text{ is purely imaginary} \\ \{\{0, i_1, i_2, \ldots, i_{l-1}, 0\}\}, & \text{if } m+n \text{ is even and } \xi \text{ is not purely real, or } m+n \text{ is odd and } \xi \text{ is not purely imaginary} \end{cases}$$

Step 10c: Determine the *admissible* strings $\mathcal{I} \in A_{k_p}$ such that the following equality holds:

$$n - (l(M(s)) - r(M(s))) =$$

$$\begin{cases} \frac{1}{2}\{i_0 \cdot (-1)^{l-1} + 2\sum_{r=1}^{l-1} i_r \cdot (-1)^{l-1-r} - i_l\} \cdot \text{sgn}[q(\infty,\ k_p)], \\ \text{if } m+n \text{ is even and } \xi \text{ is purely real, or } m+n \text{ is odd and } \xi \text{ is purely imaginary} \\ \frac{1}{2}\{2\sum_{r=1}^{l-1} i_r \cdot (-1)^{l-1-r}\} \cdot \text{sgn}[q(\infty,\ k_p)], \\ \text{if } m+n \text{ is even and } \xi \text{ is not purely real, or } m+n \text{ is odd and } \xi \text{ is not purely imaginary} \end{cases} \tag{12.30}$$

We now give some application examples of PID performance using the complex stabilization algorithm.

12.5.2 *Synthesis of H_∞ PID Controllers*

First let us consider the problem of synthesizing PID controllers for which the closed-loop system is internally stable and the H_∞-norm of a certain closed-loop transfer function is less than a prescribed level. In particular, the following closed-loop transfer functions are considered:

- The sensitivity function:

$$S(s) = \frac{1}{1 + C(s)G(s)}. \tag{12.31}$$

- The complementary sensitivity function:

$$T(s) = \frac{C(s)G(s)}{1 + C(s)G(s)}. \tag{12.32}$$

- The input sensitivity function:

$$U(s) = \frac{C(s)}{1 + C(s)G(s)}. \tag{12.33}$$

Various performance and robustness specifications can be captured by using the H_∞-norm of weighted versions of the transfer functions (12.31)–(12.33). It can be verified that when $C(s)$ is a PID controller, the transfer functions (12.31)–(12.33) can all be represented in the following general form:

$$T_{cl}(s,\ k_p,\ k_i,\ k_d) \quad = \quad \frac{A(s) + (k_d s^2 + k_p s + k_i)B(s)}{sD(s) + (k_d s^2 + k_p s + k_i)N(s)} \tag{12.34}$$

where $A(s)$ and $B(s)$ are some real polynomials. For the transfer function $T_{cl}(s,\ k_p,\ k_i,\ k_d)$ and a given number $\gamma > 0$, the standard H_∞ performance specification usually takes the form:

$$\|W(s)T_{cl}(s,\ k_p,\ k_i,\ k_d)\|_\infty < \gamma \tag{12.35}$$

where $W(s)$ is a stable frequency-dependent weighting function that is selected to capture the desired design objectives at hand. Suppose the weighting function $W(s) = \frac{W_n(s)}{W_d(s)}$, where $W_n(s)$ and $W_d(s)$ are coprime polynomials and $W_d(s)$ is Hurwitz. Define the polynomials $\delta(s, k_p, k_i, k_d)$ and $\phi(s, k_p, k_i, k_d, \gamma, \theta)$ as follows:

$$\delta(s, k_p, k_i, k_d) \quad \triangleq \quad sD(s) + (k_i + k_p s + k_d s^2)N(s)$$

and

$$\begin{aligned}\phi(s, k_p, k_i, k_d, \gamma, \theta) \quad &\triangleq \quad \left[sW_d(s)D(s) + \frac{1}{\gamma}e^{j\theta}W_n(s)A(s)\right] \\ &\quad +(k_d s^2 + k_p s + k_i)\left[W_d(s)N(s)\right. \\ &\quad \left. +\frac{1}{\gamma}e^{j\theta}W_n(s)B(s)\right].\end{aligned}$$

Then we can establish the following relationship between H_∞ synthesis using PID controllers and simultaneous stabilization of a complex polynomial family:

For a given $\gamma > 0$, there exist PID gain values $(k_d,\ k_p,\ k_i)$ such that $\|W(s)T_{cl}(s, k_p, k_i, k_d)\|_\infty < \gamma$ if and only if the following conditions hold:

(1) $\delta(s, k_p, k_i, k_d)$ is Hurwitz;

(2) $\phi(s, k_p, k_i, k_d, \gamma, \theta)$ is Hurwitz for all θ in $[0,\ 2\pi)$;

(3) $|W(\infty)T_{cl}(\infty, k_p, k_i, k_d)| < \gamma$.

The above equivalence can be used to determine stabilizing (k_p, k_i, k_d) values such that the H_∞-norm of a certain closed-loop transfer function is less than a prescribed level. This is illustrated using the following example.

Example 12.4 *Consider the plant* $G(s) = \frac{N(s)}{D(s)}$ *given by*

$$\begin{aligned} N(s) &= s-1 \\ D(s) &= s^2+0.8s-0.2 \end{aligned}$$

and the PID controller

$$C(s) = \frac{k_d s^2 + k_p s + k_i}{s}.$$

In this example, we consider the problem of determining all stabilizing PID gain values for which $\|W(s)T(s,k_p,k_i,k_d)\|_\infty < 1$, *where* $T(s,k_p,k_i,k_d)$ *is the complementary sensitivity function:*

$$T(s,\ k_p,\ k_i,\ k_d) = \frac{(k_d s^2+k_p s+k_i)(s-1)}{s(s^2+0.8s-0.2)+(k_d s^2+k_p s+k_i)(s-1)}$$

and the weight $W(s)$ *is chosen as a high pass transfer function:*

$$W(s) = \frac{s+0.1}{s+1}.$$

We know that (k_p, k_i, k_d) *values meeting the* H_∞ *performance constraint exist if and only if the following conditions hold:*

(1) $\delta(s,\ k_p,\ k_i,\ k_d) = s(s^2+0.8s-0.2)+(k_d s^2+k_p s+k_i)(s-1)$ *is Hurwitz;*

(2) $\phi(s,\ k_p,\ k_i,\ k_d,\ 1,\ \theta) = s(s+1)(s^2+0.8s-0.2)+(k_d s^2+k_p s+k_i)[(s+1)(s-1)+e^{j\theta}(s+0.1)(s-1)]$ *is Hurwitz for all* θ *in* $[0,\ 2\pi)$;

(3) $|W(\infty)T(\infty,\ k_p,\ k_i,\ k_d)| = |\frac{k_d}{k_d+1}| < 1.$

The set of all (k_p, k_i, k_d) *values for which the* H_∞ *performance specifications are met are precisely the values of* k_p, k_i, k_d *for which conditions (1), (2), and (3) are satisfied. To search for such values of* (k_p, k_i, k_d), *we fix* k_p *and determine all the values of* (k_i, k_d) *for which conditions (1), (2), and (3) hold.*

For the condition (1) with a fixed k_p, *for instance* $k_p = -0.35$, *by setting* $L(s) = s(s^2+0.8s-0.2)$ *and* $M(s) = s-1$, *and using the algorithm of Section 12.2, we obtain the set of* (k_i, k_d) *values for which the closed-loop system is stable. This set is denoted by* $S_{(1,\ -0.35)}$ *and is sketched in Fig. 12.10.*

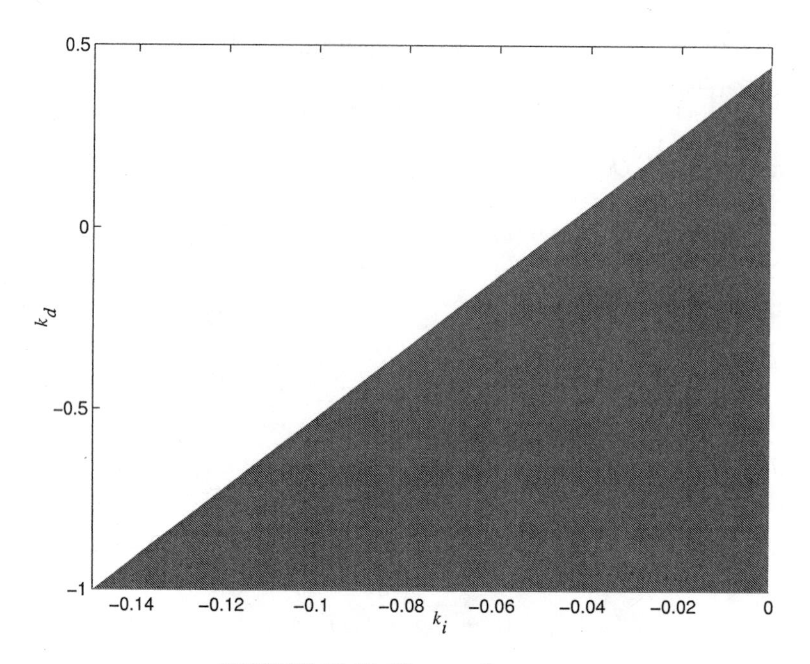

FIGURE 12.10. The set $\mathcal{S}_{(1,\ -0.35)}$.

Fixing $k_p = -0.35$ *and any fixed* $\theta \in [0, 2\pi)$, *by setting* $L(s) = s(s+1)(s^2+0.8s-0.2)$ *and* $M(s,\ \theta) = (s+1)(s-1) + e^{j\theta}(s+0.1)(s-1)$ *and using the complex stabilization algorithm of Section 12.5.1 again we can solve a linear programming problem to determine the set of* (k_i, k_d) *values. Let this set be denoted by* $\mathcal{S}_{(2,\ -0.35,\ \theta)}$. *By keeping* k_p *fixed, sweeping over* $\theta \in [0, 2\pi)$, *and using the complex stabilization algorithm of Section 12.5.1 at each stage, we can determine the set of* (k_i, k_d) *values for which condition (2) is satisfied. This set is denoted by* $\mathcal{S}_{(2,\ -0.35)}$ *and is given by*

$$\mathcal{S}_{(2,\ -0.35)} = \cap_{\theta \in [0,\ 2\pi)} \mathcal{S}_{(2,\ -0.35,\ \theta)}.$$

The set $\mathcal{S}_{(2,\ -0.35)}$ *is sketched in Fig. 12.11.*

Let $\mathcal{S}_{(3,\ -0.35)}$ *be the set of* $(k_i,\ k_d)$ *values satisfying condition (3) and this set is given by*

$$\mathcal{S}_{(3,\ -0.35)} = \{(k_i,\ k_d) |\ k_i \in \mathcal{R},\ k_d > -0.5\}.$$

Then for $k_p = -0.35$, *the set of* $(k_i,\ k_d)$ *values for which*

$$\|W(s)T(s,\ k_p,\ k_i,\ k_d)\|_\infty < 1$$

is denoted by $\mathcal{S}_{(-0.35)}$ *and is given by*

$$\mathcal{S}_{(-0.35)} = \cap_{i=1,2,3} \mathcal{S}_{(i,\ -0.35)}.$$

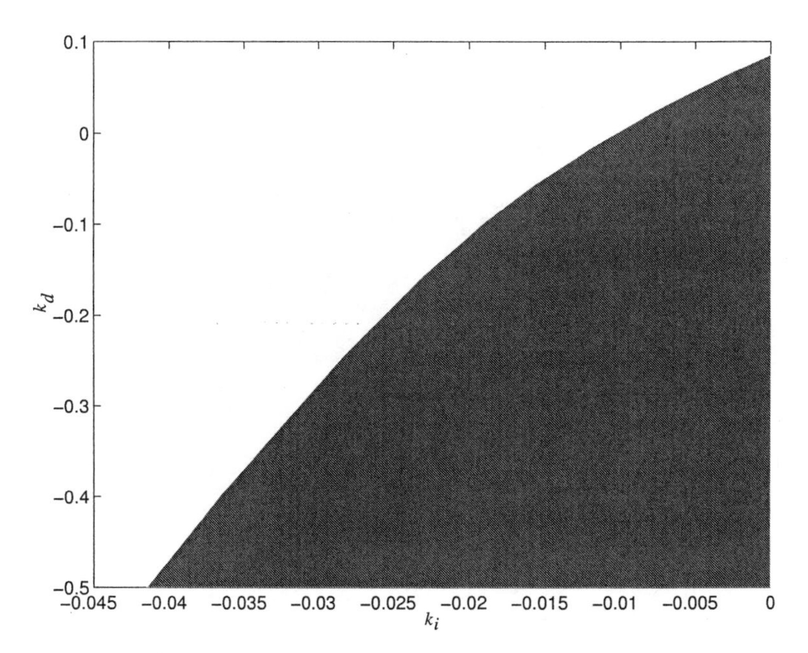

FIGURE 12.11. The set $\mathcal{S}_{(2,\ -0.35)} = \cap_{\theta \in [0,\ 2\pi)} \mathcal{S}_{(2,\ -0.35,\ \theta)}$.

In this case, we have $\mathcal{S}_{(-0.35)} = \mathcal{S}_{(2,\ -0.35)}$. Using root loci, it was determined that a necessary condition for the existence of stabilizing (k_i, k_d) values is that $k_p \in (-0.5566,\ -0.2197)$. Then, by sweeping over $k_p \in (-0.5566,\ -0.2197)$ and repeating the above procedure, we obtained the stabilizing set of (k_p, k_i, k_d) values for which $\|W(s)T(s, k_p, k_i, k_d)\|_\infty < 1$. This set is sketched in Fig. 12.12. $\triangle$

12.5.3 PID Controller Design for Robust Performance

This subsection is devoted to the problem of synthesizing PID controllers for robust performance. In particular, we focus on the following robust performance specification:

$$\| |W_1(s)S(s)| + |W_2(s)T(s)| \|_\infty < 1, \tag{12.36}$$

where $W_1(s) = \frac{N_{W1}(s)}{D_{W1}(s)}$ and $W_2(s) = \frac{N_{W2}(s)}{D_{W2}(s)}$ are stable weighting functions, and $S(s)$ and $T(s)$ are the sensitivity and the complementary sensitivity functions, respectively.

As before, let $\delta(s, k_p, k_i, k_d)$ denote the closed-loop characteristic polynomial

$$\delta(s, k_p, k_i, k_d) \ \stackrel{\Delta}{=}\ sD(s) + (k_i + k_p s + k_d s^2)N(s).$$

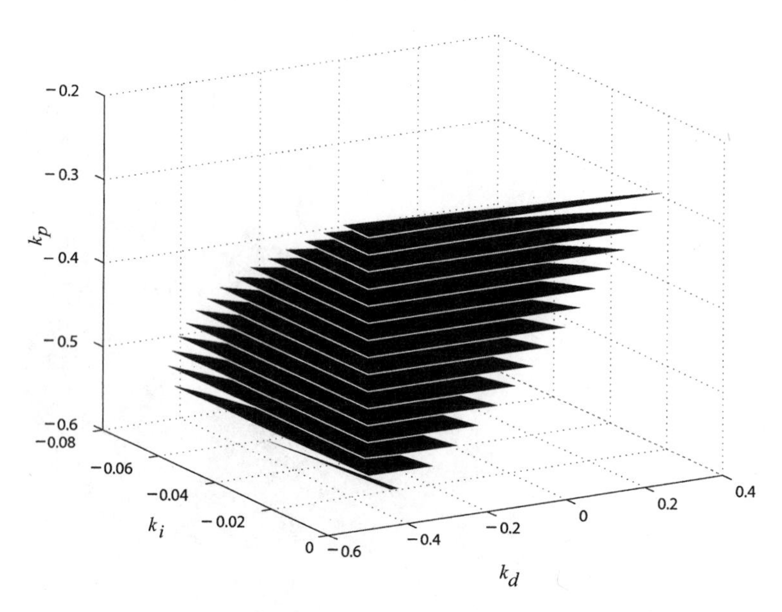

FIGURE 12.12. The set of stabilizing (k_p, k_i, k_d) values for which $\|W(s) T(s, k_p, k_i, k_d)\|_\infty < 1$.

We define the complex polynomial $\psi(s, k_p, k_i, k_d, \theta, \phi)$ by

$$\begin{aligned}\psi(s, k_p, k_i, k_d, \theta, \phi) \triangleq\ & sD_{W1}(s)D_{W2}(s)D(s) \\ & +e^{j\theta}sN_{W1}(s)D_{W2}(s)D(s) \\ & +(k_d s^2 + k_p s + k_i)[D_{W1}(s)D_{W2}(s)N(s) \\ & +e^{j\phi}D_{W1}(s)N_{W2}(s)N(s)].\end{aligned}$$

The problem of synthesizing PID controllers for robust performance can be converted into the problem of determining values of (k_p, k_i, k_d) for which the following conditions hold:

(1) $\delta(s, k_p, k_i, k_d)$ is Hurwitz;

(2) $\psi(s, k_p, k_i, k_d, \theta, \phi)$ is Hurwitz for all $\theta \in [0,\ 2\pi)$ and for all $\phi \in [0,\ 2\pi)$;

(3) $|W_1(\infty)S(\infty)| + |W_2(\infty)T(\infty)| < 1$.

The following example shows how the above conditions can be used to determine the set of stabilizing gains (k_p, k_i, k_d) for which the robust performance specification (12.36) is met.

Example 12.5 *Consider the plant* $G(s) = \frac{N(s)}{D(s)}$ *where*

$$\begin{aligned}N(s) &= s - 15 \\ D(s) &= s^2 + s - 1.\end{aligned}$$

Then the sensitivity function and complementary sensitivity function are

$$S(s,k_p,k_i,k_d) = \frac{s(s^2+s-1)}{s(s^2+s-1)+(k_ds^2+k_ps+k_i)(s-15)},$$
$$T(s,k_p,k_i,k_d) = \frac{(k_ds^2+k_ps+k_i)(s-15)}{s(s^2+s-1)+(k_ds^2+k_ps+k_i)(s-15)}.$$

The weighting functions are chosen as $W_1(s) = \frac{0.2}{s+0.2}$ *and* $W_2(s) = \frac{s+0.1}{s+1}$. *We know that stabilizing* (k_p, k_i, k_d) *values meeting the performance specification (12.36) exist if and only if the following conditions hold:*

(1) $\delta(s,k_p,k_i,k_d) = s(s^2+s-1)+(k_ds^2+k_ps+k_i)(s-15)$ *is Hurwitz;*

(2) $\psi(s,k_p,k_i,k_d,\theta,\phi) = s(s+0.2)(s+1)(s^2+s-1)+e^{j\theta}s(0.2)(s+1)(s^2+s-1)+(k_ds^2+k_ps+k_i)[(s+0.2)(s+1)(s-15)+e^{j\phi}(s+0.2)(s+0.1)(s-15)]$ *is Hurwitz for all* $\theta \in [0,\ 2\pi)$ *and for all* $\phi \in [0,\ 2\pi)$;

(3) $|W_1(\infty)S(\infty,k_p,k_i,k_d)|+|W_2(\infty)T(\infty,k_p,k_i,k_d)| = |\frac{k_d}{k_d+1}| < 1.$

The procedure for determining the set of (k_p, k_i, k_d) *values satisfying conditions (1), (2), and (3) is similar to that presented in the previous example. First using root loci, it was determined that a necessary condition for the existence of stabilizing* (k_i, k_d) *values is that* $k_p \in (-0.5079,\ -0.1155)$. *For any fixed* $k_p \in (-0.5079,\ -0.1155)$, *we use the algorithm of Section 12.5.1 to determine the set of* (k_i, k_d) *values satisfying conditions (1) and (2). The condition (3) gives that the admissible set of* (k_i, k_d) *is* $\{(k_i,k_d) |\ k_i \in \mathcal{R},\ k_d > -0.5\}$. *Then for a fixed* k_p, *we obtain the set of all* (k_i, k_d) *values for which* $\| |W_1(s)S(s,k_p,k_i,k_d)| + |W_2(s)T(s,k_p,k_i,k_d)| \|_\infty < 1$ *by taking the intersection of the set of* (k_i, k_d) *values satisfying conditions (1), (2), and (3). Thus by sweeping over* $k_p \in (-0.5079,\ -0.1155)$ *and repeating the above procedure, we obtain the set of* (k_p, k_i, k_d) *values for which* $\| |W_1(s)S(s,k_p,k_i,k_d)| + |W_2(s)T(s,k_p,k_i,k_d)| \|_\infty < 1$. *This set is sketched in Fig. 12.13.* △

12.5.4 PID Controller Design with Guaranteed Gain and Phase Margins

In this subsection, we consider the problem of designing PID controllers that achieve prespecified gain and phase margins for a given plant. To this end, let A_m and θ_m denote the desired upper gain and phase margins, respectively. From the definitions of the upper gain and phase margins, it follows that the PID gain values (k_p, k_i, k_d) achieving gain margin A_m and phase margin θ_m must satisfy the following conditions:

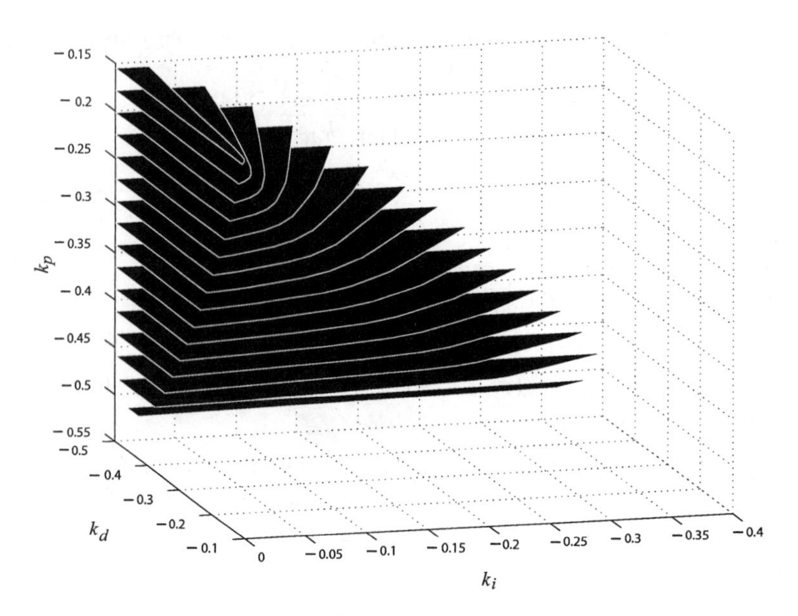

FIGURE 12.13. The set of (k_p, k_i, k_d) values for which $\||W_1(s)S(s,k_p,k_i,k_d)| + |W_2(s)T(s,k_p,k_i,k_d)|\|_\infty < 1$.

(1) $sD(s) + A(k_d s^2 + k_p s + k_i)N(s)$ is Hurwitz for all $A \in [1,\ A_m]$; and

(2) $sD(s) + e^{-j\theta}(k_d s^2 + k_p s + k_i)N(s)$ is Hurwitz for all $\theta \in [0,\ \theta_m]$.

Thus the problem to be solved is reduced to the problem of simultaneous stabilization of two families of polynomials. The algorithm of Section 12.5.1 can now be used to solve these simultaneous stabilization problems. The following example illustrates the procedure.

Example 12.6 *Consider the plant* $G(s) = \frac{N(s)}{D(s)}$ *where*

$$\begin{aligned} N(s) &= 2s - 1 \\ D(s) &= s^4 + 3s^3 + 4s^2 + 7s + 9. \end{aligned}$$

In this example, we consider the problem of determining all (k_p, k_i, k_d) *gain values that provide a gain margin* $A_m \geq 3.0$ *and a phase margin* $\theta_m \geq 40°$. *A given set of* (k_p, k_i, k_d) *values will meet these specifications if and only if the following conditions hold:*

(1) $s(s^4 + 3s^3 + 4s^2 + 7s + 9) + A(k_d s^2 + k_p s + k_i)(2s - 1)$ *is Hurwitz for all* $A \in [1,\ 3.0]$;

(2) $s(s^4 + 3s^3 + 4s^2 + 7s + 9) + e^{-j\theta}(k_d s^2 + k_p s + k_i)(2s - 1)$ *is Hurwitz for all* $\theta \in [0°,\ 40°]$.

Again, the procedure for determining the set of (k_p, k_i, k_d) values is similar to that presented in Section 12.5.2 and, therefore, a detailed description is omitted. The resulting set is sketched in Fig. 12.14.

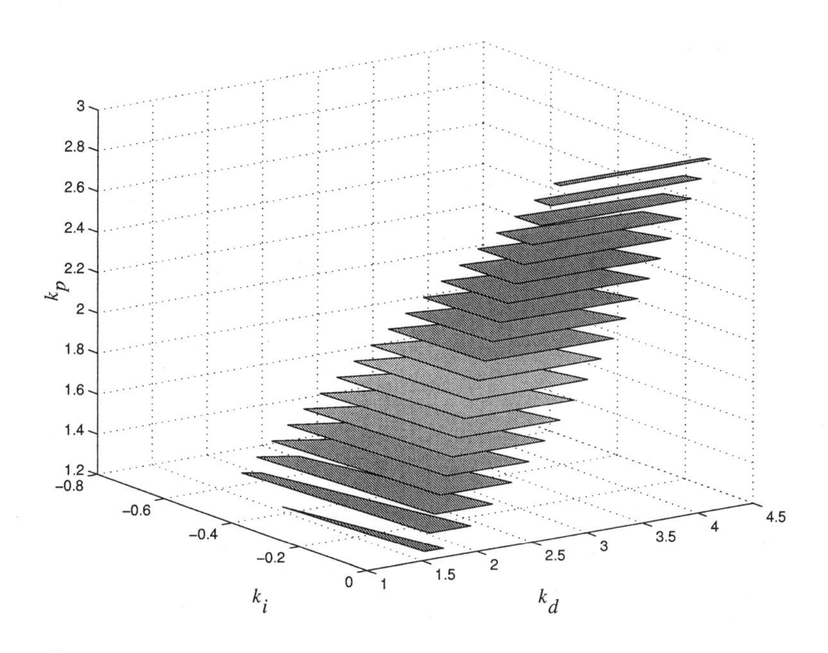

FIGURE 12.14. The set of (k_p, k_i, k_d) values for which the resulting closed-loop system achieves a gain margin $A_m \geq 3.0$ and a phase margin $\theta_m \geq 40°$.

△

12.6 Notes and References

The material presented in this chapter is based on Ho et al. [23]. The results presented in Section 12.5 were developed by Ho [21] and Ho and Lin [22]. For a detailed description of various performance and robustness specifications that can be captured by using the H_∞-norm, the reader is referred to [12].

A

Proof of Lemmas 8.3, 8.4, and 8.5

A.1 Preliminary Results

We begin by making the following observations, which follow from the proof of Lemma 8.1.

Remark A.1 *As we can see from Figs. 8.2 through 8.4(a), for* $k_p \in (-\frac{1}{k}, k_u)$*, the odd roots of (8.7), i.e.,* z_j *where* $j = 1, 3, 5, ...$*, are getting closer to* $(j-1)\pi$ *as* j *increases. So in the limit for odd values of* j *we have*

$$\lim_{j \to \infty} \cos(z_j) = 1 \ .$$

Moreover, since the cosine function is monotonically decreasing between $(j-1)\pi$ *and* $j\pi$ *for odd values of* j*, in view of the previous observation we have*

$$\cos(z_1) < \cos(z_3) < \cos(z_5) < \cdots \ .$$

Remark A.2 *From Fig. 8.2 and Fig. 8.4(a) we see that for* $k_p \in (-\frac{1}{k}, \frac{1}{k}) \cup (\frac{1}{k}, k_u)$*, the even roots of (8.7), i.e.,* z_j *where* $j = 2, 4, 6, ...$*, are getting closer to* $(j-1)\pi$ *as* j *increases. So in the limit for even values of* j *we have*

$$\lim_{j \to \infty} \cos(z_j) = -1 \ .$$

We also see in Fig. 8.2 that these roots approach $(j-1)\pi$ *from the right whereas in Fig. 8.4(a) we see that they approach* $(j-1)\pi$ *from the left. Since the cosine function is monotonically decreasing between* $(j-2)\pi$ *and*

$(j-1)\pi$ $(j = 2, 4, 6, \ldots)$ and is monotonically increasing between $(j-1)\pi$ and $j\pi$ $(j = 2, 4, 6, \ldots)$, we have

$$\cos(z_2) > \cos(z_4) > \cos(z_6) > \cdots$$

for $k_p \in (-\frac{1}{k}, \frac{1}{k}) \cup (\frac{1}{k}, k_u)$. In the particular case of Fig. 8.3, i.e., $k_p = \frac{1}{k}$, we see that $\cos(z_2) = \cos(z_4) = \cos(z_6) = \cdots = -1$.

Before proving Lemmas 8.3, 8.4, and 8.5, we first state and prove the following technical lemmas that will simplify the subsequent analysis.

Lemma A.1 *Consider the function $E_1 : \mathcal{Z}^+ \times \mathcal{Z}^+ \to \mathcal{R}$ defined by*

$$E_1(m,n) \triangleq b_m - b_n$$

where m and n are natural numbers and b_j, $j = m, n$ are as defined in (8.15). Then, for z_m, $z_n \neq l\pi$, $l = 0, 1, 2, \ldots$, $E_1(m,n)$ can be equivalently expressed as

$$E_1(m,n) = \frac{L^2}{kT} \frac{[1-(kk_p)^2][\cos(z_m) - \cos(z_n)]}{z_m z_n \sin(z_m)\sin(z_n)} .$$

Proof. We will first show that for $z_j \neq l\pi$, $j = 1, 2, 3, \ldots$, the following identity holds:

$$\sin(z_j) + \frac{T}{L} z_j \cos(z_j) = \frac{1 + kk_p \cos(z_j)}{\sin(z_j)} . \tag{A.1}$$

For $z_j \neq l\pi$, from (8.7) we obtain

$$\begin{aligned}
\sin(z_j) + \frac{T}{L} z_j \cos(z_j) &= \sin(z_j) + \left[\frac{kk_p + \cos(z_j)}{\sin(z_j)}\right] \cos(z_j) \\
&= \frac{1 + kk_p \cos(z_j)}{\sin(z_j)} .
\end{aligned}$$

From (8.15) we can rewrite $E_1(m,n)$ as follows:

$$\begin{aligned}
E_1(m,n) &= -\frac{L}{kz_m}\left[\sin(z_m) + \frac{T}{L} z_m \cos(z_m)\right] \\
&\quad + \frac{L}{kz_n}\left[\sin(z_n) + \frac{T}{L} z_n \cos(z_n)\right] \\
\Rightarrow -\frac{k}{L} E_1(m,n) &= \frac{1 + kk_p \cos(z_m)}{z_m \sin(z_m)} - \frac{1 + kk_p \cos(z_n)}{z_n \sin(z_n)} \quad \text{[using (A.1)]} \\
&= \frac{z_n \sin(z_n)[1 + kk_p \cos(z_m)]}{z_m z_n \sin(z_m)\sin(z_n)} \\
&\quad - \frac{z_m \sin(z_m)[1 + kk_p \cos(z_n)]}{z_m z_n \sin(z_m)\sin(z_n)} .
\end{aligned}$$

Since z_j, $j = 1, 2, 3, \ldots$, satisfy (8.7), we can rewrite the previous expression as follows:

$$
\begin{aligned}
-\frac{kT}{L^2} E_1(m,n) &= \frac{[kk_p + \cos(z_n)][1 + kk_p \cos(z_m)]}{z_m z_n \sin(z_m)\sin(z_n)} \\
&\quad - \frac{[kk_p + \cos(z_m)][1 + kk_p \cos(z_n)]}{z_m z_n \sin(z_m)\sin(z_n)} \\
&= \frac{[(kk_p)^2 - 1][\cos(z_m) - \cos(z_n)]}{z_m z_n \sin(z_m)\sin(z_n)} \,.
\end{aligned}
$$

Thus we finally obtain

$$
E_1(m,n) = \frac{L^2}{kT} \frac{[1 - (kk_p)^2][\cos(z_m) - \cos(z_n)]}{z_m z_n \sin(z_m)\sin(z_n)} \,.
$$

■

Before stating the next lemma, we recall here for convenience the standard signum function sgn : $\mathcal{R} \to \{-1, 0, 1\}$ already introduced in Chapter 2:

$$
\operatorname{sgn}[x] = \begin{cases} -1 & \text{if } x < 0 \\ 0 & \text{if } x = 0 \\ 1 & \text{if } x > 0 \,. \end{cases}
$$

Lemma A.2 *Consider the function* $E_2 : \mathcal{Z}^+ \times \mathcal{Z}^+ \to R$ *defined by*

$$
E_2(m,n) \triangleq v_m - v_n
$$

where m *and* n *are natural numbers and* v_j, $j = m, n$ *are as defined in (8.18). If* $k_p \neq \frac{1}{k}$ *and* z_m, $z_n \neq l\pi$, $l = 1, 2, 3, \ldots$, *then*

$$
\operatorname{sgn}[E_2(m,n)] = \operatorname{sgn}[T] \cdot \operatorname{sgn}[\cos(z_m) - \cos(z_n)] \,.
$$

Proof. First, since z_j, $j = 1, 2, 3, \ldots$, satisfies (8.7), we can rewrite v_j as follows:

$$
\begin{aligned}
v_j &= \frac{z_j}{kL}\left[\sin(z_j) + \frac{kk_p + \cos(z_j)}{\sin(z_j)}(\cos(z_j) - 1)\right] \quad [\text{since } z_j \neq l\pi] \\
\Rightarrow v_j &= \frac{z_j}{kL} \frac{(1 - kk_p)[1 - \cos(z_j)]}{\sin(z_j)} \,.
\end{aligned} \tag{A.2}
$$

Using (A.2) the function $E_2(m,n)$ can be equivalently expressed as

$$
\begin{aligned}
E_2(m,n) &= \frac{z_m}{kL} \frac{(1 - kk_p)[1 - \cos(z_m)]}{\sin(z_m)} \\
&\quad - \frac{z_n}{kL} \frac{(1 - kk_p)[1 - \cos(z_n)]}{\sin(z_n)} \\
\Rightarrow \frac{kL}{1 - kk_p} E_2(m,n) &= z_m \frac{[1 - \cos(z_m)]}{\sin(z_m)} - z_n \frac{[1 - \cos(z_n)]}{\sin(z_n)} \,.
\end{aligned}
$$

Once more we use the fact that z_j, $j = 1, 2, 3, \ldots$, satisfies (8.7):

$$\begin{aligned}
\frac{kL}{1-kk_p} E_2(m,n) &= \frac{[kk_p + \cos(z_m)][1-\cos(z_m)]}{\frac{T}{L}\sin^2(z_m)} \\
&\quad - \frac{[kk_p + \cos(z_n)][1-\cos(z_n)]}{\frac{T}{L}\sin^2(z_n)} \\
\Rightarrow \frac{kT}{1-kk_p} E_2(m,n) &= \frac{kk_p + \cos(z_m)}{1+\cos(z_m)} - \frac{kk_p + \cos(z_n)}{1+\cos(z_n)} \\
&\quad [\text{since } \sin^2(x) = 1 - \cos^2(x)] \\
&= \frac{(1-kk_p)[\cos(z_m) - \cos(z_n)]}{[1+\cos(z_m)][1+\cos(z_n)]} .
\end{aligned}$$

Thus the function $E_2(m,n)$ is given by

$$E_2(m,n) = \frac{(1-kk_p)^2[\cos(z_m) - \cos(z_n)]}{kT[1+\cos(z_m)][1+\cos(z_n)]} .$$

Since $k_p \neq \frac{1}{k}$, then $(1-kk_p)^2 > 0$. Also, since z_m, $z_n \neq l\pi$, $l = 1, 2, 3, \ldots$, then $1+\cos(z_m) > 0$ and $1+\cos(z_n) > 0$. Thus from the previous expression for $E_2(m,n)$ it is clear that

$$\text{sgn}[E_2(m,n)] = \text{sgn}[T] \cdot \text{sgn}[\cos(z_m) - \cos(z_n)] .$$

This completes the proof of the lemma. ∎

Lemma A.3 *Consider the function $E_3 : \mathcal{Z}^+ \times \mathcal{Z}^+ \to R$ defined by*

$$E_3(m,n) \triangleq w_m - w_n$$

where m and n are natural numbers and w_j, $j = m, n$ are as defined in (8.19). If $k_p \neq -\frac{1}{k}$ and z_m, $z_n \neq l\pi$, $l = 1, 2, 3, \ldots$, then

$$\text{sgn}[E_3(m,n)] = \text{sgn}[T] \cdot \text{sgn}[\cos(z_m) - \cos(z_n)] .$$

Proof. As in the previous proof, we use the fact that z_j, $j = 1, 2, 3, \ldots$, satisfies (8.7). Thus, w_j can be rewritten as follows:

$$\begin{aligned}
w_j &= \frac{z_j}{kL}\left[\sin(z_j) + \frac{kk_p + \cos(z_j)}{\sin(z_j)}(\cos(z_j)+1)\right] \quad [\text{since } z_j \neq l\pi] \\
\Rightarrow w_j &= \frac{z_j}{kL}\frac{(1+kk_p)[1+\cos(z_j)]}{\sin(z_j)} . \qquad \text{(A.3)}
\end{aligned}$$

Following the same procedure used in the proof of Lemma A.2 we obtain

$$E_3(m,n) = \frac{(1+kk_p)^2[\cos(z_m) - \cos(z_n)]}{kT[1-\cos(z_m)][1-\cos(z_n)]} .$$

Since $k_p \neq -\frac{1}{k}$, then $(1+kk_p)^2 > 0$. Also, since z_m, $z_n \neq l\pi$, $l = 1, 2, 3, ...$, then $1-\cos(z_m) > 0$ and $1-\cos(z_n) > 0$. Thus, from the previous expression for $E_3(m,n)$ it is clear that

$$\operatorname{sgn}[E_3(m,n)] = \operatorname{sgn}[T] \cdot \operatorname{sgn}[\cos(z_m) - \cos(z_n)] \,.$$

This completes the proof of the lemma. ■

A.2 Proof of Lemma 8.3

(i) First we show that $b_j < -\frac{T}{k}$ for odd values of j. Recall from Fig. 8.2 that z_j is either in the first or second quadrant for odd values of j. Thus $\sin(z_j) > 0$ for $j = 1, 3, 5, ...$. For $-\frac{1}{k} < k_p < \frac{1}{k}$ and $\cos(z_j) < 1$ we have

$$\begin{aligned}
\cos(z_j)(kk_p - 1) &> kk_p - 1 \\
\Rightarrow 1 + kk_p \cos(z_j) &> \frac{T}{L} z_j \sin(z_j) \;\; \text{[using (8.7)]} \\
\Rightarrow \frac{1 + kk_p \cos(z_j)}{\sin(z_j)} &> \frac{T}{L} z_j \;\; [\text{since } \sin(z_j) > 0] \\
\Rightarrow \sin(z_j) + \frac{T}{L} z_j \cos(z_j) &> \frac{T}{L} z_j \;\; \text{[using (A.1)]} \\
\Rightarrow b_j = -\frac{L}{kz_j}\left[\sin(z_j) + \frac{T}{L} z_j \cos(z_j)\right] &< -\frac{T}{k} \,.
\end{aligned}$$

Next we show that $b_j < b_{j+2}$ for odd values of j. Since in this case, i.e., for $k_p \in (-\frac{1}{k}, \frac{1}{k})$, $z_j \neq l\pi$ for odd values of j, from Lemma A.1 we have

$$\begin{aligned}
E_1(j, j+2) &:= b_j - b_{j+2} \\
&= \frac{L^2}{kT} \frac{[1 - (kk_p)^2][\cos(z_j) - \cos(z_{j+2})]}{z_j z_{j+2} \sin(z_j) \sin(z_{j+2})} \,.
\end{aligned}$$

Since $-\frac{1}{k} < k_p < \frac{1}{k}$ then $1 - (kk_p)^2 > 0$. We also know that $z_j > 0$ and $\sin(z_j) > 0$ for odd values of j. Then from the previous expression for $E_1(j, j+2)$ and recalling that $T > 0$, we have

$$\operatorname{sgn}[E_1(j, j+2)] = \operatorname{sgn}[\cos(z_j) - \cos(z_{j+2})] \,.$$

From Remark A.1 we know that $\cos(z_j) < \cos(z_{j+2})$. Then,

$$\begin{aligned}
\operatorname{sgn}[E_1(j, j+2)] &= -1 \\
\Rightarrow E_1(j, j+2) = b_j - b_{j+2} &< 0
\end{aligned}$$

and $b_j < b_{j+2}$ for odd values of j. Thus we have shown that

$$b_j < b_{j+2} < -\frac{T}{k} \;\; \text{for odd values of } j.$$

(ii) We now show that $b_j > \frac{T}{k}$ for even values of j. From Fig. 8.2 we see that z_j is either in the third or fourth quadrant for even values of j. Thus $\sin(z_j) < 0$ in this case. Since $\cos(z_j) > -1$ and $-1 < kk_p < 1$ we have

$$\begin{aligned}
\cos(z_j)(1+kk_p) &> -(1+kk_p) \\
\Rightarrow 1 + kk_p\cos(z_j) &> -\frac{T}{L}z_j\sin(z_j) \quad \text{[using (8.7)]} \\
\Rightarrow \frac{1+kk_p\cos(z_j)}{\sin(z_j)} &< -\frac{T}{L}z_j \quad [\text{since } \sin(z_j) < 0] \\
\Rightarrow \sin(z_j) + \frac{T}{L}z_j\cos(z_j) &< -\frac{T}{L}z_j \quad \text{[using (A.1)]} \\
\Rightarrow b_j = -\frac{L}{kz_j}\left[\sin(z_j) + \frac{T}{L}z_j\cos(z_j)\right] &> \frac{T}{k}\,.
\end{aligned}$$

Note from Fig. 8.2 that as $j \to \infty$, $z_j \to (j-1)\pi$. Then $b_j \to \frac{T}{k}$.
(iii) It only remains for us to show the properties of the parameter v_j when j takes on odd values. From (A.2) we have

$$v_j = \frac{z_j}{kL}\frac{(1-kk_p)[1-\cos(z_j)]}{\sin(z_j)}\,.$$

Since $-1 < kk_p < 1$ then $1 - kk_p > 0$. Also note that $1 - \cos(z_j) > 0$. Moreover, when j takes on odd values then $\sin(z_j) > 0$. Thus we conclude that $v_j > 0$ for odd values of j. We now make use of Lemma A.2 to determine the sign of the quantity

$$E_2(j, j+2) := v_j - v_{j+2}\,.$$

Since $k_p \neq \frac{1}{k}$ and $z_j \neq l\pi$ for odd values of j, the conditions in Lemma A.2 are satisfied and we obtain

$$\text{sgn}[E_2(j, j+2)] = \text{sgn}[T] \cdot \text{sgn}[\cos(z_j) - \cos(z_{j+2})]\,.$$

We mentioned earlier that for odd values of j we have $\cos(z_j) < \cos(z_{j+2})$ and we also have $T > 0$ for an open-loop stable plant. Then $\text{sgn}[E_2(j, j+2)] = -1$, so that $E_2(j, j+2) = v_j - v_{j+2} < 0$. Thus we conclude that

$$0 < v_j < v_{j+2} \text{ for odd values of } j.$$

This completes the proof of the lemma. ■

A.3 Proof of Lemma 8.4

(i) We first consider the case of odd values of j. The proof follows from substituting (A.1) into (8.15) since $z_j \neq l\pi$, for odd values of j:

$$b_j = -\frac{L}{kz_j}\left[\frac{1+kk_p\cos(z_j)}{\sin(z_j)}\right] \quad \text{[using (A.1)]}$$

$$= -\frac{L}{kz_j}\left[\frac{1+\cos(z_j)}{\sin(z_j)}\right] \quad \text{[since } kk_p = 1\text{]}$$

$$= -\frac{T}{k} \quad \text{[using (8.7)]}.$$

(ii) For even values of j from Fig. 8.3 we see that $z_j = (j-1)\pi$. Then $\sin(z_j) = 0$ and $\cos(z_j) = -1$ in this case. Thus from (8.15) we conclude that $b_j = \frac{T}{k}$ for even values of j. This completes the proof of this lemma. ■

A.4 Proof of Lemma 8.5

First we make a general observation regarding the roots z_j, $j = 1, 2, 3, ...$ when the parameter k_p is inside the interval $\left(\frac{1}{k}, \frac{1}{k}\left[\frac{T}{L}\alpha_1 \sin(\alpha_1) - \cos(\alpha_1)\right]\right)$. From Fig. 8.4(a) we see that these roots lie either in the first or second quadrant. Then

$$\sin(z_j) > 0 \quad \text{for } j = 1, 2, 3, \ldots . \tag{A.4}$$

(i) We now consider the case of odd values of j. Since $k_p > \frac{1}{k}$ and $\cos(z_j) < 1$ we have

$$\begin{aligned}
\cos(z_j)(kk_p - 1) &< kk_p - 1 \\
\Rightarrow 1 + kk_p \cos(z_j) &< \frac{T}{L} z_j \sin(z_j) \quad \text{[using (8.7)]} \\
\Rightarrow \sin(z_j) + \frac{T}{L} z_j \cos(z_j) &< \frac{T}{L} z_j \quad \text{[using (A.1) and (A.4)]} \\
\Rightarrow b_j = -\frac{L}{kz_j}\left[\sin(z_j) + \frac{T}{L} z_j \cos(z_j)\right] &> -\frac{T}{k}.
\end{aligned}$$

We now show that $b_j > b_{j+2}$. Since $z_j \neq l\pi$ for odd values of j, from Lemma A.1 we have

$$\begin{aligned}
E_1(j, j+2) &= b_j - b_{j+2} \\
&= \frac{L^2}{kT} \frac{[1 - (kk_p)^2][\cos(z_j) - \cos(z_{j+2})]}{z_j z_{j+2} \sin(z_j) \sin(z_{j+2})}.
\end{aligned}$$

Since $k_p > \frac{1}{k}$ then $1 - (kk_p)^2 < 0$. We also know that $z_j > 0$ and $\sin(z_j) > 0$. Then from the previous expression for $E_1(j, j+2)$ we have

$$\text{sgn}[E_1(j, j+2)] = -\text{sgn}[\cos(z_j) - \cos(z_{j+2})].$$

From Remark A.1 we have that $\cos(z_j) < \cos(z_{j+2})$. Then

$$\begin{aligned}
\text{sgn}[E_1(j, j+2)] &= 1 \\
\Rightarrow E_1(j, j+2) = b_j - b_{j+2} &> 0
\end{aligned}$$

and $b_j > b_{j+2}$ for odd values of j. Thus we have shown that

$$b_j > b_{j+2} > -\frac{T}{k} \quad \text{for odd values of } j.$$

(ii) We now consider the case of even values of the parameter j. Since $\cos(z_j) > -1$ and $k_p > \frac{1}{k}$ we have

$$\begin{aligned}
[\cos(z_j) + 1]\,(1 + kk_p) &> 0 \\
\Rightarrow 1 + kk_p \cos(z_j) &> -\frac{T}{L} z_j \sin(z_j) \;\; \text{[using (8.7)]} \\
\Rightarrow \sin(z_j) + \frac{T}{L} z_j \cos(z_j) &> -\frac{T}{L} z_j \;\; \text{[using (A.1) and (A.4)]} \\
\Rightarrow b_j = -\frac{L}{kz_j}\left[\sin(z_j) + \frac{T}{L} z_j \cos(z_j)\right] &< \frac{T}{k}\,.
\end{aligned}$$

We now show that $b_j < b_{j+2}$ for this case. We know that $z_j \neq l\pi$ for even values of j. Then from Lemma A.1 we have

$$\begin{aligned}
E_1(j, j+2) &= b_j - b_{j+2} \\
&= \frac{L^2}{kT} \frac{[1 - (kk_p)^2][\cos(z_j) - \cos(z_{j+2})]}{z_j z_{j+2} \sin(z_j) \sin(z_{j+2})}\,.
\end{aligned}$$

Once more, since $k_p > \frac{1}{k}$ then $1 - (kk_p)^2 < 0$. We also know that $z_j > 0$ and $\sin(z_j) > 0$. Then from the previous expression for $E_1(j, j+2)$ we have

$$\operatorname{sgn}[E_1(j, j+2)] = -\operatorname{sgn}[\cos(z_j) - \cos(z_{j+2})]\,.$$

From Remark A.2 we have that $\cos(z_j) > \cos(z_{j+2})$ for even values of the parameter j. Using this fact we obtain

$$\begin{aligned}
\operatorname{sgn}[E_1(j, j+2)] &= -1 \\
\Rightarrow E_1(j, j+2) = b_j - b_{j+2} &< 0
\end{aligned}$$

and $b_j < b_{j+2}$ for even values of j. Thus we have shown that

$$b_j < b_{j+2} < \frac{T}{k} \quad \text{for even values of } j.$$

(iii) We now consider the properties of the parameter w_j. From (A.3) we have

$$w_j = \frac{z_j}{kL} \frac{(1 + kk_p)[1 + \cos(z_j)]}{\sin(z_j)}\,.$$

Clearly, since $k_p > \frac{1}{k}$ then $1 + kk_p > 0$. Also notice that $1 + \cos(z_j) > 0$. Thus since $\sin(z_j) > 0$ we conclude that $w_j > 0$ for even values of the parameter j. We now invoke Lemma A.3 and evaluate the function $E_3(m, n)$ at $m = j$, $n = j + 2$:

$$E_3(j, j+2) = w_j - w_{j+2}\,.$$

Since $k_p \neq -\frac{1}{k}$ and $z_j \neq l\pi$ for even values of j, then we have

$$\operatorname{sgn}[E_3(j,j+2)] = \operatorname{sgn}[T] \cdot \operatorname{sgn}[\cos(z_j) - \cos(z_{j+2})] \ .$$

We know from Remark A.2 that $\cos(z_j) > \cos(z_{j+2})$ for even values of j, and also that $T > 0$. Then, $\operatorname{sgn}[E_3(j,j+2)] = 1$, so that $E_3(j,j+2) = w_j - w_{j+2} > 0$. Thus we have shown that

$$w_j > w_{j+2} > 0 \ \text{ for even values of } j.$$

(iv) We show first that $b_1 < b_2$. Since z_1, $z_2 \neq l\pi$, from Lemma A.1 we have

$$\begin{aligned} E_1(1,2) &= b_1 - b_2 \\ &= \frac{L^2}{kT} \frac{[1-(kk_p)^2][\cos(z_1) - \cos(z_2)]}{z_1 z_2 \sin(z_1)\sin(z_2)} \ . \end{aligned}$$

We know that $\sin(z_1) > 0$ and $\sin(z_2) > 0$. Moreover, since $k_p > \frac{1}{k}$ we obtain the following:

$$\operatorname{sgn}[E_1(1,2)] = -\operatorname{sgn}[\cos(z_1) - \cos(z_2)] \ .$$

As we can see from Fig. 8.4(a), both z_1 and z_2 are in the interval $(0,\pi)$ and $z_1 < z_2$. Since the cosine function is monotonically decreasing in $(0,\pi)$ then $\cos(z_1) > \cos(z_2)$. Thus

$$\begin{aligned} \operatorname{sgn}[E_1(1,2)] &= -1 \\ \Rightarrow E_1(1,2) = b_1 - b_2 &< 0 \ . \end{aligned}$$

Hence we have $b_1 < b_2$. Finally we show that $w_1 > w_2$. To do so, we invoke Lemma A.3 and evaluate the function $E_3(m,n)$ at $m = 1$, $n = 2$:

$$E_3(1,2) = w_1 - w_2 \ .$$

Since $k_p \neq -\frac{1}{k}$ and z_1, $z_2 \notin \{0,\pi\}$, the conditions in Lemma A.3 are satisfied and we obtain

$$\operatorname{sgn}[E_3(1,2)] = \operatorname{sgn}[T] \cdot \operatorname{sgn}[\cos(z_1) - \cos(z_2)] \ .$$

We already pointed out that $\cos(z_1) > \cos(z_2)$ and since $T > 0$ we have

$$\begin{aligned} \operatorname{sgn}[E_3(1,2)] &= 1 \\ \Rightarrow E_3(1,2) = w_1 - w_2 &> 0 \ . \end{aligned}$$

Thus $w_1 > w_2$ and this completes the proof of the lemma. ∎

B

Proof of Lemmas 8.7 and 8.9

B.1 Proof of Lemma 8.7

Let us define the function $f : (0, \pi) \times \mathcal{R} \to \mathcal{R}$ by

$$f(z, k_p) \triangleq \frac{kk_p + \cos(z)}{\sin(z)} .$$

Consider $k_{p1}, k_{p2} \in \mathcal{R}$ such that $k_{p1} < k_{p2}$. Then for any $z \in (0, \pi)$ we have

$$\begin{aligned} kk_{p1} + \cos(z) &< kk_{p2} + \cos(z) \\ \Rightarrow \frac{kk_{p1} + \cos(z)}{\sin(z)} &< \frac{kk_{p2} + \cos(z)}{\sin(z)} \quad [\text{since } \sin(z) > 0] \\ \Rightarrow f(z, k_{p1}) &< f(z, k_{p2}) . \end{aligned}$$

Thus for any fixed $z \in (0, \pi)$, $f(z, k_p)$ is monotonically increasing with respect to k_p. Hence for $k_p < -\frac{1}{k}$ we have

$$f(z, k_p) < f\left(z, -\frac{1}{k}\right) \quad \forall z \in (0, \pi).$$

This means that if the line $\frac{T}{L}z$ does not intersect the curve $f(z, -\frac{1}{k})$ in $z \in (0, \pi)$, then it will not intersect any other curve $f(z, k_p)$ in $z \in (0, \pi)$. Observe that $\forall z \in (0, \pi)$

$$\begin{aligned} f\left(z, -\frac{1}{k}\right) &= \frac{-1 + \cos(z)}{\sin(z)} \\ &= -\tan\left(\frac{z}{2}\right) . \end{aligned}$$

Accordingly, define a continuous extension of $f(z, -\frac{1}{k})$ to $[0, \pi)$ by

$$f_1\left(z, -\frac{1}{k}\right) = -\tan\left(\frac{z}{2}\right) .$$

Clearly, the curve $f_1(z, -\frac{1}{k})$ intersects the line $\frac{T}{L}z$ at $z = 0$. This is depicted in Fig. B.1. Also note that the slope of the tangent to $f_1(z, -\frac{1}{k})$ at $z = 0$ is given by

$$\begin{aligned} \left.\frac{df_1}{dz}\right|_{z=0} &= \left.-\frac{1}{2}\sec^2\left(\frac{z}{2}\right)\right|_{z=0} \\ &= -\frac{1}{2} . \end{aligned}$$

If this slope is less than or equal to $\frac{T}{L}$ then we are guaranteed that no further intersections will take place in the interval $(0, \pi)$. Since $f_1(z, -\frac{1}{k}) = f(z, -\frac{1}{k})$ on $(0, \pi)$, it follows that if $-0.5 \leq \frac{T}{L}$, then the curve $f(z, -\frac{1}{k})$ will not intersect the line $\frac{T}{L}z$ in the interval $(0, \pi)$. This completes the proof. ∎

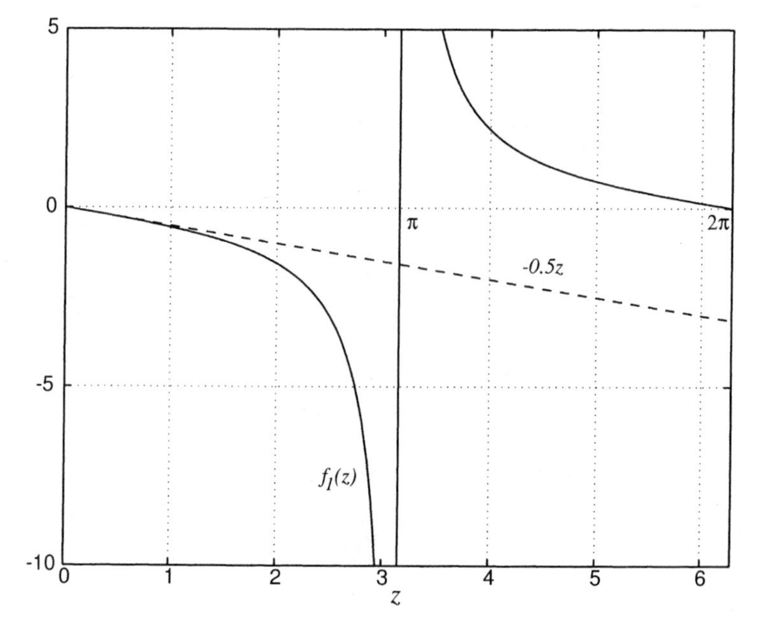

FIGURE B.1. Plot of the curve $f_1(z, -\frac{1}{k})$ and the line $\frac{T}{L}z$.

B.2 Proof of Lemma 8.9

We begin by making the following observations, which follow from the proof of Lemma 8.6.

Remark B.1 *From Fig. 8.11(a) we see that the odd roots of (8.27), i.e., z_j where $j = 1, 3, 5, ...$, are getting closer to $(j-1)\pi$ as j increases. Since the cosine function is monotonically decreasing between $(j-1)\pi$ and $j\pi$ for odd values of j, in view of the previous observation we have*

$$\cos(z_1) < \cos(z_3) < \cos(z_5) < \cdots .$$

Remark B.2 *From Fig. 8.11(a) we see that the even roots of (8.27), i.e., z_j where $j = 2, 4, 6, ...$, are getting closer to $(j-1)\pi$ as j increases. Since the cosine function is monotonically decreasing between $(j-2)\pi$ and $(j-1)\pi$ for even values of j, and because of the previous observation we have*

$$\cos(z_2) > \cos(z_4) > \cos(z_6) > \cdots .$$

From Fig. 8.11(a) we see that the roots z_j, $j = 1, 2, 3, ...$ lie either in the first or in the second quadrant when the parameter k_p is inside the interval $(k_l, -\frac{1}{k})$. Thus

$$\sin(z_j) > 0 \quad \text{for } j = 1, 2, 3, \ldots . \tag{B.1}$$

(i) First we analyze the case of odd values of j. Since $k_p < -\frac{1}{k}$ and $\cos(z_j) < 1$, we have

$$\begin{aligned}
\cos(z_j)(kk_p - 1) &> kk_p - 1 \\
\Rightarrow 1 + kk_p \cos(z_j) &> \frac{T}{L} z_j \sin(z_j) \quad \text{[using (8.27)]} \\
\Rightarrow \sin(z_j) + \frac{T}{L} z_j \cos(z_j) &> \frac{T}{L} z_j \quad \text{[using (A.1) and (B.1)]} \\
\Rightarrow b_j = -\frac{L}{kz_j}\left[\sin(z_j) + \frac{T}{L} z_j \cos(z_j)\right] &< -\frac{T}{k} .
\end{aligned}$$

We now show that $b_j < b_{j+2}$. Since $z_j \neq l\pi$ for odd values of j, from Lemma A.1 we have

$$\begin{aligned}
E_1(j, j+2) &= b_j - b_{j+2} \\
&= \frac{L^2}{kT} \frac{[1 - (kk_p)^2][\cos(z_j) - \cos(z_{j+2})]}{z_j z_{j+2} \sin(z_j) \sin(z_{j+2})} .
\end{aligned}$$

Since $k_p < -\frac{1}{k}$ then $1 - (kk_p)^2 < 0$. We also know that $z_j > 0$, $\sin(z_j) > 0$, and $T < 0$. Then from the previous expression for $E_1(j, j+2)$ we have

$$\text{sgn}[E_1(j, j+2)] = \text{sgn}[\cos(z_j) - \cos(z_{j+2})] .$$

From Remark B.1 we have $\cos(z_j) < \cos(z_{j+2})$. Thus

$$\begin{aligned}
\text{sgn}[E_1(j, j+2)] &= -1 \\
\Rightarrow E_1(j, j+2) = b_j - b_{j+2} &< 0
\end{aligned}$$

and $b_j < b_{j+2}$ for odd values of j. Thus we have shown that

$$b_j < b_{j+2} < -\frac{T}{k} \quad \text{for odd values of } j.$$

(ii) We now consider the case of even values of the parameter j. Since $k_p < -\frac{1}{k}$ and $\cos(z_j) > -1$, we have

$$\begin{aligned}
\cos(z_j)(kk_p+1) &< -(kk_p+1) \\
\Rightarrow 1 + kk_p\cos(z_j) &< -\frac{T}{L}z_j\sin(z_j) \quad \text{[using (8.27)]} \\
\Rightarrow \sin(z_j) + \frac{T}{L}z_j\cos(z_j) &< -\frac{T}{L}z_j \quad \text{[using (A.1) and (B.1)]} \\
\Rightarrow b_j = -\frac{L}{kz_j}\left[\sin(z_j) + \frac{T}{L}z_j\cos(z_j)\right] &> \frac{T}{k}.
\end{aligned}$$

We now show that $b_j > b_{j+2}$ for this case. Again from Lemma A.1, since $z_j \neq l\pi$ for even values of j, we have

$$\begin{aligned}
E_1(j, j+2) &= b_j - b_{j+2} \\
&= \frac{L^2}{kT}\frac{[1-(kk_p)^2][\cos(z_j) - \cos(z_{j+2})]}{z_j z_{j+2}\sin(z_j)\sin(z_{j+2})}.
\end{aligned}$$

Since $k_p < -\frac{1}{k}$ then $1-(kk_p)^2 < 0$. We also know that $z_j > 0$, $\sin(z_j) > 0$, and $T < 0$. Then from the previous expression for $E_1(j, j+2)$ we have

$$\operatorname{sgn}[E_1(j, j+2)] = \operatorname{sgn}[\cos(z_j) - \cos(z_{j+2})].$$

From Remark B.2 we have $\cos(z_j) > \cos(z_{j+2})$ for even values of the parameter j. Using this fact we obtain

$$\begin{aligned}
\operatorname{sgn}[E_1(j, j+2)] &= 1 \\
\Rightarrow E_1(j, j+2) = b_j - b_{j+2} &> 0
\end{aligned}$$

and $b_j > b_{j+2}$ for even values of j. Thus we have shown that

$$b_j > b_{j+2} > \frac{T}{k} \quad \text{for even values of } j.$$

(iii) We will now study the properties of the parameter w_j. From (A.3) we have

$$w_j = \frac{z_j}{kL}\frac{(1+kk_p)[1+\cos(z_j)]}{\sin(z_j)}.$$

Since $k_p < -\frac{1}{k}$ then $1 + kk_p < 0$. Moreover, we know that $z_j > 0$, $\cos(z_j) > -1$, and $\sin(z_j) > 0$. Thus we conclude that $w_j < 0$ for even values of the parameter j.

We now evaluate the function $E_3(m,n)$ defined in Lemma A.3 at $m = j$, $n = j+2$:

$$E_3(j, j+2) = w_j - w_{j+2} \, .$$

Since $k_p \neq -\frac{1}{k}$ and $z_j \neq l\pi$ for even values of j, then using Lemma A.3, we have

$$\mathrm{sgn}[E_3(j, j+2)] = \mathrm{sgn}[T] \cdot \mathrm{sgn}[\cos(z_j) - \cos(z_{j+2})] \, .$$

We know from Remark B.2 that $\cos(z_j) > \cos(z_{j+2})$ for even values of j. Since $T < 0$, it follows that $\mathrm{sgn}[E_3(j, j+2)] = -1$, i.e., $E_3(j, j+2) = w_j - w_{j+2} < 0$. Thus, we have shown that

$$w_j < w_{j+2} < 0 \ \text{ for even values of } j.$$

(iv) First we show that $b_1 > b_2$. Since z_1, $z_2 \neq l\pi$, we have from Lemma A.1

$$\begin{aligned} E_1(1,2) &= b_1 - b_2 \\ &= \frac{L^2}{kT} \frac{[1 - (kk_p)^2][\cos(z_1) - \cos(z_2)]}{z_1 z_2 \sin(z_1)\sin(z_2)} \, . \end{aligned}$$

We know that $\sin(z_1) > 0$ and $\sin(z_2) > 0$. Moreover, since $k_p < -\frac{1}{k}$ and $T < 0$, we obtain the following:

$$\mathrm{sgn}[E_1(1,2)] = \mathrm{sgn}[\cos(z_1) - \cos(z_2)] \, .$$

From Fig. 8.11(a), it is clear that both z_1 and z_2 are in the interval $(0, \pi)$ and $z_1 < z_2$. Since the cosine function is monotonically decreasing in $(0, \pi)$ then $\cos(z_1) > \cos(z_2)$, and we get

$$\begin{aligned} \mathrm{sgn}[E_1(1,2)] &= 1 \\ \Rightarrow E_1(1,2) = b_1 - b_2 &> 0 \, . \end{aligned}$$

Hence, we have $b_1 > b_2$.

We finally show that $w_1 < w_2$ by evaluating $E_3(m,n)$ at $m = 1$, $n = 2$:

$$E_3(1,2) = w_1 - w_2 \, .$$

Since $k_p \neq -\frac{1}{k}$ and z_1, $z_2 \neq l\pi, l = 0, 1$, from Lemma A.3 we obtain

$$\mathrm{sgn}[E_3(1,2)] = \mathrm{sgn}[T] \cdot \mathrm{sgn}[\cos(z_1) - \cos(z_2)] \, .$$

We already pointed out that $\cos(z_1) > \cos(z_2)$. Since $T < 0$ we have

$$\begin{aligned} \mathrm{sgn}[E_3(1,2)] &= -1 \\ \Rightarrow E_3(1,2) = w_1 - w_2 &< 0 \, . \end{aligned}$$

Thus $w_1 < w_2$ and this completes the proof. ■

C

Detailed Analysis of Example 11.4

For the first-order plant

$$R_0(s) = \frac{N(s)}{sD(s)} = \frac{k}{Ts^2 + s}$$

and for a fixed k_p, we have

$$M(\omega) = \frac{1}{|R_0(j\omega)|^2} - (k_p\omega)^2 = \frac{T^2\omega^4 + (1 - k^2k_p^2)\omega^2}{k^2}.$$

For $M(\omega) \geq 0$, we must have $T^2\omega^2 + (1 - k^2k_p^2) \geq 0$.

- When $1 - k^2k_p^2 \geq 0$, i.e., $|k_p| \leq 1/k$, all ω satisfy the requirement, which means we need to consider all $\omega > 0$.
- When $1 - k^2k_p^2 < 0$, i.e., $|k_p| > 1/k$. In this case, we only need to consider $\omega \geq \omega_s$, where $\omega_s = \sqrt{k^2k_p^2 - 1}/|T|$ and $M(\omega_s) = 0$.

Let us consider $T > 0$. Now we have two cases to consider.
Case 1: $k_i - k_d\omega^2 = \sqrt{M(\omega)}$. In this case,

$$\begin{aligned} L(\omega) &= \frac{\pi + \arg\left[\left(\frac{\omega}{k}\sqrt{T^2\omega^2 + 1 - k^2k_p^2} + jk_p\omega\right) \cdot \frac{k}{-T\omega^2 + j\omega}\right]}{\omega} \\ &=: \frac{\alpha^+(\omega)}{\omega} \end{aligned}$$

where $\alpha^+(\omega) \in [0, 2\pi)$.

First, let us check $L(\omega)$. Define

$$\begin{aligned}
\alpha_1^+(\omega) &:= \arg\left(\frac{\omega}{k}\sqrt{T^2\omega^2 + 1 - k^2k_p^2} + jk_p\omega\right) \\
&= \tan^{-1}\frac{kk_p}{\sqrt{T^2\omega^2 + 1 - k^2k_p^2}} \\
\alpha_2^+(\omega) &:= \pi + \arg\left[\frac{k}{-T\omega^2 + j\omega}\right] = \tan^{-1}\frac{1}{T\omega},
\end{aligned}$$

where $\alpha_1^+(\omega) \in (-\pi/2, \pi/2)$ and $\alpha_2^+(\omega) \in (0, \pi/2)$.

- For $k_p \geq 0$, $\alpha_1^+(\omega) \in [0, \pi/2)$, thus $\alpha_1^+(\omega) + \alpha_2^+(\omega) \in (0, \pi) \subset [0, 2\pi)$.
- For $-\frac{1}{k} < k_p < 0$, $\alpha_1^+(\omega) \in (-\pi/2, 0)$ and $|\alpha_1^+(\omega)| < |\alpha_2^+(\omega)|$, thus $\alpha_1^+(\omega) + \alpha_2^+(\omega) \in (0, \pi/2) \subset [0, 2\pi)$.

Thus $L(\omega)$ can be decomposed as

$$L(\omega) = \frac{\alpha^+(\omega)}{\omega} = \frac{\alpha_1^+(\omega) + \alpha_2^+(\omega)}{\omega}. \tag{C.1}$$

Furthermore,

- For $k_p \geq 0$, $\alpha_1^+(\omega)$ and $\alpha_2^+(\omega)$ are decreasing functions of ω. So $L(\omega)$ is also a decreasing function of ω.
- For $-\frac{1}{k} < k_p < 0$, let us consider

$$\begin{aligned}
\tan[\alpha_1^+(\omega) + \alpha_2^+(\omega)] &= \frac{\tan\alpha_1^+(\omega) + \tan\alpha_2^+(\omega)}{1 - \tan\alpha_1^+(\omega)\tan\alpha_2^+(\omega)} \\
&= \frac{kk_pT\omega + \sqrt{T^2\omega^2 + 1 - k^2k_p^2}}{T\omega\sqrt{T^2\omega^2 + 1 - k^2k_p^2} - kk_p}.
\end{aligned} \tag{C.2}$$

Taking its derivative, we obtain

$$\begin{aligned}
D(\omega) &= \frac{\mathrm{d}\{\tan[\alpha_1^+(\omega) + \alpha_2^+(\omega)]\}}{\mathrm{d}\omega} \\
&= \frac{T(1 + T^2\omega^2)\left(-kk_pT\omega - \sqrt{T^2\omega^2 + 1 - k^2k_p^2}\right)}{\left(T\omega\sqrt{T^2\omega^2 + 1 - k^2k_p^2} - kk_p\right)^2\sqrt{T^2\omega^2 + 1 - k^2k_p^2}} \\
&< \frac{T(1 + T^2\omega^2)\left(T\omega - \sqrt{T^2\omega^2 + 1 - k^2k_p^2}\right)}{\left(T\omega\sqrt{T^2\omega^2 + 1 - k^2k_p^2} - kk_p\right)^2\sqrt{T^2\omega^2 + 1 - k^2k_p^2}} \\
&< 0.
\end{aligned}$$

Since $\alpha_1^+(\omega) + \alpha_2^+(\omega) \in (0, \pi/2)$, we have that $\alpha_1^+(\omega) + \alpha_2^+(\omega)$ is a monotonically decreasing function of ω. So $L(\omega)$ is also a monotonically decreasing function of ω.

From the above analysis, we know that for *any given* k_p in $\mathcal{S}_1$, $L(\omega)$ is a monotonically decreasing function of ω. This implies that there is only at most *one* ω that satisfies $L(\omega) = L_0$. We denote this ω, when it exists, by ω_1^+ (see Figs. C.1, C.2, and C.3). The quantity ω_1^+ along with the quantity ω_s, defined earlier, enables us to characterize Ω^+:

- For $-\frac{1}{k} < k_p \leq \frac{1}{k}$, $\Omega^+ = [\omega_1^+, +\infty)$
- For $k_p > \frac{1}{k}$ and $L_0 \leq L(\omega_s)$, $\Omega^+ = [\omega_1^+, +\infty)$
- For $k_p > \frac{1}{k}$ and $L_0 > L(\omega_s)$, $\Omega^+ = [\omega_s, +\infty)$.

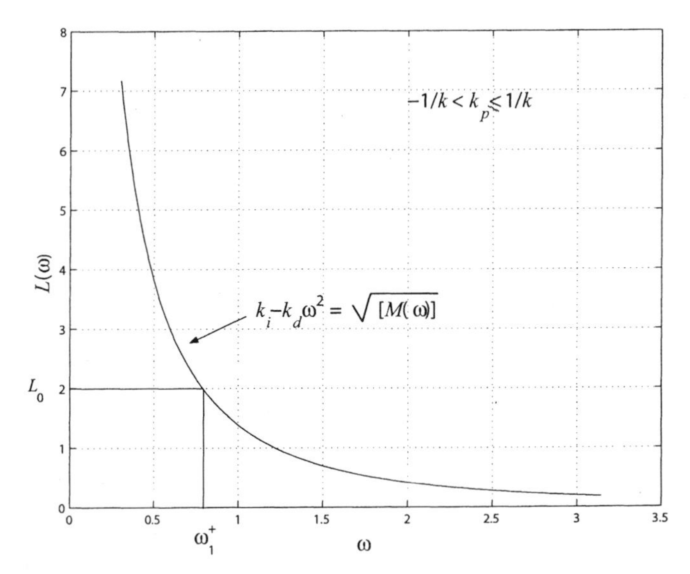

FIGURE C.1. First-order plant: $L(\omega)$ versus ω with $k_i - k_d\omega^2 = \sqrt{M(\omega)}$.

Let us check the straight lines defined by $k_i - k_d\omega^2 = \sqrt{M(\omega)}$ in the (k_i, k_d) plane. The straight line

$$k_i = \omega^2 k_d + \frac{\omega\sqrt{T^2\omega^2 + 1 - k^2k_p^2}}{k}$$

intersects the lines $k_d = \frac{T}{k}$ and $k_d = -\frac{T}{k}$ at $(k_{i,\omega}^+, \frac{T}{k})$ and $(k_{i,\omega}^-, -\frac{T}{k})$, respectively, where

$$k_{i,\omega}^+ = \frac{\omega}{k}\left(\sqrt{T^2\omega^2 + 1 - k^2k_p^2} + T\omega\right) \tag{C.3}$$

$$k_{i,\omega}^- = \frac{\omega}{k}\left(\sqrt{T^2\omega^2 + 1 - k^2k_p^2} - T\omega\right). \tag{C.4}$$

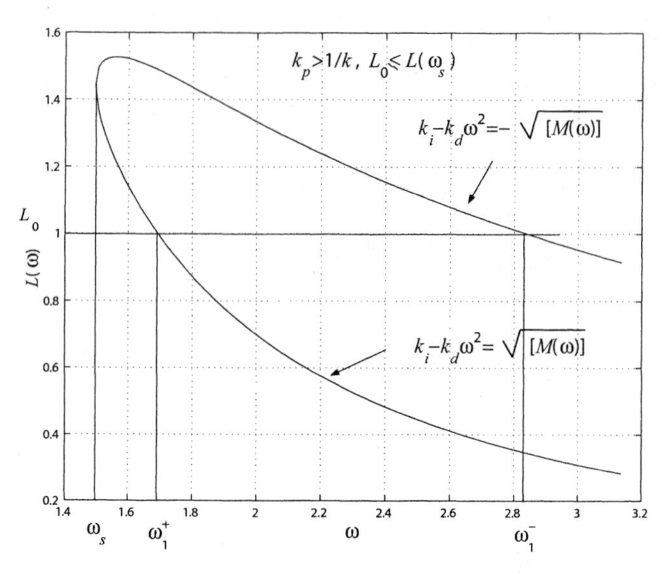

FIGURE C.2. First-order plant: $L(\omega)$ versus ω for $k_p > \frac{1}{k}$ and $L_0 \leq L(\omega_s)$.

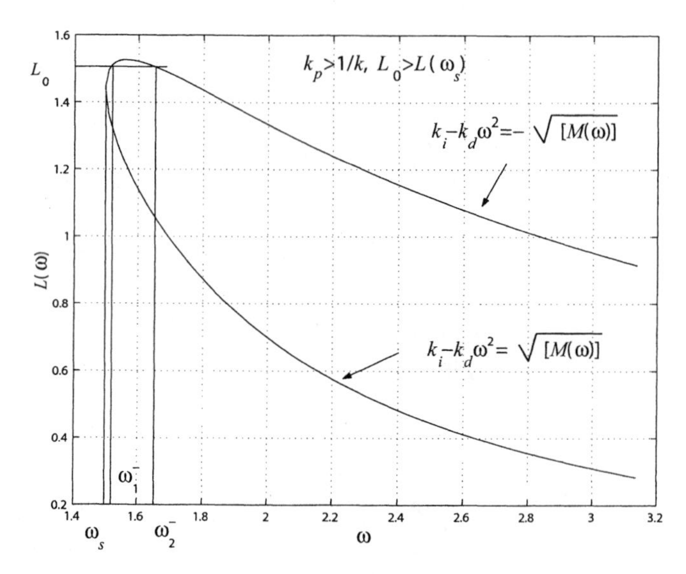

FIGURE C.3. First-order plant: $L(\omega)$ versus ω for $k_p > \frac{1}{k}$ and $L_0 > L(\omega_s)$.

The derivative of $k_{i,\omega}^-$ is

$$\begin{aligned}\frac{\mathrm{d}k_{i,\omega}^-}{\mathrm{d}\omega} &= \frac{\left(\sqrt{T^2\omega^2+1-k^2k_p^2}-T\omega\right)^2}{k\sqrt{T^2\omega^2+1-k^2k_p^2}} \\ &\geq 0\,.\end{aligned}$$

From (C.3) and (C.4), it follows that $\frac{dk_{i,\omega}^+}{d\omega}$ is also non-negative. Thus $k_{i,\omega}^-$ and $k_{i,\omega}^+$ are both monotonically increasing functions of ω. From this, it follows that the set

$$\mathcal{S}_{L,k_p}^+ \triangleq \left\{(k_i,k_d)|k_i-k_d\omega^2=\sqrt{M(\omega)},\omega\in\Omega^+\right\}\cap\mathcal{S}_{1,k_p}$$

can be described as follows, corresponding to the different values of k_p and L_0:

- For $-\frac{1}{k}<k_p\leq\frac{1}{k}$,

$$\mathcal{S}_{L,k_p}^+ = \left\{(k_i,k_d)|k_i\geq k_d(\omega_1^+)^2+\sqrt{M(\omega_1^+)}\right\}\cap\mathcal{S}_{1,k_p}\,. \tag{C.5}$$

- For $k_p>\frac{1}{k}$ and $L_0\leq L(\omega_s)$,

$$\mathcal{S}_{L,k_p}^+ = \left\{(k_i,k_d)|k_i\geq k_d(\omega_1^+)^2+\sqrt{M(\omega_1^+)}\right\}\cap\mathcal{S}_{1,k_p}\,. \tag{C.6}$$

- For $k_p>\frac{1}{k}$ and $L_0>L(\omega_s)$,

$$\begin{aligned}\mathcal{S}_{L,k_p}^+ &= \left\{(k_i,k_d)|k_i\geq k_d(\omega_s)^2+\sqrt{M(\omega_s)}\right\}\cap\mathcal{S}_{1,k_p} \\ &= \{(k_i,k_d)|k_i\geq k_d(\omega_s)^2\}\cap\mathcal{S}_{1,k_p} \\ &\quad \text{(since } M(\omega_s)=0\text{, by definition)}\;.\end{aligned} \tag{C.7}$$

Case 2: $k_i-k_d\omega^2=-\sqrt{M(\omega)}$. Here we first check the positions of these lines. They intersect $k_d=\frac{T}{k}$ and $k_d=-\frac{T}{k}$ at $(k_{i,\omega}^+,\frac{T}{k})$ and $(k_{i,\omega}^-,-\frac{T}{k})$, respectively, where

$$\begin{aligned}k_{i,\omega}^+ &= \frac{\omega}{k}\left(-\sqrt{T^2\omega^2+1-k^2k_p^2}+T\omega\right) \\ k_{i,\omega}^- &= \frac{\omega}{k}\left(-\sqrt{T^2\omega^2+1-k^2k_p^2}-T\omega\right)\,.\end{aligned}$$

Here, for $-\frac{1}{k}<k_p\leq\frac{1}{k}$, $k_{i,\omega}^+\leq 0$, which means that the lines $k_i-k_d\omega^2=-\sqrt{M(\omega)}$ lie outside $\mathcal{S}_{1,k_p}$. So, $\mathcal{S}_{L,k_p}^-=\emptyset$ for these k_p values.

On the other hand, for $k_p > \frac{1}{k}$, $k^+_{i,\omega} > 0$, i.e., the lines $k_i - k_d\omega^2 = -\sqrt{M(\omega)}$ have a nonempty intersection with $\mathcal{S}_{1,k_p}$ and, therefore, affect the set of all stabilizing PID controllers for the system with time delay.

We next proceed to determine this intersection. The derivative of $k^+_{i,\omega}$ is

$$\begin{aligned}\frac{\mathrm{d}k^+_{i,\omega}}{\mathrm{d}\omega} &= -\frac{\left(\sqrt{T^2\omega^2+1-k^2k_p^2}-T\omega\right)^2}{k\sqrt{T^2\omega^2+1-k^2k_p^2}} \\ &\leq 0.\end{aligned}$$

So $k^+_{i,\omega}$ and $k^-_{i,\omega}$ are monotonically decreasing functions of ω and $k^+_{i,\omega}$ tends to zero as $\omega \to \infty$. This result will be used to determine $\mathcal{S}^-_{L,k_p}$. In order to do that, we also need to examine $L(\omega)$ when $k_p > \frac{1}{k}$. In this case,

$$L(\omega) = \frac{\pi + \arg\left[\left(-\frac{\omega}{k}\sqrt{T^2\omega^2+1-k^2k_p^2}+jk_p\omega\right)\cdot\frac{k}{-T\omega^2+j\omega}\right]}{\omega} =: \frac{\alpha^-(\omega)}{\omega}$$

where $\alpha^-(\omega) \in [0, 2\pi)$. Define

$$\begin{aligned}\alpha_1^-(\omega) &= \arg\left(-\frac{\omega}{k}\sqrt{T^2\omega^2+1-k^2k_p^2}+jk_p\omega\right) \\ &= \pi - \tan^{-1}\frac{kk_p}{\sqrt{T^2\omega^2+1-k^2k_p^2}} \qquad (C.8)\\ &= \pi - \alpha_1^+(\omega) \\ \alpha_2^-(\omega) &= \pi + \arg\left[\frac{k}{-T\omega^2+j\omega}\right] \\ &= \tan^{-1}\frac{1}{T\omega} \qquad (C.9)\\ &= \alpha_2^+(\omega)\end{aligned}$$

where $\alpha_1^-(\omega) \in (\pi/2, \pi)$ and $\alpha_2^-(\omega) \in (0, \pi/2)$ for $k_p > \frac{1}{k}$. Thus $\alpha_1^-(\omega) + \alpha_2^-(\omega) \in (\pi/2, 3\pi/2) \subset [0, 2\pi)$, so that $L(\omega)$ can be decomposed as

$$L(\omega) = \frac{\alpha^-(\omega)}{\omega} = \frac{\alpha_1^-(\omega)+\alpha_2^-(\omega)}{\omega}. \qquad (C.10)$$

We first evaluate $\tan[\alpha_1^-(\omega)+\alpha_2^-(\omega)]$:

$$\tan[\alpha_1^-(\omega)+\alpha_2^-(\omega)] = \frac{\sqrt{T^2\omega^2+1-k^2k_p^2}-kk_pT\omega}{T\omega\sqrt{T^2\omega^2+1-k^2k_p^2}+kk_p}$$

and its derivative:

$$D(\omega) = \frac{\mathrm{d}\tan[\alpha_1^-(\omega)+\alpha_2^-(\omega)]}{\mathrm{d}\omega}$$

$$
\begin{aligned}
&= \frac{T(1+T^2\omega^2)\left(kk_pT\omega - \sqrt{T^2\omega^2+1-k^2k_p^2}\right)}{\left(T\omega\sqrt{T^2\omega^2+1-k^2k_p^2}+kk_p\right)^2\sqrt{T^2\omega^2+1-k^2k_p^2}} \\
&> \frac{T(1+T^2\omega^2)\left(T\omega - \sqrt{T^2\omega^2+1-k^2k_p^2}\right)}{\left(T\omega\sqrt{T^2\omega^2+1-k^2k_p^2}+kk_p\right)^2\sqrt{T^2\omega^2+1-k^2k_p^2}} \\
&\quad (\text{since } kk_p > 1) \\
&> 0\,.
\end{aligned}
$$

Since $\alpha_1^-(\omega)+\alpha_2^-(\omega) \in (\pi/2, 3\pi/2)$, $\alpha_1^-(\omega)+\alpha_2^-(\omega)$ is a monotonically increasing function of ω. Next we evaluate the derivative of $L(\omega)$.

$$
\begin{aligned}
\frac{\mathrm{d}L(\omega)}{\mathrm{d}\omega} &= \frac{\mathrm{d}}{\mathrm{d}\omega}\left[\frac{\alpha_1^-(\omega)+\alpha_2^-(\omega)}{\omega}\right] \\
&= \frac{1}{\omega^2}\left[\frac{kk_pT^2\omega^2}{(1+T^2\omega^2)\sqrt{T^2\omega^2+1-k^2k_p^2}} - \frac{T\omega}{1+T^2\omega^2}\right. \\
&\quad \left. - \left(\pi - \tan^{-1}\frac{kk_p}{\sqrt{T^2\omega^2+1-k^2k_p^2}} + \tan^{-1}\frac{1}{T\omega}\right)\right] \\
&= \frac{1}{\omega^2}\left\{\frac{T\omega}{1+T^2\omega^2}\left(\frac{kk_pT\omega}{\sqrt{T^2\omega^2+1-k^2k_p^2}} - 1\right) - \right. \\
&\quad \left. [\alpha_1^-(\omega)+\alpha_2^-(\omega)]\right\} \\
&= \frac{1}{\omega^2}\{\beta(\omega) - [\alpha_1^-(\omega)+\alpha_2^-(\omega)]\}
\end{aligned}
$$

where

$$
\beta(\omega) := \frac{T\omega}{1+T^2\omega^2}\left(\frac{kk_pT\omega}{\sqrt{T^2\omega^2+1-k^2k_p^2}} - 1\right).
$$

For $\omega \le 1/T$,

$$
\begin{aligned}
\frac{\mathrm{d}\beta(\omega)}{\mathrm{d}\omega} &= \frac{T}{(1+T^2\omega^2)^2}\cdot \\
&\quad \left[kk_p(1+T^2\omega^2)\frac{T\omega}{\sqrt{T^2\omega^2+1-k^2k_p^2}}\left(1 - \frac{T^2\omega^2}{T^2\omega^2+1-k^2k_p^2}\right) + \right. \\
&\quad \left. (1-T^2\omega^2)\left(\frac{kk_pT\omega}{\sqrt{T^2\omega^2+1-k^2k_p^2}} - 1\right)\right]
\end{aligned}
$$

$$< \frac{T}{(1+T^2\omega^2)^2}\left[\left(1-\frac{T^2\omega^2}{T^2\omega^2+1-k^2k_p^2}\right)+\right.$$

$$\left.\left(\frac{kk_pT\omega}{\sqrt{T^2\omega^2+1-k^2k_p^2}}-1\right)\right] \quad \text{(using } \omega T \leq 1 \text{ and } kk_p > 1)$$

$$= \frac{T^2\omega}{(1+T^2\omega^2)^2(T^2\omega^2+1-k^2k_p^2)}$$

$$\left(kk_p\sqrt{T^2\omega^2+1-k^2k_p^2}-T\omega\right).$$

Since

$$\begin{aligned}\left(kk_p\sqrt{T^2\omega^2+1-k^2k_p^2}\right)^2-(T\omega)^2 &= k^2k_p^2(1-k^2k_p^2)+k^2k_p^2T^2\omega^2- \\ &\quad T^2\omega^2 \\ &= (k^2k_p^2-T^2\omega^2)(1-k^2k_p^2) \\ &< 0,\end{aligned}$$

we have

$$\frac{\mathrm{d}\beta(\omega)}{\mathrm{d}\omega} < 0.$$

For $\omega > 1/T$, $\frac{T\omega}{1+T^2\omega^2}$ and $\frac{kk_pT\omega}{\sqrt{T^2\omega^2+1-k^2k_p^2}}-1$ are both positive while their derivatives are both negative, so that when $\omega > 1/T$, we have $\mathrm{d}\beta(\omega)/\mathrm{d}\omega < 0$. Thus for all values of ω, $\beta(\omega)$ is a monotonically decreasing function of ω. At $\omega = \omega_s$,

$$\beta(\omega_s) - [\alpha_1^-(\omega_s)+\alpha_2^-(\omega_s)] = \infty - \left(\frac{\pi}{2}+\tan^{-1}\frac{1}{T\omega_s}\right) = \infty > 0,$$

and at $\omega = \infty$,

$$\beta(\infty) - [\alpha_1^-(\infty)+\alpha_2^-(\infty)] = 0-(\pi+0) = -\pi < 0.$$

As already shown, $\alpha_1^-(\omega)+\alpha_2^-(\omega)$ is a monotonically increasing function of ω. So there is only one finite solution for the equation

$$\beta(\omega) - [\alpha_1^-(\omega)+\alpha_2^-(\omega)] = 0$$

in the interval (ω_s, ∞). The above analysis suggests that $\mathrm{d}L(\omega)/\mathrm{d}\omega$ has only one finite zero, which indicates only *one* maximum point for $L(\omega)$ (see Figs. C.2 and Fig. C.3). Depending on the value of L_0, the sets S_{L,k_p}^- can be characterized as follows:

- For $L_0 \le L(\omega_s)$, there is only one solution for $L(\omega) = L_0$, denoted by ω_1^- and $\Omega^- = [\omega_1^-, +\infty)$. With the knowledge about the positions of $k_i - k_d\omega^2 = -\sqrt{M(\omega)}$ that we acquired earlier (recall the monotonicity property of $k_{i,\omega}^+$ and $k_{i,\omega}^-$), we have

$$\begin{aligned} \mathcal{S}_{L,k_p}^- &= \left\{(k_i, k_d) | k_i - k_d\omega^2 = -\sqrt{M(\omega)}, \omega \in \Omega^-\right\} \cap \mathcal{S}_{1,k_p} \\ &= \left\{(k_i, k_d) | k_i \le k_d(\omega_1^-)^2 - \sqrt{M(\omega_1^-)}\right\} \cap \mathcal{S}_{1,k_p} . \quad \text{(C.11)} \end{aligned}$$

- For $L(\omega_s) < L_0 < \max_{\omega \in (\omega_s, \infty)} L(\omega)$, there are two solutions for $L(\omega) = L_0$, denoted as ω_1^- and ω_2^- with $\omega_1^- < \omega_2^-$. So $\Omega^- = [\omega_s, \omega_1^-] \cup [\omega_2^-, +\infty)$, and

$$\begin{aligned} \mathcal{S}_{L,k_p}^- = &\left\{(k_i, k_d) | k_d(\omega_1^-)^2 - \sqrt{M(\omega_1^-)} \le k_i \le k_d(\omega_s)^2 \text{ or} \right. \\ &\left. k_i \le k_d(\omega_2^-)^2 - \sqrt{M(\omega_2^-)}\right\} \cap \mathcal{S}_{1,k_p}. \quad \text{(C.12)} \end{aligned}$$

- For $L_0 > \max_{\omega \in (\omega_s, \infty)} L(\omega)$, there is no solution for $L(\omega) = L_0$ and we have $\Omega^- = [\omega_s, +\infty)$ and

$$\mathcal{S}_{L,k_p}^- = \{(k_i, k_d) | k_i \le k_d(\omega_s)^2\} \cap \mathcal{S}_{1,k_p} . \quad \text{(C.13)}$$

We can now compute $\mathcal{S}_{R,k_p} = \mathcal{S}_{1,k_p} \backslash (\mathcal{S}_{L,k_p}^+ \cup \mathcal{S}_{L,k_p}^-)$.

- For $-\frac{1}{k} < k_p \le \frac{1}{k}$, $\mathcal{S}_{R,k_p}$ is defined by

$$\begin{aligned} k_i &> 0 \\ -\frac{T}{k} &< k_d < \frac{T}{k} \\ k_i &< (\omega_1^+)^2 k_d + \sqrt{M(\omega_1^+)} \text{ (using (C.6))} \end{aligned}$$

  where ω_1^+ satisfies

$$L_0 = [\alpha_1^+(\omega_1^+) + \alpha_2^+(\omega_1^+)]/\omega_1^+ \text{ (see (C.1))} .$$

  This region $\mathcal{S}_{R,k_p}$ is a trapezoid (see Fig. 11.12(a)).

- For $k_p > \frac{1}{k}$ and $L_0 \le L(\omega_s) = (\frac{\pi}{2} + \tan^{-1} \frac{1}{T\omega_s})/\omega_s$ (see (C.8), (C.9), and (C.10)) $\mathcal{S}_{R,k_p}$ is given by

$$\begin{aligned} k_i &> 0 \\ k_d &< \frac{T}{k} \\ k_i &< (\omega_1^+)^2 k_d + \sqrt{M(\omega_1^+)} \text{ (using (C.6))} \\ k_i &> (\omega_1^-)^2 k_d - \sqrt{M(\omega_1^-)}, \text{ (using (C.11))} \end{aligned}$$

where ω_1^+ and ω_1^- satisfy

$$L_0 = [\alpha_1^+(\omega_1^+) + \alpha_2^+(\omega_1^+)]/\omega_1^+$$

and

$$L_0 = [\alpha_1^-(\omega_1^-) + \alpha_2^-(\omega_1^-)]/\omega_1^-,$$

respectively. This set $\mathcal{S}_{R,k_p}$ is a quadrilateral (see Fig. 11.12(b)). Note that the inequality $k_d > -\frac{T}{k}$ is redundant in this case since for $k_p > \frac{1}{k}$ and $\omega > 0$, $\frac{\sqrt{M(\omega)}}{\omega^2} < \frac{T}{k}$, so that in particular

$$\frac{-\sqrt{M(\omega_1^+)}}{(\omega_1^+)^2} > -\frac{T}{k}.$$

- For $k_p > \frac{1}{k}$ and $L(\omega_s) < L_0 < \max_{\omega \in (\omega_s, \infty)}[\frac{\alpha_1^-(\omega)+\alpha_2^-(\omega)}{\omega}]$, $\mathcal{S}_{R,k_p}$ is given by

$$\begin{aligned} k_i &> 0 \\ k_d &< \frac{T}{k} \\ k_i &< (\omega_1^-)^2 k_d - \sqrt{M(\omega_1^-)} \text{ (using (C.7) and (C.12))} \\ k_i &> (\omega_2^-)^2 k_d - \sqrt{M(\omega_2^-)}, \text{ (using (C.12))} \end{aligned}$$

where $\omega_1^- < \omega_2^-$ are solutions of the equation

$$L_0 = \frac{[\alpha_1^-(\omega) + \alpha_2^-(\omega)]}{\omega}.$$

This set $\mathcal{S}_{R,k_p}$ is also a quadrilateral (see Fig. 11.12(c)).

- For $k_p > \frac{1}{k}$ and $L_0 > \max \frac{\alpha_1^-(\omega)+\alpha_2^-(\omega)}{\omega}$, $\mathcal{S}_{R,k_p} = \emptyset$ (using (C.7) and (C.13)).

As for the case of an open-loop unstable plant, i.e., $T < 0$, the procedure to obtain the stabilizing regions is similar to the case when $T > 0$ and $k_p > \frac{1}{k}$.

References

[1] Astrom, K. J., and Hagglund, T. "Automatic Tuning of Simple Regulators with Specifications on Phase and Amplitude Margins," *Automatica*, Vol. 20, pp. 645–651, 1984.

[2] Astrom, K. J., and Hagglund, T. *PID Controllers: Theory, Design and Tuning*, Instrument Society of America, Research Triangle Park, NC, 1995.

[3] Atherton, D. P., and Majhi, S. "Limitations of PID Controllers," *Proceedings of the American Control Conference*, pp. 3843–3847, June 1999.

[4] Bellman, R., and Cooke, K. L. *Differential-Difference Equations*, Academic Press, London, UK, 1963.

[5] Bhattacharyya, S. P., Chapellat, H., and Keel, L. H. *Robust Control: The Parametric Approach*, Prentice-Hall, Upper Saddle River, NJ, 1995.

[6] Bialkowski, W. L. "Control of the Pulp and Paper Making Process," in *The Control Handbook*, W. S. Levine, Ed., pp. 1219–1242, IEEE Press, New York, 1996.

[7] Chien, K. L., Hrones, J. A., and Reswick, J. B. "On the Automatic Control of Generalized Passive Systems," *Transactions of the American Society of Mechanical Engineers*, Vol. 74, pp. 175–185, 1952.

[8] Choksy, N. H. "Time-Lag Systems—A Bibliography," *IEEE Transactions on Automatic Control*, Vol. 5, pp. 66–70, 1960.

[9] Cohen, G. H., and Coon, G. A. "Theoretical Consideration of Retarded Control," *Transactions of the American Society of Mechanical Engineers*, Vol. 76, pp. 827–834, 1953.

[10] Datta, A., Ho, M. T., and Bhattacharyya, S. P. *Structure and Synthesis of PID Controllers*, Springer-Verlag, London, UK, 2000.

[11] Desoer, C. A., and Wu, M. Y. "Stability of Linear Time-Invariant Systems," *IEEE Transactions on Circuit Theory*, Vol. 15, pp. 245–250, 1968.

[12] Doyle, J. C., Francis, B. A., and Tannenbaum, A. R. *Feedback Control Theory*, Macmillan, New York, 1992.

[13] Gantmacher, F. R. *The Theory of Matrices*, Chelsea, New York, 1959.

[14] Grimble, M. J., and Johnson, M. A. "Algorithm for PID Controller Tuning Using LQG Cost Minimization," *Proceedings of the American Control Conference*, pp. 3922–3928, June 1997.

[15] Gu, K., Kharitonov, V. L., and Chen, J. *Stability of Time-Delay Systems*, Birkhäuser, Boston, 2003.

[16] Ho, K. W., and Chan, W. K. "Stability of Third-Order Control Systems with Time Lag Using Pontryagin's Method for Zero Determination," *Proceedings of IEE*, Vol. 116, No. 12, pp. 2063–2068, 1969.

[17] Ho, M. T., Datta, A., and Bhattacharyya, S. P. "Generalizations of the Hermite-Biehler Theorem," *Linear Algebra and its Applications*, Vol. 302-303, pp. 135–153, 1999.

[18] Ho, M. T., Datta, A., and Bhattacharyya, S. P. "Control System Design Using Low Order Controllers: Constant Gain, PI and PID," *Proceedings of the American Control Conference*, pp. 571–578, June 1997.

[19] Ho, M. T., Datta, A., and Bhattacharyya, S. P. "A Linear Programming Characterization of All Stabilizing PID Controllers," *Proceedings of the American Control Conference*, pp. 3922–3928, June 1997.

[20] Ho, M. T., Datta, A., and Bhattacharyya, S. P. "Robust and Non-Fragile PID Controller Design," *International Journal of Robust and Nonlinear Control*, Vol. 11, pp. 681–708, 2001.

[21] Ho, M. T. "Synthesis of H_∞ PID Controllers," *Proceedings of the 40th IEEE Conference on Decision and Control*, pp. 255–260, December 2001.

[22] Ho, M. T., and Lin, C. Y. "PID Controller Design for Robust Performance," *Proceedings of the 41st IEEE Conference on Decision and Control*, pp. 1063–1067, December 2002.

[23] Ho, M. T., Silva, G. J., Datta, A., and Bhattacharyya, S. P. "Real and Complex Stabilization: Stability and Performance," *Proceedings of the American Control Conference*, pp. 4126–4138, July 2004.

[24] Ho, W., Hang, C., and Zhou, J. "Self-Tuning PID Control for a Plant with Underdamped Response with Specification on Gain and Phase Margins," *IEEE Transactions on Control, Systems and Technology*, Vol. 5, No. 4, pp. 446–452, 1997.

[25] Karmarkar, J. S., and Siljak, D. D. "Stability Analysis of Systems with Time Delay," *Proceedings of IEE*, Vol. 117, No. 7, pp. 1421–1424, 1970.

[26] Keel, L. H., and Bhattacharyya, S. P. "Robust, Fragile or Optimal?" *IEEE Transactions on Automatic Control*, Vol. 42, pp. 1098–1105, 1997.

[27] Khalil, H. K. *Nonlinear Systems*, Macmillan, New York, 1992.

[28] Kharitonov, V. L., and Zhabko, A. P. "Robust Stability of Time-Delay Systems," *IEEE Transactions on Automatic Control*, Vol. 39, pp. 2388–2397, 1994.

[29] Marshall, J. E., Gorecki, H., Korytowski, A., and Walton, K. *Time-Delay Systems: Stability and Performance Criteria with Applications*, Ellis Horwood, New York, 1992.

[30] Marshall, J. E. *Control of Time Delay Systems*, Peter Peregrinus, London, UK, 1979.

[31] Morari, M., and Zafiriou, E. *Robust Process Control*, Prentice-Hall, Englewood Cliffs, NJ, 1989.

[32] Newton, G., Gould, L. A., and Kaiser, J. F. *Analytical Design of Linear Feedback Controls*, John Wiley, New York, 1957.

[33] Ogata, K. *Discrete-Time Control Systems*, Prentice-Hall, Englewood Cliffs, NJ, 1995.

[34] Ozguler, A. B., and Kocan, A. A. "An Analytic Determination of Stabilizing Feedback Gains," Report 321, Institut für Dynamische Systeme, Universitat Bremen, Sept. 1994.

[35] Panagopoulos, H., Astrom, K. J., and Hagglund, T. "Design of PID Controllers Based on Constrained Optimization," *Proceedings of the American Control Conference*, pp. 3858–3862, June 1999.

[36] Pessen, D. W. "A New Look at PID-controller Tuning," *Transactions of the American Society of Mechanical Engineers, Journal of Dynamic Systems, Measurement and Control*, Vol. 116, pp. 553–557, 1994.

[37] Pontryagin, L. S. "On the Zeros of Some Elementary Transcendental Function," *American Mathematical Society Translation*, Vol. 2, pp. 95–110, 1955.

[38] Silva, G. J., Datta, A., and Bhattacharyya, S. P. "PI Stabilization of First-Order Systems with Time-Delay," *Automatica*, Vol. 37, pp. 2025–2031, 2001.

[39] Silva, G. J., Datta, A., and Bhattacharyya, S. P. "New Results on the Synthesis of PID Controllers," *IEEE Transactions on Automatic Control*, Vol. 47, pp. 241–252, 2002.

[40] Silva, G. J., Datta, A., and Bhattacharyya, S. P. "Controller Design Via Padé Approximation Can Lead to Instability," *Proceedings of the 40th IEEE Conference on Decision and Control*, pp. 4733–4737, December 2001.

[41] Silva, G. J., Datta, A., and Bhattacharyya, S. P. "Stabilization of Time Delay Systems," *Proceedings of the American Control Conference*, pp. 963–970, June 2000.

[42] Silva, G. J., Datta, A., and Bhattacharyya, S. P. "Determination of Stabilizing Feedback Gains for Second-Order Systems with Time Delay," *Proceedings of the American Control Conference*, pp. 4658–4663, June 2001.

[43] Silva, G. J., Datta, A., and Bhattacharyya, S. P. "On the Stability and Controller Robustness of Some Popular PID Tuning Rules," *IEEE Transactions on Automatic Control*, Vol. 48, pp. 1638–1641, 2003.

[44] Silva, G. J., Datta, A., and Bhattacharyya, S. P. "Robust Control Design Using the PID Controller," *Proceedings of the 41st IEEE Conference on Decision and Control*, pp. 1313–1318, December 2002.

[45] Silva, G. J., Datta, A., and Bhattacharyya, S. P. "PID Tuning Revisited: Guaranteed Stability and Non-Fragility," *Proceedings of the American Control Conference*, pp. 5000–5006, May 2002.

[46] Thompson, W. M., Vacroux, A. G., and Hoffman, C. H. "Application of Pontryagin's Time Lag Stability Criterion to Force-Reflecting Servomechanisms," *9th Joint Automatic Control Conference*, pp. 432–443, 1968.

[47] Tsypkin, Y. Z. "Stability of Systems with Delayed Feedback," English Translation in *Frequency-Response Methods in Control Systems*, pp. 45–56, A. G. J. Mac Farlane, IEEE Press, Piscataway, NJ, 1979.

[48] Voda, A. A., and Landau, I. D. "A Method of the Auto-Calibration of PID Controllers," *Automatica*, Vol. 31, No. 1, pp. 41–53, 1995.

[49] Walton, K., and Marshall, J. E. "Direct Method for TDS Stability Analysis," *Proceedings of IEE, Part D*, Vol. 134, pp. 101–107, 1987.

[50] Xu, H., Datta, A., and Bhattcharyya, S. P. "Computation of All Stabilizing PID Gains for Digital Control Systems," *IEEE Transactions on Automatic Control*, Vol. 46, pp. 647–652, 2001.

[51] Xu, H., Datta, A., and Bhattcharyya, S. P. "PID Stabilization of LTI Plants with Time-Delay," *Proceedings of the 42nd IEEE Conference on Decision and Control*, pp. 4038–4043, December 2003.

[52] Yamamoto, S., and Hashimoto, I. "Present Status and Future Needs: The View from Japanese Industry," *Proceedings of the 4th International Conference on Chemical Process Control*, pp. 512–521, 1991.

[53] Zhuang, M., and Atherton, D. P. "Automatic Tuning of Optimum PID Controllers," *IEE Proceedings-D*, Vol. 140, No. 3, pp. 216–224, 1993.

[54] Ziegler, J. G., and Nichols, N. B. "Optimum Settings for Automatic Controllers," *Transactions of the American Society of Mechanical Engineers*, Vol. 64, pp. 759–768, 1942.